Lecture Notes in Computational Science and Engineering **123**

More information about this series at http://www.springer.com/series/3527

Jochen Garcke • Dirk Pflüger •
Clayton G. Webster • Guannan Zhang
Editors

Sparse Grids and Applications – Miami 2016

Springer

Editors
Jochen Garcke
Institute for Numerical Simulation
University of Bonn
Bonn, Germany

Fraunhofer SCAI
Sankt Augustin, Germany

Dirk Pflüger
Institute for Parallel and Distributed Systems (IPVS)
University of Stuttgart
Stuttgart, Germany

Clayton G. Webster
Department of Mathematics
University of Tennessee
Knoxville, TN, USA

Oak Ridge National Laboratory
Oak Ridge, TN, USA

Guannan Zhang
Oak Ridge National Laboratory
Oak Ridge, TN, USA

Department of Mathematics and Statistics
Auburn University
Auburn, AL, USA

ISSN 1439-7358 ISSN 2197-7100 (electronic)
Lecture Notes in Computational Science and Engineering
ISBN 978-3-030-09227-6 ISBN 978-3-319-75426-0 (eBook)
https://doi.org/10.1007/978-3-319-75426-0

Mathematics Subject Classification (2010): 65D99, 65M12, 65N99, 65Y20, 65N12, 62H99

Printed on acid-free paper

This Springer imprint is published by the registered company Springer International Publishing AG part of Springer Nature.
The registered company address is: Gewerbestrasse 11, 6330 Cham, Switzerland

Preface

Sparse grids are a popular approach for the numerical treatment of high-dimensional problems. Where classical numerical discretization schemes fail in more than three or four dimensions, sparse grids, in their different flavors, are frequently the method of choice, be it spatially adaptive in the hierarchical basis or via the dimensionally adaptive combination technique.

The Fourth Workshop on Sparse Grids and Applications (SGA 2016), which took place in Miami, Florida, USA, from October 4 to 7 in 2016, demonstrated once again the importance of this numerical discretization scheme. Organized by Hans-Joachim Bungartz, Jochen Garcke, Michael Griebel, Markus Hegland, Dirk Pflüger, Clayton Webster, and Guannan Zhang, almost 50 participants from six different countries have presented and discussed the current state of the art of sparse grids and their applications. Thirty-seven talks covered their numerical analysis as well as efficient data structures and new forms of adaptivity and a range of applications from clustering and model order reduction to uncertainty quantification settings and optimization. As a novelty, the topic high-performance computing covered several talks, targeting exascale computing and related tasks. Besides data structures and communication patterns with excellent parallel scalability, fault tolerance was introduced to the SGA series, the hierarchical approach providing novel approaches to the treatment of hardware failures without checkpoint restart. This volume of LNCSE collects selected contributions from attendees of the workshop.

We thank the U.S. Department of Energy Office of Science and the Oak Ridge National Laboratory for their financial support. Furthermore, we thank Kasi Arnold and Lora Wolfe for their assistance with the local organization.

Bonn, Germany — Jochen Garcke
Stuttgart, Germany — Dirk Pflüger
Knoxville, TN, USA — Clayton G. Webster
Oak Ridge, TN, USA — Guannan Zhang
February 2018

Contents

Contributors

Gustavo Avila Chemistry Department, Queen's University, Kingston, ON, Canada

Bastian Bohn Institute for Numerical Simulation, University of Bonn, Bonn, Germany

Hans-Joachim Bungartz Technical University of Munich, Garching, Germany

Tucker Carrington Jr. Chemistry Department, Queen's University, Kingston, ON, Canada

Ionuţ-Gabriel Farcaş Technical University of Munich, Garching, Germany

Fabian Franzelin Institute for Parallel and Distributed Systems, University of Stuttgart, Stuttgart, Germany

Alfredo Parra Hinojosa Chair of Scientific Computing, Technische Universität München, München, Germany

Peter Jantsch University of Tennessee, Knoxville, TN, USA

Tobias Neckel Technical University of Munich, Garching, Germany

Jens Oettershagen Institute for Numerical Simulation, University of Bonn, Bonn, Germany

Dirk Pflüger Institute for Parallel and Distributed Systems, University of Stuttgart, Stuttgart, Germany

Paul Cristian Sârbu Technical University of Munich, Garching, Germany

Peter Schober Goethe University Frankfurt, Frankfurt am Main, Germany

Miroslav Stoyanov Oak Ridge National Laboratory, Oak Ridge, TN, USA

Raúl Tempone King Abdullah University of Science and Technology (KAUST), Thuwal, Kingdom of Saudi Arabia

Benjamin Uekermann Technical University of Munich, Garching, Germany

Julian Valentin Simulation of Large Systems (SGS), Institute for Parallel and Distributed Systems (IPVS), University of Stuttgart, Stuttgart, Germany

Clayton G. Webster University of Tennessee, Knoxville, TN, USA

Oak Ridge National Laboratory, Oak Ridge, TN, USA

Sören Wolfers King Abdullah University of Science and Technology (KAUST), Thuwal, Kingdom of Saudi Arabia

Comparing Nested Sequences of Leja and PseudoGauss Points to Interpolate in 1D and Solve the Schroedinger Equation in 9D

Gustavo Avila, Jens Oettershagen, and Tucker Carrington Jr.

Abstract In this article, we use nested sets of weighted Leja points, which have previously been studied as interpolation points, as collocation points to solve a 9D vibrational Schroedinger equation. Collocation has the advantage that it obviates the need to compute integrals with quadrature. A multi-dimension sparse grid is built from the Leja points and Hermite-type basis functions by restricting sparse grid levels i_c using $\sum_c g^c(i_c) \leq H$, where $g^c(i_c)$ is a non-decreasing function and H is a parameter that controls the accuracy. Results obtained with Leja points are compared to those obtained with PseudoGauss points. PseudoGauss points are also nested. They are chosen to improve the accuracy of the Gram matrix. With both Leja and PseudoGauss points it is possible to add one point per level. We also compare Lebesgue constants for weighted Leja and PseudoGauss points.

1 Introduction

For several years, PseudoGauss nested sets of points [2] have been used to build sparse grids which are then employed with collocation [6, 7, 14] to compute vibrational energy levels of molecules. In this article, we define PseudoGauss nested point sets and compare results computed with them to those obtained from sets of new Leja-type points.

A molecule is composed of nuclei and electrons. Most properties of a molecule can be determined by solving the Schroedinger equation,

$$(K_e + K_n + V_{ee} + V_{nn} + V_{en})\Theta_t(q, x) = E_t \Theta_t(q, x) , \tag{1}$$

G. Avila · T. Carrington Jr. (✉)
Chemistry Department, Queen's University, Kingston, ON, Canada
e-mail: Tucker.Carrington@queensu.ca

J. Oettershagen
Institute for Numerical Simulation, University of Bonn, Bonn, Germany

J. Garcke et al. (eds.), *Sparse Grids and Applications – Miami 2016*,
Lecture Notes in Computational Science and Engineering 123,
https://doi.org/10.1007/978-3-319-75426-0_1

where K_e and K_n are kinetic energy operators for the electrons and the nuclei and V_{ee}, V_{nn}, and V_{en} are the Coulomb potentials for the interaction between electrons, between nuclei, and between electrons and nuclei. Because electrons are much lighter than nuclei it is often an excellent approximation to use the Born-Oppenheimer approximation [22]. Vibrational energy levels E_n are eigenvalues of the vibrational Schroedinger equation,

$$\hat{H}\psi_n = E_n\psi_n \,, \tag{2}$$

for which the Hamiltonian operator, $\hat{H}$, is composed of two terms, $\hat{H} = \hat{K} + \hat{V}$, and $\hat{V}$, the potential, is an eigenvalue of the electronic Schroedinger equation. $\hat{K}$ has the general form

$$\hat{K} = \sum_{k,l}^{3N-6} \left(G_{kl}({}^1x, \cdots, {}^{3N-6}x) \frac{\partial}{\partial^k x} \frac{\partial}{\partial^l x} \right) + \sum_{k}^{3N-6} \left(H_k({}^1x, \cdots, {}^{3N-6}x) \frac{\partial}{\partial^k x} \right) + W({}^1x, \cdots, {}^{3N-6}x) \,, \tag{3}$$

where $G_{kl}({}^1x, \cdots, {}^{3N-6}x)$, $H_k({}^1x, \cdots, {}^{3N-6}x)$, and $W({}^1x, \cdots, {}^{3N-6}x)$ are functions of the coordinates ${}^c x, c = 1, 2, \cdots, 3N-6$ and N is the number of atoms in the molecule. $\hat{V}$ is also a function of the coordinates. There are $3N-6$ because this is the number of vibrational degrees of freedom (there are three translation and three rotation coordinates). To solve Eq. (2), one writes ψ_n as a linear combination of basis functions

$$\psi_n(\mathbf{x}) = \sum_k c_{kn} \ \varphi_k(\mathbf{x}) \tag{4}$$

and solves for the coefficients [13]. The sum is over all the basis functions. $\mathbf{x}$ represents $({}^1x, \cdots, {}^{3N-6}x)$. The coefficients are usually determined with a variational (Galerkin) method, by substituting Eq. (4) into Eq. (2), multiplying on the left by $\varphi_{k'}(\mathbf{x})$ and integrating. This yields the generalized eigenvalue problem,

$$\mathbf{HU} = \mathbf{SUE} \,. \tag{5}$$

The $(k'k)$th element of $\mathbf{H}$ is $\langle\varphi_{k'}|\hat{H}|\varphi_k\rangle$. The $(k'k)$th element of $\mathbf{S}$ is $\langle\varphi_{k'}|\varphi_k\rangle$. The $(k'k)$th element of $\mathbf{V}$ is $\langle\varphi_{k'}|\hat{V}|\varphi_k\rangle$. The size of the matrices is the number of basis functions. For more detail see [13]. The potential $\hat{V}$ is smooth, but may be a complicated function. For many molecules, delocalised basis functions are a good choice. Usually, orthogonal functions are used. It is often not possible to calculate the matrix elements of $\mathbf{V}$ exactly and they are therefore obtained by quadrature. When $D = 3N-6 \geq 12$, it is extremely costly to use a direct product quadrature grid. One way to deal with the problem is to use Smolyak quadrature [2, 19, 23, 32, 38]. The Smolyak grid is built from nested 1D quadrature rules. In some coordinates $\hat{K}$ is simple enough that its matrix elements can be obtained from algebraic formulae. When this is not possible, calculating matrix elements of $\hat{K}$ is expensive. The matrix eigenvalue problem is solved with a Lanczos algorithm [31]. It requires evaluating matrix-vector products (MVPs). It would be impossible to build and costly to use the **H** and **S** matrices. Instead MVPs are computed by doing sums sequentially [3–5]. For decades, similar ideas have been used without quadrature (when exact matrix elements are known) [43] and with direct product bases and quadratures [12, 24, 27, 28, 36, 44, 46].

Quadrature-based methods work, however, they have several disadvantages. (1) The quadrature must be good enough that **S** matrix elements computed with it are exact; otherwise it is necessary to solve a generalized eigenvalue problem and iterative (Krylov space) algorithms for generalized eigenvalue problems are much more costly. (2) Determining a good quadrature scheme requires finding not only points but also weights. (3) If $\hat{K}$ is fairly simple, it is possible to choose basis functions so that it is not necessary to use quadrature to compute its matrix elements, but this limits one's ability to use the best possible basis functions. (4) If $\hat{K}$ is so complicated that one is forced to use quadrature to compute its matrix elements, the calculation is very costly.

Due to these disadvantages, we have also developed a collocation-based method [6, 7, 14]. Eigenvalues are computed with an Arnoldi algorithm [25]. MVPs are evaluated by doing sums sequentially. To use collocation, one introduces an interpolant for the ψ_n one wishes to compute and solves an eigenvalue problem in which elements of the eigenvectors are values of the wavefunctions at collocation points. There are no integrals and therefore no need for quadrature. It is necessary to choose collocation points, but if the best possible representation of the wavefunctions in the basis used to make the interpolant is excellent then energy levels are not sensitive to the choice of the points [11]. If a good basis set is at hand, this makes it possible to obtain accurate solutions to Eq. (2) without developing an accurate multi-dimensional quadrature. If many of the G_{kl} (or H_k) coefficients of $\hat{K}$ are complicated, collocation has the advantage that they need only be evaluated at points (they do not appear in integrands of integrals). In our calculations, it is imperative that the interpolation and collocation points be the same. Interpolation points are used to make an interpolant for ψ_n; collocation points are the points at which we require that Eq. (2) be satisfied. As interpolation and collocation points, we have used points on a nested sparse grid built from 1D PseudoGauss (PG) rules. In recent years, it has become common to use nested Leja points for interpolation. Are Leja points better 1D interpolation points than PG points? If PG points are as good, then they are an alternative to Leja points. Are Leja points better collocation points? Is it obvious that better interpolation points will also be better collocation points?

2 Interpolation

It is possible to make a multi-dimensional sparse grid interpolant by combining 1D interpolation operators, U^{i_c}, constructed from 1D Lagrange functions [10]. The 1D interpolation operators are used to make 1D interpolants

$$\overline{f}(x) = U^m f(x) = \sum_{k=1}^{m} a_k(x) f(x_k) , \tag{6}$$

where $f(x)$ is a 1D function to be interpolated at the points $x_1, \ldots, x_m$. This sparse grid interpolation method can be generalized [7] by using Lagrange-type functions (LTF) that span the same space as a basis, $\varphi_n(x),\ n = 0, 1, \cdots, m-1$,

$$a_k(x) = \sum_{n=0}^{m-1} B_{k,n}\varphi_n(x),\ k = 1, 2, \cdots, m . \tag{7}$$

In this paper we start k with one and n with zero. $B_{k,n}$ is an element of the inverse of the matrix whose elements are $\varphi_n(x_k)$, where $x_k, k = 1, 2, \cdots, m$ are interpolation points. Each $a_k(x)$ is equal to one at one x_k and zero at all the other x_k. Note that we are not using localized or hat basis functions.

The multi-dimensional interpolation operator is [30, 38]

$$I(D, H) = \sum_{g(i_1,i_2,\cdots,i_D)\leq H} \Delta U^{i_1} \otimes \Delta U^{i_2} \otimes \cdots \otimes \Delta U^{i_D}, \tag{8}$$

where $c = 1, 2, \cdots, D$ and

$$\Delta U^{i_c} = U^{i_c} - U^{i_c-1} . \tag{9}$$

i_c labels the 1D level for coordinate c. $g(i_1, i_2, \cdots, i_D)$ is a condition that determines what combinations of levels contribute. The corresponding set must be downward closed. In Eq. (8), there is a different U^{i_c} for each i_c and the accuracy of the associated 1D interpolant is determined by the number of points $m_c(i_c)$ used to make U^{i_c},

$$U^{i_c} f({}^c x) = \sum_{k_c=1}^{m_c(i_c)} a_{i_c,k_c}({}^c x) f({}^c x_{k_c}), \tag{10}$$

where $a_{i_c,k_c}({}^c x)$ are the Lagrange-type functions,

$$\begin{aligned} a_{i_c,k_c}({}^c x) = 1\ , \quad {}^c x = {}^c x_{k_c} \\ a_{i_c,k_c}({}^c x) = 0\ , \quad {}^c x = {}^c x_{k_c'}, \quad k_c' \neq k_c\ . \end{aligned} \tag{11}$$

In this article, we shall set $m_c(i_c) = i_c$, for the sake of simplicity. In papers [6, 7, 14], we use $\varphi_{n_c}({}^c x)$ that are eigenfunctions of 1D operators extracted from the full $\hat{H}$ by removing all derivative terms in $\hat{K}$ except those that involve ${}^c x$ and setting ${}^{c'} x = 0$ for $c' \neq c$, in $\hat{V}$ and $\hat{K}$.

3 The Importance of Nesting

It is important to use nested sets of points in Eq. (8). In a nested set of points, the points used to make U^{i_c} are also used for U^{i_c+1}. Non-nested sets of points are easier to make, but they have significant disadvantages. First, the number of sparse grid points increases less quickly with the maximum number of 1D basis functions per coordinate when nested sequences are used. This is especially important when the maximum number of 1D basis functions per coordinate is large. In our calculations, the maximum number of 1D basis functions per coordinate is typically about 25, because to compute many accurate ψ_n we need many basis functions. Second, the MVPs we must compute are much less costly if we use a nested grid. For example, the transformation of a vector labeled by grid indices to a vector labeled by basis indices is efficient if nested sequences are used [7, 37].

What makes a good nested sequence of points? It is best to have a sequence in which the number of points increases slowly as i_c is increased. One well-known nested sequence is obtained from Chebyshev (type 1) points,

$$x_k = \cos\left(\frac{2k-1}{2N_i}\pi\right), k = 1, 2, \cdots, N_i \, . \tag{12}$$

The sequence with $N_i = 1, 3, 9, 27, 81 \cdots$ points is nested. Another well-known nested sequence is obtained from the Clenshaw–Curtis points. If $N_i > 1$ they are

$$x_k = \cos\left(\frac{k-1}{N_i - 1}\pi\right), k = 1, 2, \cdots, N_i \, , \tag{13}$$

and if $N_i = 1$ the point is $x_1 = 0$. The sequence with $N_i = 1, 3, 5, 9, 17, 33, 65, 129, \cdots$ points is nested. Both the Chebyshev and Clenshaw–Curtis sequences have the crucial disadvantage that the number of points increases exponentially with i. The problem can be mitigated by using delay [32]. When delay is used the same number of points is used in several successive levels. However, delay only really helps if the maximum number of basis functions per coordinate is small. Otherwise, even with delay the sparse grid is large. The grid also is determined by choosing $g(i_1, i_2, \cdots, i_D)$ and H. They must be chosen so that the grid includes the points required to represent functions we wish to compute. With Chebyshev or Clenshaw–Curtis sequences, the grid is excessively large. It is therefore important to find nested sequences for which the number of points increases more slowly with i. Ideal would be to have nested sequences in which U^{i_c} uses one (or maybe two) fewer point than U^{i_c+1}. PseudoGauss points are one option. Leja points are another.

3.1 PseudoGauss Nested Points

PseudoGauss (PG) points and weights are determined, together [2], so that 1D overlap integrals (elements of the Gram matrix) associated with the $\varphi_m(x)$ basis, when computed by quadrature, are as exact as possible. The points and weights are linked. To do collocation no weights are required, but we nonetheless expect PG points to be good collocation, and maybe good interpolation, points. A set of PG points with N points is designed to maximize the accuracy of the overlap integrals

$$L_{n,m} = \int \varphi_n(x)\varphi_m(x)dx \tag{14}$$

with $0 \leq n+m \leq 2N-1$. They are called PseudoGauss because Gauss points and weights satisfy this requirement. We do not use Gauss points because they are not nested.

To make PG point sequences, we start with a one-point Gauss quadrature and denote the point by x_1. Let Q^N denote the set of points and weights for the quadrature with N points. To make Q^N from Q^{N-1} we add a point, x_N, and use DGLES in the LAPACK library [1] to compute a least-squares solution for an overdetermined system of linear equations which gives us N weights. The system of linear equations is **PW** = **L**, where **W** is a vector of weights, **W** $=(w_1, \cdots w_N)^T$ and

$$P_{n,m;i} = \varphi_n(x_i)\varphi_m(x_i), \tag{15}$$

and n, m are all n, m pairs with $n \leq m$ and $n+m \leq (2N-1)$ (e.g. for $N = 2$ n, m = 0,0; 0,1; 1,1; 0,2; 1,2; 0,3). **PW** is a quadrature approximation for the integrals in **L**. We then vary x_N to minimize

$$R(x_N) = \sum_{m=0}^{2N-1} \sum_{n=m}^{2N-1-m} \mid \delta_{m,n} - \sum_{i=1}^{N} w_i \varphi_n(x_i)\varphi_m(x_i) \mid, \tag{16}$$

(similar results are obtained by squaring the difference). Because the $\varphi_n(x)$ are orthogonal, the exact overlap matrix is an identity matrix. We use the subplex optimization procedure [35] to minimize Eq. (16). Subplex solves an unconstrained optimization problems using a simplex method on subspaces. In general, different initial values of x_N may give different minimized $R(x_N)$. We use several starting values and choose the one with the smallest $R(x_N)$. Differences between different approximate solutions are usually very small.

In this article, we compare the quality of PG and Leja interpolation points using a basis, $\varphi_m(x) = h_m^{-1/2} e^{-x^2/2} H_m(x)$, where $H_m(x)$ is a Hermite polynomial and h_m is the normalisation constant. Because the basis functions are either even or odd, we add points to the list of PG points in groups of two. The second new point is obtained from the first by changing its sign. In this case we determine only $Q^1, Q^3, Q^5, \cdots$ For example, the PG Q^5 quadrature rule is determined by adding two new points x_4 and $x_5 = -x_4$ to the set $(x_1, x_2, x_3 = -x_2)$. The vector of weights $(w_1, w_2, w_3, w_4, w_5)$ is calculated by solving $\mathbf{PW} = \mathbf{L}$ with values of $(n, m) = (0, 0), (0, 1), (0, 2), (0, 3), (0, 4), (0, 5), (1, 1), (1, 2), (1, 3), (1, 4), (2, 2), (2, 3)$, i.e. all the indices restricted by $n + m \leq 5$. As is the case for the points, $w_2 = w_3, w_4 = w_5$, etc. The value of x_4 is optimized to reduce the error between the exact overlap matrix elements (the identity $\mathbf{I}$ in this case) and the approximate $\mathbf{L}$ matrix elements calculated with the approximate quadrature Q^5.

3.2 Leja Nested Points

It is becoming popular to use Leja points as interpolation points [16, 21, 26, 29]. The classical Leja points are determined as follows. Choose the first point, x_1, to be any point in the relevant domain $[a, b]$. The kth point is then chosen so that

$$x_k = \arg\max_{x \in [a,b]} \prod_{i=1}^{k-1} |x - x_i|, \; k > 1 . \tag{17}$$

The resulting Leja points $x_1, x_2, \ldots$ have favorable properties when using a polynomial basis $1, x, x^2, \ldots$ to interpolate functions on $[a, b]$ with the error measured in the standard $L^\infty([a, b])$-norm, i.e. $\|f\|_{L^\infty([a,b])} = \sup_{x \in [a,b]} |f(x)|$. Moreover, their distribution follows the distribution of Gauss-Legendre points [29].

Sometimes, one is interested in interpolating functions where the error is measured in a υ-weighted supremum norm, i.e. $\|f\|_{L^\infty([a,b],\upsilon)} = \sup_{x \in [a,b]} \upsilon(x)|f(x)|$. If υ is given by $\upsilon(x) = \sqrt{w(x)}$, where $w : [a, b] \to \mathbb{R}_+$ is a positive weight function, then weighted Leja points can be defined as [21, 29]

$$x_k = \arg\max_{x \in [a,b]} \left\{ \sqrt{w(x)} \prod_{i=1}^{k-1} |x - x_i| \right\}, \; k > 1 . \tag{18}$$

Their distribution often follows the distribution of the Gaussian quadrature points associated with the weight function w [29]. The same Leja points can be used for interpolation with any basis composed of the functions $b_0 = \sqrt{w(x)} P_0$, $b_1 = \sqrt{w(x)} P_1, b_2 = \sqrt{w(x)} P_2, \cdots$, where P_j is the orthogonal polynomial of degree j associated with the weight function $w(x)$ and the interval $[a, b]$. In this case, Eq. (18) is equivalent to

$$x_k = \arg\max_{x\in[a,b]}\left\{\sqrt{w(x)}P_k(x) - U^{k-1}(x_1, x_2, \cdots, x_{k-1})[\sqrt{w(x)}P_k(x)]\right\}, \quad (19)$$

where $U^{k-1}(x_1, x_2, \cdots, x_{k-1})[\sqrt{w(x)}P_k(x)]$ is the interpolant of $\sqrt{w(x)}P_k(x)$ made using the first $k-1$ basis functions and points. This is true because the residual $\left\{\sqrt{w(x)}P_k(x) - U^{k-1}(x_1, x_2, \cdots, x_{k-1})[\sqrt{w(x)}P_k(x)]\right\}$ is both, (1) a linear combination of the functions $\{b_0, b_1, \cdots b_{k-1}\}$ and hence equal to a product of $\sqrt{w(x)}$ and a polynomial of degree $k-1$; and (2) equal to zero at $x_1, x_2, \cdots x_{k-1}$. Similarly, for basis functions, $\varphi_0, \varphi_1, \cdots$ that are linear combinations of $\sqrt{w(x)}P_k(x)$ (with a common weight function), the kth point is chosen so that

$$x_k = \arg\max_{x\in[a,b]}\left\{\varphi_k(x) - U^{k-1}(x_1, x_2, \cdots, x_{k-1})[\varphi_k(x)]\right\}. \quad (20)$$

The point used to make U^k that is not also used to make U^{k-1}, is the point at which the difference between the basis function, φ_k and its the interpolation with U^{k-1} is maximum.

In this article, we increase the number of points in the Leja set two at a time, because we are using basis functions each of which is even or odd. The first point is the quadrature point for the Gauss-Hermite quadrature rule with one point, i.e. $x_1 = 0$. The second and third points are constrained to satisfy $x_3 = -x_2$ and x_2 is determined by maximizing the function ($[a, b] = [-\infty, +\infty]$)

$$(\varphi_1(x) - C_{1,0}\varphi_0(x))^2 + (\varphi_2(x) - C_{2,0}\varphi_0(x))^2, \quad (21)$$

where $C_{1,0}\varphi_0(x)$ is the interpolant for $\varphi_1(x)$ computed with the single basis function $\varphi_0(x)$ and the single point x_1 and $C_{2,0}\varphi_0(x)$ is the interpolant for $\varphi_2(x)$ computed with the single basis function $\varphi_0(x)$ and the single point x_1. The new two points $x_5 = -x_4$ and x_4 are determined by maximizing the function

$$\begin{aligned}(\varphi_3(x) - C_{3,0}\varphi_0(x) - C_{3,1}\varphi_1(x) - C_{3,2}\varphi_2(x))^2 +\\ (\varphi_4(x) - C_{4,0}\varphi_0(x) - C_{4,1}\varphi_1(x) - C_{4,2}\varphi_2(x))^2,\end{aligned} \quad (22)$$

where $C_{3,0}\varphi_0(x)+C_{3,1}\varphi_1(x)+C_{3,2}\varphi_2(x)$ is the interpolant for $\varphi_3(x)$ made with the basis $\{\varphi_0(x), \varphi_1(x), \varphi_2(x)\}$ and the points $\{x_1, x_2, x_3\}$ and $C_{4,0}\varphi_0(x)+C_{4,1}\varphi_1(x)+C_{4,2}\varphi_2(x)$ is the interpolant for $\varphi_4(x)$ made with the basis $\{\varphi_0(x), \varphi_1(x), \varphi_2(x)\}$ and the points $\{x_1, x_2, x_3\}$ The same procedure is applied to find $(x_6, x_7), (x_8, x_9)\cdots$, where $x_6 = -x_7, x_8 = -x_9\cdots$. We remark that there are many different variants and modifications of the original idea of Leja points. A comprehensive overview is given in [39, Sec. 3], where so-called Leja-odd points are defined. Leja-odd point sets are not symmetric as are the point sets we construct. Rather, they are groupings of the regular Leja points. Moreover, there are $\Re$-Leja points, which are symmetric if grouped as $N = 2k - 1$. They are, however, tailored to interpolation on the complex unit disk and, by projection, also on the interval $[-1, 1]$. Therefore, they are not relevant in our setting.

4 Lebesgue Constants

It is common to use the Lebesgue constant to assess the quality of a set of interpolation points [34, 41]. The Lebesgue constant, Λ_N, bounds the interpolation error,

$$||f(x) - X(f(x))|| \leq (\Lambda_N + 1)||f(x) - X^*(f(x))|| . \tag{23}$$

In this equation, $X(f(x))$ is the interpolant for $f(x)$ constructed with a set of N basis functions, $\varphi_k(x), k = 0, 1, \cdots N - 1$, and n points in an interval $[a, b]$, and $X^*(f(x))$ is the best possible approximation that can be made with the same basis functions. We use the maximum norm. The best points clearly depend on the basis. The Lebesgue constant for the case $\varphi_k(x) = x^k$ is discussed in many books [34, 41]. It is well known that when $\varphi_k(x) = x^k$, the Lebesgue constant of a set equally spaced points increases exponentially with n, but that the Lebesgue constant of a set of Chebyshev points increases only logarithmically [17, 20]. For this reason, it is often stated that Chebyshev points are good interpolation points. For many functions, there are bases much better than $\varphi_k(x) = x^k, k = 0, 1, 2 \ldots$. Owing to the fact that the functions we compute decrease exponentially, if $|x|$ is large enough, we want (nested) interpolation points that work well with $\varphi_k(x) = h_m^{-1/2} e^{-x^2/2} H_k(x)$. What is a good set of interpolation points for a general $\{\varphi_k(x)\}_{k=0}^{N-1}$ basis? Lebesgue constants have previously been calculated for bases $\sqrt{w(x)}P_k(x)$, with $P_k(x)$ as in Sect. 3.2 [15, 40, 42].

When interpolating with $\varphi_k(x) = x^k$, the Lebesgue constant Λ_N can be calculated from Lagrange functions $a_j(x)$

$$a_j(x) = \prod_{\substack{i=1 \\ j \neq i}}^{N} \frac{x - x_i}{x_j - x_i} \tag{24}$$

by finding the maximum value of

$$\lambda_N(x) = \sum_{j=1}^{N} |a_j(x)|. \tag{25}$$

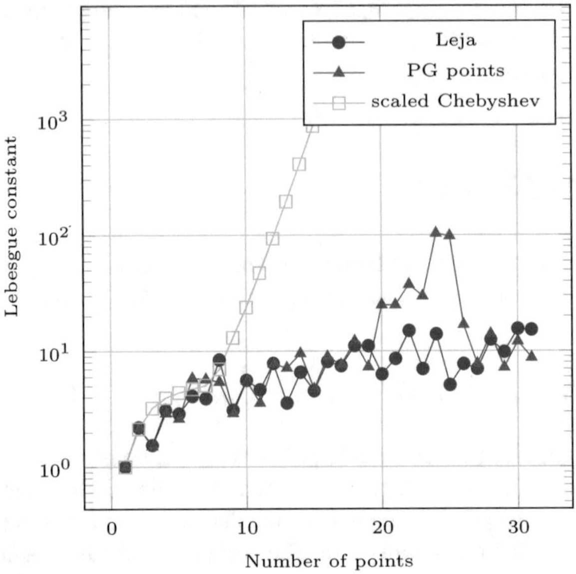

Fig. 1 Semi-logarithmic plot of the Lebesgue constants computed with Leja points, PG points and scaled Chebyshev points

When interpolating with a general basis, $\{\varphi_k(x)\}_{k=0}^{N-1}$, one can define Lagrange-type functions (LTF) $\{a_j\}_{j=1}^{N}$ as in Eq. (10) and compute the Lebesgue constant from

$$\Lambda_N(T) = \max_{x\in[a,b]} \sum_{j=1}^{N} \mid a_j(x) \mid . \tag{26}$$

In this article, we use the LTFs made from $\varphi_k(x) = h_m^{-1/2} e^{-x^2/2} H_k(x)$ to compute Lebesgue constants with both Leja and PseudoGauss points. They are compared to Lebesgue constants for scaled Chebyshev points, which are known to have favorable properties for interpolation with $\varphi_k(x) = x^k$ [20].

To make a nested set of Chebyshev point sequences requires using levels with $m = 1, 3, 9, 27, 81, \cdots$ points. The Smolyak grids build from Chebyshev points are therefore much larger than their Leja and PseudoGauss counterparts. The interval for the Chebyshev point set with N points is defined by setting a equal to the smallest Gauss-Hermite quadrature rule with N points and b equal to the largest Gauss-Hermite quadrature rule with N points. Therefore, for $N = 2$, $[a, b] = [-0.707106781186548, 0.707106781186547]$ and the Chebyshev points are 0.5 and -0.50, and for $N = 3$, $[a, b] = [-1.22474487139159, 1.22474487139159]$ and the Chebyshev points are 1.06066017177982, 0 and -1.06066017177982. The Lebesgue constants are reported in Fig. 1 and Table 1. Note that the Lebesgue constants of the scaled Chebyshev points increase exponentially, whereas the

Table 1 Lebesgue constants for Leja, PseudoGauss, and Chebyshev points

N	Λ_{Leja}	Λ_{PG}	Λ_{CH}
1	1.00	1.00	1.00
2	2.21	2.21	2.14
3	1.54	1.50	3.24
4	3.12	2.92	3.96
5	2.90	2.63	4.39
6	4.10	5.84	4.72
7	3.92	5.71	5.07
8	8.42	5.43	6.99
9	3.11	2.92	13.05
10	5.61	5.55	23.71
11	4.65	3.58	47.14
12	7.83	7.75	93.33
13	3.57	7.14	193.93
14	6.57	9.51	405.86
15	4.55	4.70	870.12
16	8.11	8.94	1891.65
17	7.45	7.58	4151.81
18	11.09	12.13	9279.35
19	11.05	7.29	20,745.74
20	6.31	24.83	47,354.43
21	8.57	24.80	107,471.08
22	14.91	37.34	249,422.18
23	7.00	29.53	573,197.21
24	13.95	102.87	1,348,297.50
25	5.10	97.73	3,131,662.87
26	7.72	16.77	7,448,633.91
27	7.01	7.72	17,459,802.45
28	12.49	13.94	41,929,091.80
29	9.80	7.20	99,085,230.50
30	15.49	12.07	239,713,640.92
31	15.21	8.71	570,620,030.02

Lebesgue constants of both Leja and PG points increase at most algebraically. Moreover, we note that the Lebesgue constant for $N = 2k$ is almost always larger than the Lebesgue constant for $2k - 1$, $k = 1, 2, \cdots$. This is because pairs of points are determined to be used together, but, if N is even only the positive member of a pair is used to make the interpolant. When N is even, the points do not have the correct symmetry.

The Leja and PseudoGauss points appear to be of similar quality. Not surprisingly, Chebyshev points are poor.

5 Comparison Between Leja Points and PseudoGauss Points in Collocation Calculations

PseudoGauss points ought to be good collocation points. They might be better collocation points than interpolation points. The best possible vibrational energies will be obtained, from a given basis set, by exactly calculating all of the Hamiltonian and overlap matrix elements. This is often impossible, for two reasons. First, for some potentials and bases there are no algebraic expressions for the matrix elements. Second, if the potential does not have a special form (and the corresponding matrix is therefore not sparse) then to efficiently evaluate MVPs it is imperative that we impose structure on the matrix representation of the potential by using quadrature. This structure makes it possible to do sums sequentially [3–5]. When it is not possible to calculate *all* of the Hamiltonian and overlap matrix elements exactly then one must choose between (1) computing *all* of them with quadrature or (2) computing only those for which exact matrix elements are not known. For example, if orthogonal basis functions are used then one might choose to replace **S** in Eq. (5) with an identity matrix. Although it might seem obvious that one should use exact matrix elements when they are available, Boys showed it is best to be consistent and to use the same quadrature for *all* matrix elements [11]. For a 1D Hamiltonian with a unit mass, computing all matrix elements by quadrature means solving

$$[\mathbf{TWK} + \mathbf{TWVT^T}]\tilde{\mathbf{U}} = [\mathbf{TWT^T}]\tilde{\mathbf{U}}\tilde{\mathbf{E}}\,, \tag{27}$$

where $\tilde{\mathbf{U}}$ is a matrix of eigenvectors and $\tilde{\mathbf{E}}$ is a diagonal matrix whose diagonal elements are eigenvalues. **T** is a matrix whose elements are

$$T_{j,\alpha} = \varphi_j(x_\alpha)\,, \tag{28}$$

where x_α and w_α are quadrature points and weights. The range of α and j is the number of quadrature points (equal to the number of basis functions). **K** is a matrix whose elements are

$$K_{\alpha,j} = \frac{-1}{2}\frac{d^2}{dx^2}\varphi_j(x)|_{x=x_\alpha}\,. \tag{29}$$

V is a matrix whose elements are

$$V_{\beta\alpha} = \delta_{\beta,\alpha}V(x_\alpha)\,. \tag{30}$$

W is a matrix whose elements are

$$W_{\beta,\alpha} = \delta_{\beta,\alpha}w_\alpha\,. \tag{31}$$

Equation (27) is equivalent to

$$[\mathbf{K} + \mathbf{V}\mathbf{T}^{\mathbf{T}}]\tilde{\mathbf{U}} = \mathbf{T}^{\mathbf{T}}\tilde{\mathbf{U}}\tilde{\mathbf{E}}\,, \tag{32}$$

which is the collocation equation. Therefore, if we can find points and weights that accurately evaluate each of the three terms in Eq. (27) then the points will be good collocation points. It is in general not possible to choose points to optimise the accuracy of the kinetic and potential matrix elements, however, it is possible to choose points to optimise the accuracy of $\mathbf{S} \approx \mathbf{TWT}^{\mathbf{T}}$. They are PseudoGauss points.

The 100 lowest levels of the CH_2NH (a 9D problem) were calculated with a sparse grid collocation method [6, 7, 14], using both Leja and PseudoGauss points. We use normal coordinates [45] and

$$\hat{K} = -\sum_{c=1}^{3N-6} \left(\frac{\omega_c}{2} \frac{\partial^2}{\partial {}^c x^2} \right) . \tag{33}$$

The potential is a Taylor series. We use the interpretation of [18] of the potential of Pouchan and Zaki [33]. The basis functions are, $\varphi_k({}^c x) = h_m^{-1/2} e^{-x^2/2} H_k({}^c x)$. We use $m_c(i_c) = i_c$. The restriction function that defines the basis is [5, 8]

$$g(i_1, i_2, \cdots i_d) = g^1(H, n_1) + g^2(H, n_2) + \cdots + g^D(H, n_D) \leq H\,. \tag{34}$$

Note that these 1D functions depend, in general, on H. It is not possible to use the same 1D functions for all H without including many unnecessary functions in the basis as H is increased. In the past, we have used step functions for the 1D functions. For the three normal mode coordinates with the largest frequencies, $c = 1, 2, 3$, we use 1D functions that are independent of H:

$$\begin{aligned} g^c(0) = 0,\, g^c(1) = 17,\, g^c(2) = 26,\, g^c(3) = 35,\, g^c(4) = 44,\, g^c(5) = 53,\, g^1(6) = 62, \\ g^c(7) = 71,\, g^c(8) = 79,\, g^c(9) = 89,\, g^c(10) = 99,\, g^c(11) = 109,\, g^c(12) = 120, \\ g^c(13) = 130,\, g^c(14) = 140,\, g^c(15) = 150,\, g^c(16) = 160,\, g^c(17) = 169, \\ g^c(18) = 178,\, g^c(19) = 187,\, g^c(20) = 196 \cdots \end{aligned} \tag{35}$$

For the remaining coordinates we use

$$\begin{aligned} g^c(0) = 0,\, g^c(1) = 15,\, g^c(2) = 24,\, g^c(3) = 33,\, g^c(4) = 42,\, g^c(5) = 51, \\ g^1(6) = 60,\, g^c(7) = 69,\, g^c(8) = 79,\, g^c(9) = 89,\, g^c(10) = 99, \\ g^c(11) = 110,\, g^c(12) = 120,\, g^c(13) = 130,\, g^c(14) = 139, \\ g^c(15) = 148,\, g^c(16) = 157,\, g^c(17) = 166,\, g^c(18) = 175, \\ g^c(19) = 187,\, g^c(20) = 196 \cdots \end{aligned} \tag{36}$$

Table 2 Errors for the 100 lowest levels of CH_2NH

	Leja		PG	
	Average error	Maximum error	Average error	Maximum error
Basis I	0.0026905	0.02366	0.0242918	0.06925
Basis II	0.0005770	0.00471	0.0032222	0.04544
Basis III	0.0001857	0.00062	0.0004222	0.00210
Basis IV	0.0002091	0.00057	0.0002298	0.00209

All errors are in cm^{-1}

and

$$\begin{aligned} g^c(h(H), n_c) &= g^c(n_c) + h(H), \quad 0 < n_c \leq 4; \\ g^c(h(H), n_c) &= g^c(n_c), \quad 5 \leq n_c \leq 12; \\ g^c(h(H), n_c) &= g^c(n_c) - h(H), \quad n_c \geq 13\,. \end{aligned} \tag{37}$$

When H is increased we increase also h to avoid increasing the number of unnecessary basis functions [9]. $h(H)$ is a step function. Values are specified in the next paragraph.

We report results for the 100 lowest eigenvalues of CH_2NH. The largest calculation was done for $H = 264$ and $h = 7$ with 12,684,284 pruned product basis functions. With this basis the largest relative error in the eigenvalues is 10^{-9}. Calculations with smaller basis sets were done to test the convergence: Basis I with $H = 150$ and $h = 0$ and 342,152 functions; Basis II with $H = 160$ and $h = 0$ and 586,808 basis functions; Basis III with $H = 170$ and $h = 0$ and 932,231 basis functions; Basis IV with $H = 190$ and $h = 0$ and 2,272,064 basis functions. Results are given in Table 2. Leja and PseudoGauss points are about equally good, but Leja points seem to be slightly better when the basis is small.

6 Conclusion

In this article, we present a recipe for determining PseudoGauss (PG) points and assess their usefulness for interpolating a 1D function in a basis $\varphi_k(^c x) = h_m^{-1/2} e^{-x^2/2} H_k(^c x)$ by computing Lebesgue constants. The PG Lebesgue constants are compared to Leja Lebesgue constants. PG Lebesgue constants are usually smaller than Leja Lebesgue constants, when the number of points is less than about 12. They are larger, when the number of points is larger than 12. In most cases, the PG and Leja Lebesgue constants are similar. By combining Leja and PG points with a sparse grid interpolation method, we are able to use them to compute vibrational energy levels of a 9D Hamiltonian. We have demonstrated that Leja points are slightly more accurate. It is known that Lebesgue constants increase subexponentially [21] and they appear to be promising for calculating vibrational energy levels.

Acknowledgements Research reported in this article was funded by The Natural Sciences and Engineering Research Council of Canada and the DFG via project GR 1144/21-1. We are grateful for important discussions about Leja points with Peter Jantsch.

References

1. E. Anderson, Z. Bai, C. Bischof, S. Blackford, J. Demmel, J. Dongarra, J. Du Croz, A. Greenbaum, S. Hammarling, A. McKenney, D. Sorensen, *LAPACK Users' Guide*, 3rd edn. (Society for Industrial and Applied Mathematics, Philadelphia, 1999). ISBN: 0-89871-447-8 (paperback)
2. G. Avila, T. Carrington Jr., Nonproduct quadrature grids for solving the vibrational Schrödinger equation. J. Chem. Phys. **131**, 174103 (2009)
3. G. Avila, T. Carrington Jr., Using a pruned basis, a non-product quadrature grid, and the exact Watson normal-coordinate kinetic energy operator to solve the vibrational Schrödinger equation for C_2H_4. J. Chem. Phys. **135**, 064101 (2011)
4. G. Avila, T. Carrington Jr., Using nonproduct quadrature grids to solve the vibrational Schrödinger equation in 12D. J. Chem. Phys. **134**, 054126 (2011)
5. G. Avila, T. Carrington Jr., Solving the vibrational Schrödinger equation using bases pruned to include strongly coupled functions and compatible quadratures. J. Chem. Phys. **137**, 174108 (2012)
6. G. Avila, T. Carrington Jr., Solving the Schrödinger equation using Smolyak interpolants. J. Chem. Phys. **139**, 134114 (2013)
7. G. Avila, T. Carrington Jr., A multi-dimensional Smolyak collocation method in curvilinear coordinates for computing vibrational spectra. J. Chem. Phys. **143**, 214108 (2015)
8. G. Avila, T. Carrington, Pruned bases that are compatible with iterative eigensolvers and general potentials: new results for CH_3CN. Chem. Phys. **482**, 3–8 (2017)
9. G. Avila, T. Carrington, Computing vibrational energy levels of CH_4 with a Smolyak collocation method. J. Chem. Phys. **147**, 144102 (2017)
10. V. Barthelmann, E. Novak, K. Ritter, High dimensional polynomial interpolation on sparse grids. Adv. Comput. Math. **12**, 273–288 (2000)
11. S.F. Boys, Some bilinear convergence characteristics of the solutions of dissymmetric secular equations. Proc. R. Soc. Lond. A Math. Phys. Eng. Sci. **309**, 195–208 (1969)
12. M.J. Bramley, T. Carrington Jr., A general discrete variable method to calculate vibrational energy levels of three and four atom molecules. J. Chem. Phys. **99**, 8519–8541 (1993)
13. T. Carrington Jr., Perspective: computing (ro-) vibrational spectra of molecules with more than four atoms. J. Chem. Phys. **146**(12), 120902 (2017)
14. E. Castro, G. Avila, S. Manzhos, J. Agarwal, H.F. Schaefer, T. Carrington Jr., Applying a Smolyak collocation method to Cl_2CO. Mol. Phys. **115**(15–16), 1775–1785 (2017). https://doi.org/10.1080/00268976.2016.1271153
15. S. Damelin, The weighted Lebesgue constant of Lagrange interpolation for exponential weights on $[-1, 1]$. Acta Math. Hungar. **81**, 223–240 (1998)
16. M. Griebel, J. Oettershagen, On tensor product approximation of analytic functions. J. Approx. Theory **207**, 348–379 (2016)
17. R. Gunttner, Evaluation of Lebesgue constants. SIAM J. Numer. Anal. **17**(4), 512–520 (1980)
18. T. Halverson, B. Poirier, Calculation of exact vibrational spectra for P_2O and CH_2NH using a phase space wavelet basis. J. Chem. Phys. **140**, 204112 (2014)
19. F. Heiss, V. Winschel, Likelihood approximation by numerical integration on sparse grids. J. Econ. **144**, 62–80 (2008)
20. B.A. Ibrahimoglu, Lebesgue functions and Lebesgue constants in polynomial interpolation. J. Inequal. Appl. **2016**(1), 93 (2016)

21. P. Jantsch, C.G. Webster, G. Zhang, On the Lebesgue constant of weighted Leja points for Lagrange interpolation on unbounded domains (2016). arXiv preprint arXiv:1606.07093
22. H. Koeppel, W. Domcke, L.S. Cederbaum, Multimode molecular dynamics beyond the Born Oppenheimer approximation. Adv. Chem. Phys. **57**, 59–246 (1984)
23. D. Lauvergnat, A. Nauts, Quantum dynamics with sparse grids: a combination of Smolyak scheme and cubature. Application to methanol in full dimensionality. Spectrochim. Acta A Mol. Biomol. Spectrosc. **119**, 18–25 (2014)
24. C. Leforestier, L.B. Braly, K. Liu, M.J. Elrod, R.J. Saykally, Fully coupled six-dimensional calculations of the water dimer vibration-rotation-tunneling states with a split Wigner pseudo spectral approach. J. Chem. Phys. **106**, 8527–8544 (1997)
25. R.B. Lehoucq, D.C. Sorensen, C. Yang, *ARPACK Users Guide: Solution of Large-Scale Eigenvalue Problems With Implicitly Restarted Arnoldi Methods* (SIAM, Philadelphia, 1998). http://www.caam.rice.edu/software/ARPACK
26. F. Leja, Sur certaines suites liées aux ensembles plans et leur application à la représentation conforme. Ann. Polon. Math. **4**, 8–13 (1957)
27. J.C. Light, T. Carrington Jr., Discrete-variable representations and their utilization. Adv. Chem. Phys. **114**, 263–310 (2000)
28. U. Manthe, H. Köppel, New method for calculating wave packet dynamics: strongly coupled surfaces and the adiabatic basis. J. Chem. Phys. **93**, 345–356 (1990)
29. A. Narayan, J.D. Jakeman, Adaptive Leja sparse grid constructions for stochastic collocation and high-dimensional approximation. SIAM J. Sci. Comput. **36**, A2952–A2983 (2014)
30. E. Novak, K. Ritter, High dimensional integration of smooth functions over cubes. Numer. Math. **75**, 79–97 (1996)
31. C.C. Paige, Computational variants of the Lanczos method for the eigenproblem. IMA J. Appl. Math. **10**, 373–381 (1972)
32. K. Petras, Smolyak cubature of given polynomial degree with few nodes for increasing dimension. Numer. Math. **93**, 729–753 (2003)
33. C. Pouchan, K. Zaki, Ab initio configuration interaction determination of the overtone vibrations of methyleneimine in the region 2800–3200 cm^{-1}. J. Chem. Phys. **107**, 342–345 (1997)
34. T.J. Rivlin, *An Introduction to the Approximation of Functions* (Courier Corporation, North Chelmsford, 2003)
35. T. Rowan, The subplex method for unconstrained optimization. Dissertation Ph.D. thesis, Department of Computer Sciences, University of Texas, 1990
36. P. Sarkar, N. Poulin, T. Carrington Jr., Calculating rovibrational energy levels of a triatomic molecule with a simple Lanczos method. J. Chem. Phys. **110**, 10269–10274 (1999)
37. J. Shen, H. Yu, Efficient spectral sparse grid methods and applications to high-dimensional elliptic problems. SIAM J. Sci. Comput. **32**, 3228–3250 (2010)
38. S.A. Smolyak, Quadrature and interpolation formulas for tensor products of certain classes of functions. Dokl. Akad. Nauk SSSR **4**, 123 (1963)
39. M.K. Stoyanov, C.G. Webster, A dynamically adaptive sparse grids method for quasi-optimal interpolation of multidimensional functions. Comput. Math. Appl. **71**, 2449–2465 (2016)
40. J. Szabados, Weighted Lagrange and Hermite-Fejer interpolation on the real line. J. Inequal. Appl. **1**, 99–123 (1997)
41. J. Szabados, P. Vértesi, *Interpolation of Functions* (World Scientific, Singapore, 1990)
42. P. Vértesi, On the Lebesgue function of weighted Lagrange interpolation. II. J. Aust. Math. Soc. A **65**, 145–162 (1998)
43. X.G. Wang, T. Carrington, The utility of constraining basis function indices when using the Lanczos algorithm to calculate vibrational energy levels. J. Phys. Chem. A **105**, 2575–2581 (2001)
44. X.G. Wang, T. Carrington Jr., Computing rovibrational levels of methane with curvilinear internal vibrational coordinates and an Eckart frame. J. Chem. Phys. **138**, 104106 (2013)

45. E.B. Wilson Jr., J.C. Decius, P.C. Cross, *Molecular Vibrations: The Theory of Infrared and Raman Vibrational Spectra* (Dover, New York, 2000)
46. D. Xu, R. Chen, H. Guo, Probing highly excited vibrational eigenfunctions using a modified single Lanczos propagation method: application to acetylene (HCCH). J. Chem. Phys. **118**, 7273–7282 (2003)

On the Convergence Rate of Sparse Grid Least Squares Regression

Bastian Bohn

Abstract While sparse grid least squares regression algorithms have been frequently used to tackle Big Data problems with a huge number of input data in the last 15 years, a thorough theoretical analysis of stability properties, error decay behavior and appropriate couplings between the dataset size and the grid size has not been provided yet.

In this paper, we will present a framework which will allow us to close this gap and rigorously derive upper bounds on the expected error for sparse grid least squares regression. Furthermore, we will verify that our theoretical convergence results also match the observed rates in numerical experiments.

1 Introduction

One of the most common tasks in *Big Data* applications is *function regression*. Here, we aim to approximate a function $g : \Omega \to \mathbb{R}$ defined on an open domain $\Omega \subset \mathbb{R}^m$. However, we only have access to n (possibly noisy) evaluations $(\mathbf{t}_i, g(\mathbf{t}_i) + \varepsilon_i) \in \Omega \times \mathbb{R}, i = 1, \ldots, n$ of g. Note that this is a special instance of a much more general regression or even density estimation problem, see e.g. [15].

Although many successful regression algorithms such as generalized clustering methods, radial basis function neural networks or support vector machines have been proposed over the last decades, see e.g. [1, 14, 20], one of the main problems in Big Data applications, namely the vast number n of data points, still presents a severe limitation to these so-called *data-centered* algorithms. This phenomenon usually prevents the user from applying the above mentioned methods straightforwardly because of their superlinear runtime dependence on n, i.e. the number of computational steps grows much faster than n, e.g. $\mathcal{O}(n^3)$ for applying direct solvers to the regression problem. In order to cope with this problem, several enhancements

B. Bohn (✉)
Institute for Numerical Simulation, University of Bonn, Bonn, Germany
e-mail: bohn@ins.uni-bonn.de

J. Garcke et al. (eds.), *Sparse Grids and Applications – Miami 2016*,
Lecture Notes in Computational Science and Engineering 123,
https://doi.org/10.1007/978-3-319-75426-0_2

to these algorithms, such as chunking or sparse greedy matrix approximation, have been introduced, see [20]. Furthermore, since many data-centered methods are based on *kernel representations*, we need to have access to a closed form of an appropriate kernel function. However, in many cases only infinite series expansion kernels are provided and an evaluation is not straightforward, see [12, 13].

To circumvent these issues and obtain an algorithm which naturally employs linear runtime complexity with respect to n, grid based discretizations have been proposed. Here, sparse grids are particularly well-suited since they allow to efficiently treat also higher-dimensional domains, i.e. $m > 3$, which is not possible with full tensor-product grids due to the *curse of dimensionality*. This means that—for a full grid space—the number of grid points N_k scales like $\mathcal{O}\left(2^{km}\right)$, where k denotes the grid level. In the sparse grid case, however, the scaling of N_k is only $\mathcal{O}\left(2^k k^{m-1}\right)$. Many variants of sparse grid regression algorithms can be found in e.g. [3, 5, 9, 10, 19].

Even though sparse grid regression algorithms have proven to be a good choice for many practical Big Data problems, there has not yet been a thorough theoretical justification for their good performance, i.e. the overall error convergence behavior and suitable couplings between N_k and n have yet to be determined. In this paper, we aim to close this gap for the case of (unregularized) least-squares function regression. Here, the corresponding problem is to determine

$$\arg\min_{h \in V_k} \frac{1}{n} \sum_{i=1}^{n} (h(\mathbf{t}_i) - g(\mathbf{t}_i) - \varepsilon_i)^2,$$

where V_k is the sparse grid space of level k, i.e. we search for the function $h \in V_k$ which minimizes the average squared distance between point evaluations of h and the unknown function g in the input data points $\mathbf{t}_i, i = 1, \ldots, n$. The evaluation in $\mathbf{t}_i$ is perturbed by some additive noise term ε_i. For this setting, we will derive the optimal coupling between N_k and n and present the corresponding error convergence rate. As we will see, the rate is governed mainly by the best approximation error in the sparse grid space and a sample-dependent term in which the noise variance σ will play an important role. To obtain our results, we will enhance the analysis of [7] on least-squares regression with orthonormal basis sets, which has been applied to derive convergence properties for global polynomial spaces in [6, 17, 18], to arbitrary basis sets and apply it to our sparse grid basis functions. While the choice of the particular basis is arbitrary in the orthonormal case, the quotient of the frame constants enters our estimates for non-orthonormal bases. Therefore, we use the sparse grid prewavelets since they form an L_2 Riesz frame and reveal essentially the same properties in our estimates as an orthonormal basis does in [7]. Furthermore, the prewavelets lead to sparsely populated system matrices for least-squares regression because of their compact support. Thus, our basis choice leads to a fast least-squares algorithm with quasi-optimal convergence rate in the piecewise linear case.

The remainder of this paper is structured as follows: In Sect. 2 we recapitulate the least squares regression problem and introduce the necessary notation. Then, we briefly present our sparse grid spaces and the according basis functions in Sect. 3. Our main results on the coupling and the convergence rate can be found in Sect. 4. Subsequently, we provide numerical experiments to underscore our theoretical results in Sect. 5. Finally, we conclude in Sect. 6 with a short summary and an outlook on possible future research directions.

2 Least-Squares Regression

We now define the necessary ingredients to state and analyze the least squares function regression problem. To this end, let ρ be a probability measure on the Lebesgue σ-algebra of $\Omega \subset \mathbb{R}^m$ and let $g : \Omega \to \mathbb{R}$ be a point-evaluable, bounded function, i.e. there exists an $r > 0$ such that $\|g\|_{L_{\infty,\rho}(\Omega)} \leq r$. We define a real-valued random variable $\varepsilon = \varepsilon(\mathbf{t})$, which models the noise and fulfills

$$\mathbb{E}\left[\varepsilon \mid \mathbf{t}\right] = 0 \text{ for all } \mathbf{t} \in \Omega \quad \text{and} \quad \sigma^2 := \sup_{\mathbf{t}\in\Omega} \mathbb{E}\left[\varepsilon^2 \mid \mathbf{t}\right] < \infty. \tag{1}$$

Our n input data points for the least-squares regression are then given by

$$\mathscr{Z}_n := (\mathbf{t}_i, g(\mathbf{t}_i) + \varepsilon_i)_{i=1}^n \subset \Omega \times \mathbb{R},$$

where the $\mathbf{t}_i$ are drawn i.i.d. according to ρ and the $\varepsilon_i = \varepsilon(\mathbf{t}_i)$ are instances of the random variable ε. Finally, we denote our scale of finite-dimensional search spaces, i.e. the spaces in which the solution to the regression problem will lie, by $V_k \subset L_{2,\rho}(\Omega)$ for a scale parameter $k \in \mathbb{N}$, which will be the level of our grid spaces later on. In the following we will write $N_k := \dim(V_k)$ to denote the dimension of the search space of level k. Then, as already mentioned in the introduction, we can write the least-squares regression problem as

$$\text{Determine } f_{\mathscr{Z}_n,V_k} := \arg\min_{h\in V_k} \frac{1}{n} \sum_{i=1}^{n} \left(h(\mathbf{t}_i) - g(\mathbf{t}_i) - \varepsilon_i\right)^2. \tag{2}$$

Note that a regularized version of this problem, where a penalty term is added to the above formulation, is also often considered. However, in this paper we solely focus on the unregularized case (2) and give sufficient conditions such that this problem is stably solvable also without a penalty term.

To solve (2), let $\nu_1, \ldots, \nu_{N_k}$ be an arbitrary basis of V_k. Then it is straightforward to show that the coefficients $\boldsymbol{\alpha} := \left(\alpha_1, \ldots, \alpha_{N_k}\right)^T$ of $f_{\mathscr{Z}_n,V_k} = \sum_{i=1}^{N_k} \alpha_i \nu_i$ can be computed by solving the linear system

$$nBB^T\boldsymbol{\alpha} = B\mathbf{x}, \tag{3}$$

where $B\in\mathbb{R}^{N_k\times n}$ is given by $B_{ij}:=\frac{1}{n}\nu_i(\mathbf{t}_j)$ and $\mathbf{x} := (g(\mathbf{t}_1)+\varepsilon_1,\ldots,g(\mathbf{t}_n)+\varepsilon_n)^T$. For a more detailed discussion on this system, we refer to [2, 9].

3 Full Grids and Sparse Grids

In order to solve (3) on a full grid space, i.e. $V_k = \mathcal{V}_k^{\text{full}}$ of level k, or a sparse grid space, i.e. $V_k = \mathcal{V}_k^{\text{sparse}}$ of level k, we have to define appropriate basis functions $\nu_1,\ldots,\nu_{N_k}$. To this end, we consider the so-called piecewise linear prewavelet basis, see also [11], since it forms a Riesz frame, which will be of major importance for the analysis in the subsequent section. The prewavelets are based on linear combinations of the hat functions

$$\phi_{l,i}(t) := \phi(2^l\cdot t - i)|_{[0,1]} \quad \text{with} \quad \phi(t) := \begin{cases} 1-|t| & \text{if } t\in[-1,1], \\ 0 & \text{else.}\end{cases} \tag{4}$$

The univariate prewavelet basis functions $\gamma_{l,i} : [0,1]\to\mathbb{R}$ are then defined by

$$\gamma_{0,0} := 1,\ \gamma_{0,1} := \phi_{0,1},\ \gamma_{1,1} := 2\cdot\phi_{1,1} - 1.$$

for $l \leq 1$ and by

$$\gamma_{l,i} := 2^{\frac{l}{2}}\cdot\left(\frac{1}{10}\phi_{l,i-2} - \frac{6}{10}\phi_{l,i-1} + \phi_{l,i} - \frac{6}{10}\phi_{l,i+1} + \frac{1}{10}\phi_{l,i+2}\right)$$

for $l \geq 2$ and $i \in I_l \setminus \{1, 2^l-1\}$ with $I_l := \{i\in\mathbb{N} \mid 1\leq i\leq 2^l-1,\ i \text{ odd}\}$. For the boundary cases $i \in \{1, 2^l-1\}$, we have

$$\gamma_{l,1} := 2^{\frac{l}{2}}\cdot\left(-\frac{6}{5}\phi_{l,0} + \frac{11}{10}\phi_{l,1} - \frac{3}{5}\phi_{l,2} + \frac{1}{10}\phi_{l,3}\right),\quad \gamma_{l,2^l-1}(t) := \gamma_{l,1}(1-t).$$

The m-variate prewavelet functions are defined by a simple product approach

$$\gamma_{\mathbf{l},\mathbf{i}}(\mathbf{t}) := \prod_{j=1}^{m}\gamma_{l_j,i_j}(t_j), \tag{5}$$

where $\mathbf{l} = (l_1,\ldots,l_m)$ denotes the multivariate level index and $\mathbf{i} = (i_1,\ldots,i_m)$ denotes the multivariate position index. The graph of two exemplary univariate and

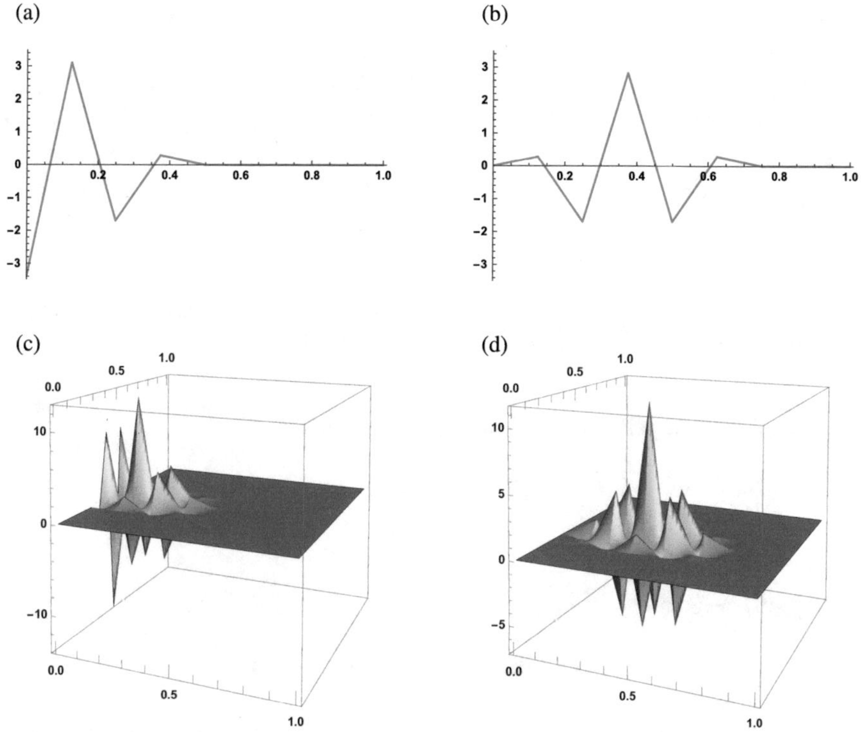

Fig. 1 Piecewise linear prewavelet examples. (**a**) $\gamma_{3,1}$. (**b**) $\gamma_{3,3}$. (**c**) $\gamma_{(3,4),(1,7)}$. (**d**) $\gamma_{(3,4),(3,7)}$

two exemplary bivariate prewavelet basis functions can be found in Fig. 1. In the multivariate case, the appropriate index sets are given by

$$\mathbf{I_l} := \left\{ \mathbf{i} \in \mathbb{N}^m \,\middle|\, \begin{matrix} 0 \le i_j \le 1, & \text{if } l_j = 0, \\ 1 \le i_j \le 2^{l_j} - 1, \; i_j \text{ odd} & \text{if } l_j > 0 \end{matrix} \quad \text{for all } 1 \le j \le m \right\},$$

which lead to the hierarchical increment spaces

$$W_{\mathbf{l}} := \operatorname{span} \left\{ \gamma_{\mathbf{l},\mathbf{i}} \mid \mathbf{i} \in \mathbf{I_l} \right\}.$$

Now, the full grid space of level $k > 0$ is defined by

$$\mathcal{V}_k^{\text{full}} := \bigoplus_{\substack{\mathbf{l} \in \mathbb{N}^m \\ |\mathbf{l}|_{\ell_\infty} \le k}} W_{\mathbf{l}},$$

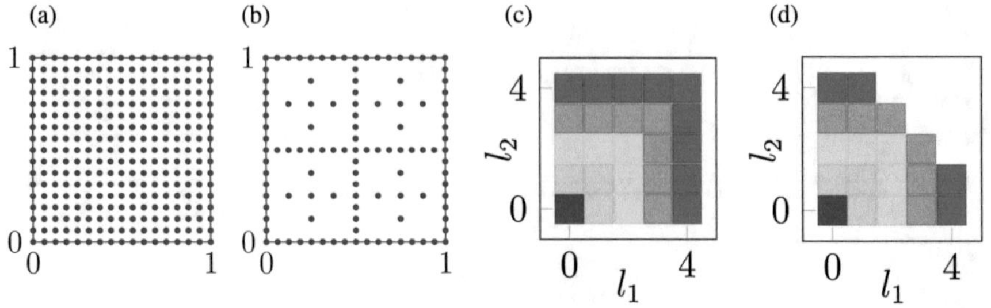

Fig. 2 Two-dimensional full grid and sparse grid and their corresponding index sets. (**a**) Full grid G_4^{full}. (**b**) Sparse grid G_4^{sparse}. (**c**) $\{\mathbf{l} \in \mathbb{N}^2 \mid |\mathbf{l}|_{\ell_\infty} \leq 4\}$. (**d**) $\{\mathbf{l} \in \mathbb{N}^2 \mid |\zeta_2(\mathbf{l}) \leq 4\}$

whereas the sparse grid space of level $k > 0$ is given by

$$\mathcal{V}_k^{\text{sparse}} := \bigoplus_{\substack{\mathbf{l} \in \mathbb{N}^m \\ \zeta_m(\mathbf{l}) \leq k}} W_{\mathbf{l}}$$

with $\zeta_m(\mathbf{0}) := 0$ and

$$\zeta_m(\mathbf{l}) := |\mathbf{l}|_{\ell_1} - m + \left|\{j \mid l_j = 0\}\right| + 1$$

for a non-zero $\mathbf{l} \in \mathbb{N}^m$. The specific choice of ζ_m guarantees that the highest resolution of a subgrid on the boundary is the same as the highest resolution of a subgrid in the interior of $[0, 1]^m$. The corresponding grids G_k^{full} and G_k^{sparse}, i.e. the centers of the support of the involved prewavelet basis functions, can be found in Fig. 2.

As we mentioned above, full grids suffer from the curse of dimensionality, i.e. the degrees of freedom grow like

$$\dim\left(\mathcal{V}_k^{\text{full}}\right) = (2^k + 1)^m = \mathcal{O}\left(2^{km}\right),$$

which depends exponentially on the dimension m of the domain. For sparse grids, it can easily be obtained that

$$\dim\left(\mathcal{V}_k^{\text{sparse}}\right) = \mathcal{O}\left(2^k k^{m-1}\right)$$

see e.g. [4] for grids in the interior of the domain and [8] for grids which are also allowed to live on the boundary. As we see, the curse of dimensionality only appears with respect to the level k instead of 2^k. Therefore, sparse grids can be used also for $m > 3$.

4 Error Analysis

After introducing the least-squares problem and our grid discretization in the previous sections, we can now present our main theorems on the stability and the error decay of a sparse grid regression algorithm. Our results are built on theorems 1 and 3 of [7] and can be seen as an extension thereof since only orthonormal bases are treated there, whereas our result also holds for arbitrary non-orthonormal bases.

4.1 Well-Posedness and Error Decay

In the following, we denote the maximum and minimum eigenvalues of a symmetric matrix X by $\lambda_{\max}(X)$ and $\lambda_{\min}(X)$. We start with a Matrix Chernoff bound, which is proven in section 5 of [21].

Theorem 1 (Chernoff Inequality for Random Matrices) *Let $D \in \mathbb{N}$ and $\delta \in [0, 1)$ be arbitrary and let $X_1, \dots, X_n \in \mathbb{R}^{D\times D}$ be independent, symmetric and positive semidefinite matrices with random entries. Let $R > 0$ be such that $\lambda_{\max}(X_i) \leq R$ holds for all $i = 1, \dots, n$. Then, it holds*

$$\mathbb{P}\left[\lambda_{\min}\left(\sum_{i=1}^{n} X_i\right) \leq (1-\delta)c_{\min}\right] \leq D\left(\frac{e^{-\delta}}{(1-\delta)^{1-\delta}}\right)^{\frac{c_{\min}}{R}}$$

and

$$\mathbb{P}\left[\lambda_{\max}\left(\sum_{i=1}^{n} X_i\right) \geq (1+\delta)c_{\max}\right] \leq D\left(\frac{e^{\delta}}{(1+\delta)^{1+\delta}}\right)^{\frac{c_{\max}}{R}}$$

with $c_{\min} := \lambda_{\min}\left(\mathbb{E}\left[\sum_{i=1}^{n} X_i\right]\right)$ and $c_{\max} := \lambda_{\max}\left(\mathbb{E}\left[\sum_{i=1}^{n} X_i\right]\right)$.

For a basis $\nu_1, \dots, \nu_{N_k}$ of V_k, we introduce the quantity

$$S(\nu_1, \dots, \nu_{N_k}) := \sup_{\mathbf{t}\in\Omega} \sum_{i=1}^{N_k} |\nu_i(\mathbf{t})|^2, \tag{6}$$

which will play a pivotal role throughout the rest of this paper. Note that this quantity is named $K(N_k)$ in [7] since it is independent of the basis choice there as the authors only deal with orthonormal bases. However, in our more general case, the quantity $S(\nu_1, \dots, \nu_{N_k})$ is highly dependent on the concrete choice of the basis of V_k.

In the following, we denote the mass matrix on level k by $M = M(\nu_1, \dots, \nu_{N_k}) \in \mathbb{R}^{N_k\times N_k}$, i.e. $M_{ij} = \langle \nu_i, \nu_j\rangle_{L_{2,\rho}(\Omega)}$. With the help of Theorem 1, we are able to prove the following stability result, which is an extension of theorem 1 of [7].

Theorem 2 (Well-Posedness) *Let* $n \geq N_k$, $c = \left|\log\left(\frac{e^{0.5}}{(1.5)^{1.5}}\right)\right| \approx 0.1082$ *and let*

$$S(\nu_1, \ldots, \nu_{N_k}) \leq c \cdot \frac{\lambda_{\min}(M)}{1+\theta} \cdot \frac{n}{\log(n)} \tag{7}$$

for a $\theta > 0$. *Then, the solution* $f_{\mathscr{Z}_n, V_k} = \sum_{j=1}^{N_k} \alpha_j \nu_j$ *of* (3) *exists, is unique and fulfills*

$$\|f_{\mathscr{Z}_n, V_k}\|_{L_{2,\rho}(\Omega)} \leq \sqrt{6} \cdot \frac{\lambda_{\max}(M)}{\lambda_{\min}(M)} \cdot \frac{1}{\sqrt{n}} \|\mathbf{x}\|_{\ell_2}$$

with probability at least $1 - 2n^{-\theta}$, *where* $\mathbf{x} := (g(\mathbf{t}_1) + \varepsilon_1, \ldots, g(\mathbf{t}_n) + \varepsilon_n)^T$.

Proof The proof follows the lines of [7] with the necessary generalizations for arbitrary basis functions. Let $X \in \mathbb{R}^{N_k \times N_k}$ be the random, positive semi-definite matrix with entries $X_{ij} := \frac{1}{n}\nu_i(\mathbf{t}) \cdot \nu_j(\mathbf{t})$, where $\mathbf{t}$ is drawn according to ρ and let $X_1, \ldots, X_n$ be n realizations of X with $\mathbf{t} = \mathbf{t}_1, \ldots, \mathbf{t}_n$ from the samples $\mathscr{Z}_n$. Then, $nBB^T = \sum_{i=1}^n X_i$ and $M = \mathbb{E}\left[\sum_{i=1}^n X_i\right]$.

Note that $\lambda_{\max}(X) \leq \frac{1}{n} S(\nu_1, \ldots, \nu_{N_k})$ almost surely since $X = nAA^T$ with $A = \frac{1}{n}\left(\nu_1(\mathbf{t}), \ldots, \nu_{N_k}(\mathbf{t})\right)^T$ and we have

$$\begin{aligned}
\lambda_{\max}(X) = n\lambda_{\max}(AA^T) = n \cdot \max_{|y|=1} \|Ay\|_{\ell_2}^2 = n \cdot \left(\frac{1}{n^2}\sum_{i=1}^{N_k} |\nu_i(\mathbf{t}) \cdot 1|^2\right) \\
\leq \frac{1}{n} S(\nu_1, \ldots, \nu_{N_k}).
\end{aligned}$$

Therefore, we can apply Theorem 1 with $D = N_k$, $R = \frac{1}{n}S(\nu_1, \ldots, \nu_{N_k})$ and $\delta = \frac{1}{2}$ to obtain

$$\begin{aligned}
P := \mathbb{P}\left[\lambda_{\min}(nBB^T) \leq \frac{\lambda_{\min}(M)}{2} \quad \text{or} \quad \lambda_{\max}(nBB^T) \geq \frac{3\lambda_{\max}(M)}{2}\right] \\
\leq N_k \left(\frac{e^{-0.5}}{0.5^{0.5}}\right)^{\frac{n\lambda_{\min}(M)}{S(\nu_1,\ldots,\nu_{N_k})}} + N_k \left(\frac{e^{0.5}}{1.5^{1.5}}\right)^{\frac{n\lambda_{\max}(M)}{S(\nu_1,\ldots,\nu_{N_k})}} \leq 2N_k \left(\frac{e^{0.5}}{1.5^{1.5}}\right)^{\frac{n\lambda_{\min}(M)}{S(\nu_1,\ldots,\nu_{N_k})}},
\end{aligned}$$

where the last inequality follows from $\lambda_{\min}(M) \leq \lambda_{\max}(M)$ and $0 < \frac{e^{-0.5}}{0.5^{0.5}} < \frac{e^{0.5}}{1.5^{1.5}} < 1$. Using (7) and the definition of c, we obtain

$$P \leq 2N_k e^{-\frac{cn\lambda_{\min}(M)}{S(\nu_1,\ldots,\nu_{N_k})}} \leq 2N_k \cdot n^{-(1+\theta)} \leq 2n^{-\theta}$$

since we assumed $N_k \leq n$. Therefore, (3) is uniquely solvable with probability at least $1 - 2n^{-\theta}$. Noting that $\|B\|^2_{\mathrm{Lin}(\mathbb{R}^n,\mathbb{R}^{N_k})} = \frac{1}{n}\|nBB^T\|_{\mathrm{Lin}(\mathbb{R}^{N_k},\mathbb{R}^{N_k})} = \frac{1}{n}\lambda_{\max}(nBB^T)$ holds for the operator norm of the linear operator B and writing the L_2 norm with the help of the mass matrix, we finally get

$$\begin{aligned}
\|f_{\mathscr{X}_n,V_k}\|^2_{L_{2,\rho}(\Omega)} &= \boldsymbol{\alpha}^T M \boldsymbol{\alpha} \stackrel{(3)}{=} \mathbf{x}^T B^T (nBB^T)^{-1} M (nBB^T)^{-1} B\mathbf{x} \\
&\leq \|\mathbf{x}\|^2_{\ell_2} \|B\|^2_{\mathrm{Lin}(\mathbb{R}^n,\mathbb{R}^{N_k})} \lambda_{\max}((nBB^T)^{-1})^2 \lambda_{\max}(M) \\
&= \frac{1}{n}\|\mathbf{x}\|^2_{\ell_2} \lambda_{\max}(nBB^T) \frac{1}{\lambda_{\min}(nBB^T)^2} \lambda_{\max}(M) \\
&\leq \frac{1}{n}\|\mathbf{x}\|^2_{\ell_2} \frac{3\lambda_{\max}(M)}{2} \frac{4}{\lambda_{\min}(M)^2} \lambda_{\max}(M) = 6\frac{\lambda_{\max}(M)^2}{\lambda_{\min}(M)^2} \cdot \frac{1}{n}\|\mathbf{x}\|^2_{\ell_2}
\end{aligned}$$

with probability at least $1 - 2n^{-\theta}$, which proves our assertion. □

Theorem 2 tells us that the regression problem with basis $\nu_1, \ldots, \nu_{N_k}$ is stably solvable for all $k \in \mathbb{N}$ with high probability if the number of samples n is large enough such that $n \geq N_k$ and (7) are fulfilled and if the fraction $\frac{\lambda_{\max}(M)}{\lambda_{\min}(M)}$, i.e. the condition number of the mass matrix, does not grow too fast with $k \to \infty$. Note that it is also possible to prove a more general version of this theorem if a (Tikhonov) regularization term is added, see [2].

Recall the L_∞ bound r on the function g from which the data $\mathscr{X}_n$ is sampled. For our error bound, we need to define the truncation operator $\tau_r : L_{\infty,\rho}(\Omega) \to L_{\infty,\rho}(\Omega)$ by $\tau_r(f)(\cdot) := P_r(f(\cdot))$, where the convex projection $P_r : \mathbb{R} \to \mathbb{R}$ is defined by

$$P_r(x) = \begin{cases} x & \text{if } |x| \leq r, \\ \frac{x}{|x|} \cdot r & \text{else.} \end{cases}$$

Note that τ_r is a non-expansive operator with respect to the $L_{2,\rho}(\Omega)$ norm, i.e. $\|\tau_r(f_1) - \tau_r(f_2)\|_{L_{2,\rho}(\Omega)} \leq \|f_1 - f_2\|_{L_{2,\rho}(\Omega)}$ for all $f_1, f_2 \in L_{\infty,\rho}(\Omega)$. Now, we can provide a theorem on the expected error behavior.

Theorem 3 (Expected Regression Error) *Let $n \geq N_k$ and let $f_{\mathscr{X}_n,V_k}$ be the solution to* (3)*—or $f_{\mathscr{X}_n,V_k} = 0$ if no unique solution to* (3) *exists. Let, furthermore, $S(\nu_1, \ldots, \nu_{N_k})$ and n fulfill* (7) *for a fixed $\theta > 0$ and for all $k \in \mathbb{N}$. Then,*

$$\begin{aligned}
\mathbb{E}\left[\|\tau_r\left(f_{\mathscr{X}_n,V_k}\right) - g\|^2_{L_{2,\rho}(\Omega)}\right] &\leq \left(1 + \frac{8c\lambda_{\max}(M)}{(1+\theta)\lambda_{\min}(M)\log(n)}\right) \inf_{f \in V_k} \|f - g\|^2_{L_{2,\rho}(\Omega)} \\
&\quad + 8r^2 n^{-\theta} + 8\sigma^2 \left(\frac{\lambda_{\max}(M)}{\lambda_{\min}(M)}\right)^2 \cdot \frac{N_k}{n} \qquad (8)
\end{aligned}$$

with c from (7)*. Here, the expectation is taken with respect to the product measure $\rho^n := \rho \times \ldots \times \rho$.*

Proof Again, the proof generalizes the one in [7], where only orthonormal bases are considered. In the following we will just write L_p for $L_{p,\rho}(\Omega)$ with $p \in [1, \infty]$. Let $\Omega^n = \Omega \times \ldots \times \Omega$ and let

$$\Omega_+^n := \left\{(\mathbf{t}_1, \ldots, \mathbf{t}_n) \in \Omega^n \mid \lambda_{\max}(nBB^T) \le \frac{3\lambda_{\max}(M)}{2} \text{ and } \lambda_{\min}(nBB^T) \ge \frac{\lambda_{\min}(M)}{2}\right\}$$

and let $\Omega_-^n := \Omega^n \setminus \Omega_+^n$. We have already shown in the proof of Theorem 2 that $\mathbb{P}(\Omega_-^n) \le 2n^{-\theta}$ since (7) holds. Let us denote $E := \mathbb{E}\left[\|\tau_r\left(f_{\mathscr{Z}_n,V_k}\right) - g\|_{L_2}^2\right]$. Since $|\tau_r(f)(\mathbf{t}) - g(\mathbf{t})| \le |\tau_r(f)(\mathbf{t})| + |g(\mathbf{t})| \le 2r$ holds for all $f \in L_\infty$ and almost every $\mathbf{t} \in \Omega$, we obtain

$$\begin{aligned}
E &= \int_{\Omega_+^n} \|\tau_r\left(f_{\mathscr{Z}_n,V_k}\right) - g\|_{L_2}^2 \, \mathrm{d}\rho^n + \int_{\Omega_-^n} \|\tau_r\left(f_{\mathscr{Z}_n,V_k}\right) - g\|_{L_2}^2 \, \mathrm{d}\rho^n \\
&\le \int_{\Omega_+^n} \|\tau_r\left(f_{\mathscr{Z}_n,V_k}\right) - g\|_{L_2}^2 \, \mathrm{d}\rho^n + \int_{\Omega_-^n} 4r^2 \, \mathrm{d}\rho^n \\
&\le \int_{\Omega_+^n} \|\tau_r\left(f_{\mathscr{Z}_n,V_k}\right) - g\|_{L_2}^2 \, \mathrm{d}\rho^n + 8r^2 n^{-\theta} \\
&\le \int_{\Omega_+^n} \|f_{\mathscr{Z}_n,V_k} - g\|_{L_2}^2 \, \mathrm{d}\rho^n + 8r^2 n^{-\theta},
\end{aligned} \tag{9}$$

where the last inequality holds since τ_r is non-expansive and $g = \tau_r(g)$ holds almost everywhere.

Next, we define the projection $P_{V_k}^n$ onto V_k by

$$P_{V_k}^n(f) := \operatorname*{arg\,min}_{h \in V_k} \frac{1}{n} \sum_{i=1}^n (h(\mathbf{t}_i) - f(\mathbf{t}_i))^2,$$

which is well-defined for point-evaluable functions f on Ω_+^n since the coefficients of $P_{V_k}^n(f)$ are given by (3) if we substitute the vector $\mathbf{x}$ by $(f(\mathbf{t}_1), \ldots, f(\mathbf{t}_n))^T$. Note that the coefficients of $f_{\mathscr{Z}_n,V_k}$ are given by $P_{V_k}^n(g + \varepsilon)$. Furthermore, we need the (standard) orthogonal L_2 projector P_{V_k} onto V_k. Obviously, it holds $P_{V_k}^n \circ P_{V_k} = P_{V_k}$. Therefore, we have

$$\begin{aligned}
\|f_{\mathscr{Z}_n,V_k} - g\|_{L_2}^2 &= \|P_{V_k}^n(g + \varepsilon) - P_{V_k}^n \circ P_{V_k}(g) + P_{V_k}(g) - g\|_{L_2}^2 \\
&= \|P_{V_k}^n\left(g - P_{V_k}(g)\right) + P_{V_k}^n(\varepsilon)\|_{L_2}^2 + \|g - P_{V_k}(g)\|_{L_2}^2 \\
&\le 2\|P_{V_k}^n\left(g - P_{V_k}(g)\right)\|_{L_2}^2 + 2\|P_{V_k}^n(\varepsilon)\|_{L_2}^2 + \|g - P_{V_k}(g)\|_{L_2}^2
\end{aligned} \tag{10}$$

since $Id - P_{V_k}$ is L_2-orthogonal on V_k. To bound (9) from above, we will now deal with each of the three summands in (10) separately.

First, note that $P^n_{V_k}\left(g - P_{V_k}(g)\right) = \sum_{i=1}^{N_k} \beta_i \nu_i$ with $\boldsymbol{\beta} = (\beta_1, \ldots, \beta_{N_k})^T$ given by $\boldsymbol{\beta} = \left(nBB^T\right)^{-1} \boldsymbol{\xi}$ with $\boldsymbol{\xi} = B\mathbf{a}$ and $a_j = g(\mathbf{t}_j) - P_{V_k}(g)(\mathbf{t}_j)$ for $j = 1, \ldots, n$. Thus, we have

$$
\begin{aligned}
\| P^n_{V_k}\left(g - P_{V_k}(g)\right) \|^2_{L_2} &= \boldsymbol{\beta}^T M \boldsymbol{\beta} = \boldsymbol{\xi}^T \left(nBB^T\right)^{-1} M \left(nBB^T\right)^{-1} \boldsymbol{\xi} \\
&\leq \lambda_{\max}(M) \frac{1}{\lambda_{\min}\left(nBB^T\right)^2} \|\boldsymbol{\xi}\|^2_{\ell_2} \leq \frac{4\lambda_{\max}(M)}{\lambda_{\min}(M)^2} \|\boldsymbol{\xi}\|^2_{\ell_2}
\end{aligned}
\tag{11}
$$

on Ω^n_+, on which nBB^T is invertible. This yields

$$
\int_{\Omega^n_+} 2\| P^n_{V_k}\left(g - P_{V_k}(g)\right) \|^2_{L_2} \, \mathrm{d}\rho^n \leq \frac{8\lambda_{\max}(M)}{\lambda_{\min}(M)^2} \mathbb{E}\left[\|\boldsymbol{\xi}\|^2_{\ell_2}\right]. \tag{12}
$$

With the independence of $\mathbf{t}_1, \ldots, \mathbf{t}_n$, we deduce

$$
\begin{aligned}
\mathbb{E}\left[\|\boldsymbol{\xi}\|^2_{\ell_2}\right] &= \int_{\Omega^n} \sum_{j=1}^{N_k} \left(\frac{1}{n}\sum_{i=1}^{n} \nu_j(\mathbf{t}_i) \cdot (g - P_{V_k}(g))(\mathbf{t}_i)\right)^2 \mathrm{d}\rho^n(\mathbf{t}_1, \ldots, \mathbf{t}_n) \\
&= \frac{1}{n^2} \sum_{j=1}^{N_k} (n^2 - n) \left(\underbrace{\int_{\Omega} \nu_j(\mathbf{t}) \cdot (g - P_{V_k}(g))(\mathbf{t}) \, \mathrm{d}\rho(\mathbf{t})}_{=0}\right)^2 \\
&\quad + \frac{1}{n^2} \sum_{j=1}^{N_k} n \int_{\Omega} \left(\nu_j(\mathbf{t}) \cdot (g - P_{V_k}(g))(\mathbf{t})\right)^2 \mathrm{d}\rho(\mathbf{t}) \\
&\overset{(6)}{\leq} \frac{1}{n} S(\nu_1, \ldots, \nu_{N_k}) \|g - P_{V_k}(g)\|^2_{L_2} \overset{(7)}{\leq} \frac{c\lambda_{\min}(M)}{(1+\theta)\log(n)} \|g - P_{V_k}(g)\|^2_{L_2}.
\end{aligned}
$$

Applying this to (12), we finally obtain

$$
\int_{\Omega^n_+} 2\| P^n_{V_k}\left(g - P_{V_k}(g)\right) \|^2_{L_2} \, \mathrm{d}\rho^n \leq \frac{8c\lambda_{\max}(M)}{(1+\theta)\lambda_{\min}(M)\log(n)} \|g - P_{V_k}(g)\|^2_{L_2}. \tag{13}
$$

For the second summand of (10), we proceed similarly. Note that $\boldsymbol{\vartheta} = \left(nBB^T\right)^{-1} \boldsymbol{\eta}$ are the coefficients of $P^n_{V_k}(\varepsilon)$ with respect to $\nu_1, \ldots, \nu_{N_k}$. Here, $\boldsymbol{\eta} = B\mathbf{b}$ with $b_i = \varepsilon(\mathbf{t}_i)$. Analogously to (11), we get

$$
\| P^n_{V_k}(\varepsilon) \|^2_{L_2} \leq \frac{4\lambda_{\max}(M)}{\lambda_{\min}(M)^2} \|\boldsymbol{\eta}\|^2_{\ell_2}
$$

on Ω_+^n. Therefore, it remains to estimate

$$\int_{\Omega_+^n} 2\|P_{V_k}^n(\varepsilon)\|_{L_2}^2 \,\mathrm{d}\rho^n \le \frac{8\lambda_{\max}(M)}{\lambda_{\min}(M)^2}\mathbb{E}\left[\|\boldsymbol{\eta}\|_{\ell_2}^2\right]. \tag{14}$$

Because of (1) we have $\mathbb{E}_\rho[\varepsilon\nu_j] = 0$ for all $j \in 1, \ldots, N_k$. Thus, we obtain

$$\begin{aligned}
\mathbb{E}_{\rho^n}\left[\|\boldsymbol{\eta}\|_{\ell_2}^2\right] &= \int_{\Omega^n} \sum_{j=1}^{N_k}\left(\frac{1}{n}\sum_{i=1}^{n} \nu_j(\mathbf{t}_i)\cdot\varepsilon(\mathbf{t}_i)\right)^2 \mathrm{d}\rho^n(\mathbf{t}_1,\ldots,\mathbf{t}_n)\\
&= \frac{1}{n^2}\sum_{j=1}^{N_k}(n^2-n)\left(\underbrace{\mathbb{E}_\rho\left[\varepsilon\nu_j\right]}_{=0}\right)^2 + \frac{1}{n^2}\sum_{j=1}^{N_k} n\mathbb{E}_\rho\left[\varepsilon^2\nu_j^2\right]\\
&= \frac{1}{n}\sum_{j=1}^{N_k}\int_\Omega \nu_j(\mathbf{t})^2\mathbb{E}_\rho[\varepsilon^2 \mid \mathbf{t}]\mathrm{d}\rho(\mathbf{t}) \overset{(1)}{\le} \frac{\sigma^2}{n}\sum_{j=1}^{N_k}\int_\Omega \nu_j(\mathbf{t})^2\mathrm{d}\rho(\mathbf{t})\\
&\le \frac{\sigma^2}{n}\sum_{j=1}^{N_k}\lambda_{\max}(M) = \frac{N_k\sigma^2}{n}\lambda_{\max}(M).
\end{aligned}$$

Plugging this into (14), we get

$$\int_{\Omega_+^n} 2\|P_{V_k}^n(\varepsilon)\|_{L_2}^2 \,\mathrm{d}\rho^n \le \frac{8\sigma^2\lambda_{\max}(M)^2}{\lambda_{\min}(M)^2}\cdot\frac{N_k}{n}. \tag{15}$$

Since the third summand of (10) is independent of the samples, we have

$$\int_{\Omega_+^n} \|g - P_{V_k}(g)\|_{L_2}^2 \le \|g - P_{V_k}(g)\|_{L_2}^2 = \inf_{f\in V_k}\|g - f\|_{L_2}^2.$$

Finally, we combine this estimate together with (13) and (15) into (9) and (10), which completes the proof. □

The first term of the expected rate from Theorem 3 depends mainly on the best approximation error in V_k and the quotient $\frac{\lambda_{\max}(M)}{\lambda_{\min}(M)}$, which can be bounded from above independently from k for Riesz bases for example. The second summand scales like $n^{-\theta}$, which resembles the decay of the error with respect to the amount of data n in the noiseless case, i.e. when $\sigma^2 = 0$ and the third summand vanishes. In the noisy case, the third summand is also present and the best possible decay rate with respect to n scales like n^{-1}.

4.2 Application to Sparse Grids

In the following, we assume that the measure ρ is the m-dimensional Lebesgue measure on $\Omega = [0,1]^m$, i.e. the data $\mathbf{t}_i, i = 1, \ldots, n$ are distributed uniformly in Ω. We now apply Theorems 2 and 3 to the regression problem on sparse grid spaces $V_k = \mathcal{V}_k^{\text{sparse}}$ and need to bound

$$S(\nu_1, \ldots, \nu_{N_k}) = \sup_{\mathbf{t}\in\Omega} \sum_{\zeta_m(\mathbf{l})\le k} \sum_{\mathbf{i}\in\mathbf{I}_\mathbf{l}} \gamma_{\mathbf{l},\mathbf{i}}(\mathbf{t})^2.$$

from above. To this end, we provide the following lemma.

Lemma 1 *For each* $\mathbf{l} \in \mathbb{N}^m$*, it holds*

$$\max_{\mathbf{t}\in[0,1]^m} \sum_{\mathbf{i}\in\mathbf{I}_\mathbf{l}} \gamma_{\mathbf{l},\mathbf{i}}(\mathbf{t})^2 \le 2^{|\mathbf{l}|_{\ell_1}} \cdot 2^{|\{j\in\{1,\ldots,m\}|\,l_j=0\}|} \cdot \left(\frac{36}{25}\right)^{|\{j\in\{1,\ldots,m\}|\,l_j>0\}|}. \tag{16}$$

Proof We first consider the univariate case $m = 1$ and define $S_l(t) := \sum_{i\in I_l} \gamma_{l,i}(t)^2$. For $l = 0$, we obtain

$$S_0(t) = \gamma_{0,0}^2(t) + \gamma_{0,1}^2(t) = 1 + t^2 \le 2$$

and for $l = 1$ we have

$$S_1(t) = \gamma_{1,1}^2(t) = (2\phi_{1,1}(t) - 1)^2 \le 1$$

with $t \in [0, 1]$. In the general case $l \ge 2$, S_l is a sum of the piecewise quadratic polynomials $\gamma_{l,i}^2(\cdot)$ with $i \in I_l$. Therefore, the quadratic term of the piecewise quadratic polynomial $S_l(\cdot)$ has a positive coefficient everywhere and the maximum of S_l over $[0, 1]$ can only reside on one of the grid points $2^{-l}i$ with $i = 0, \ldots, 2^l$. This is also illustrated in Fig. 3, where S_4 is plotted exemplarily.

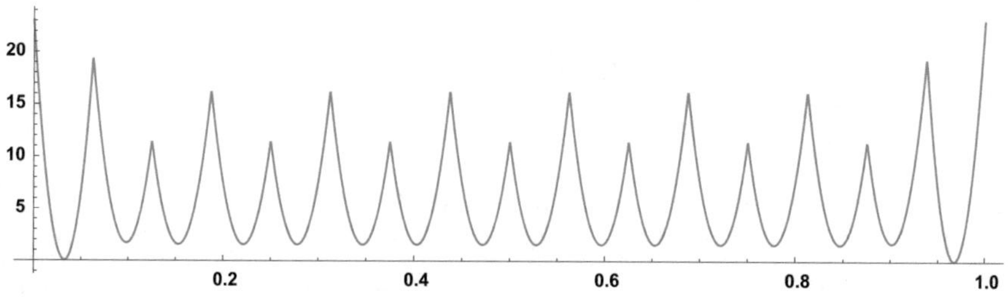

Fig. 3 The squared sum S_4 of the univariate prewavelet basis functions for $k = 4$

We now prove that the maximum of S_l is always attained at the boundary point $t = 1$. For $l = 0$ and $l = 1$, this is immediately clear. The (local) maxima of S_2 are denoted below in a mask-type notation which contains a prefactor 2^l and the nodal values at the grid points. The calculation

$$\begin{aligned} S_2(t) &= \gamma_{2,1}^2(t) + \gamma_{2,3}^2(t) \\ &= 4\,[\ \tfrac{36}{25}\ \tfrac{121}{100}\ \tfrac{9}{25}\ \tfrac{1}{100}\ 0\ ] \\ &+ 4\,[\ 0\ \tfrac{1}{100}\ \tfrac{9}{25}\ \tfrac{121}{100}\ \tfrac{36}{25}\ ] \\ &= 4\,[\ \tfrac{36}{25}\ \tfrac{61}{50}\ \tfrac{18}{25}\ \tfrac{61}{50}\ \tfrac{36}{25}\ ] \end{aligned}$$

shows that the largest value $4 \cdot \frac{36}{25}$ is attained at the boundary grid points. Analogously, we have

$$\begin{aligned} S_3(t) &= \gamma_{3,1}^2(t) + \gamma_{3,3}^2(t) + \gamma_{3,5}^2(t) + \gamma_{3,7}^2(t) \\ &= 8\,[\ \tfrac{36}{25}\ \tfrac{121}{100}\ \tfrac{9}{25}\ \tfrac{1}{100}\ 0\ 0\ 0\ 0\ 0\ ] \\ &+ 8\,[\ 0\ \tfrac{1}{100}\ \tfrac{9}{25}\ 1\ \tfrac{9}{25}\ \tfrac{1}{100}\ 0\ 0\ 0\ ] \\ &+ 8\,[\ 0\ 0\ 0\ \tfrac{1}{100}\ \tfrac{9}{25}\ 1\ \tfrac{9}{25}\ \tfrac{1}{100}\ 0\ ] \\ &+ 8\,[\ 0\ 0\ 0\ 0\ 0\ \tfrac{1}{100}\ \tfrac{9}{25}\ \tfrac{121}{100}\ \tfrac{36}{25}\ ] \\ &= 8\,[\ \tfrac{36}{25}\ \tfrac{61}{50}\ \tfrac{18}{25}\ \tfrac{51}{50}\ \tfrac{18}{25}\ \tfrac{51}{50}\ \tfrac{18}{25}\ \tfrac{61}{50}\ \tfrac{36}{25}\ ] \end{aligned}$$

for $l = 3$. Due to the local support of the basis functions, analogous calculations show that the value of S_l never exceeds $2^l \cdot \frac{36}{25}$ also for higher levels l. Therefore, the maximum of S_l is always attained for $t = 1$. If $l = 0$, the maximum value is 2 and if $l \geq 2$, it is $2^l \cdot \frac{36}{25}$. For the special case $l = 1$, we use the crude estimate $S_1(1) = 1 < 2 \cdot \frac{36}{25}$. Therefore, the assertion (16) is proven for $m = 1$.

The case $m > 1$ follows directly from the tensor product construction of the basis. To see this, let $\mathbf{t} \in [0, 1]^m$ and $\mathbf{l} \in \mathbb{N}^m$ be arbitrary. It holds

$$\sum_{\mathbf{i} \in \mathbf{I_l}} \gamma_{\mathbf{l},\mathbf{i}}(\mathbf{t})^2 = \sum_{(i_1,\ldots,i_m) \in \mathbf{I_l}} \prod_{j=1}^{m} \gamma_{l_j,i_j}(t_j)^2 = \prod_{j=1}^{m} \sum_{i_j \in I_{l_j}} \gamma_{l_j,i_j}(t_j)^2$$

due to the structure of $\mathbf{I_l}$. Therefore, the maximization of the term on the left can be split into the maximization of S_{l_j} for each direction $j \in \{1, \ldots, m\}$. Since the maximum is bounded by 2 for directions j with $l_j = 0$ and by $2^{l_j} \cdot \frac{36}{25}$ for directions j with $l_j \geq 1$, the inequality (16) follows. □

We are now able to present an upper bound on $S(\nu_1, \ldots, \nu_{N_k})$ for sparse grids.

Theorem 4 *For $V_k = \mathscr{V}_k^{sparse}$, $S(\nu_1, \ldots, \nu_{N_k})$ can be bounded by*

$$S(\nu_1, \ldots, \nu_{N_k}) \leq \left(\frac{72}{25}\right)^m (N_k + 1). \tag{17}$$

Proof In the following, we write $Z(\mathbf{l}) := |\{j \in \{1, \ldots, m\} \mid l_j = 0\}|$ for the number of zeros of a multiindex $\mathbf{l} \in \mathbb{N}^m$. Applying Lemma 1, we obtain

$$S(\nu_1, \ldots, \nu_{N_k}) \leq \sum_{|\mathbf{l}|_{\ell_1} + Z(\mathbf{l}) \leq k+m-1} 2^{|\mathbf{l}|_{\ell_1} + Z(\mathbf{l})} \cdot \left(\frac{36}{25}\right)^{m - Z(\mathbf{l})},$$

where we used $\zeta_m(\mathbf{l}) = |\mathbf{l}|_{\ell_1} - m + Z(\mathbf{l}) + 1$. Substituting $i = |\mathbf{l}|_{\ell_1} + Z(\mathbf{l})$, this becomes

$$S(\nu_1, \ldots, \nu_{N_k}) \leq \sum_{i=0}^{k+m-1} 2^i \cdot \sum_{l=0}^{m} |\{\mathbf{l} \in \mathbb{N}^m \mid |\mathbf{l}|_{\ell_1} = i - l \text{ and } Z(\mathbf{l}) = l\}| \cdot \left(\frac{36}{25}\right)^{m-l}.$$

Obviously, it holds $|\{\mathbf{l} \in \mathbb{N}^m \mid |\mathbf{l}|_{\ell_1} = i - l \text{ and } Z(\mathbf{l}) = l\}| = 0$ for all $l = 0, \ldots, m$ if $i < m$. Therefore, we can begin the summation over i from m. If $i \geq m$ holds, a simple combinatorial argument, see also [4], leads to

$$\begin{aligned} |\{\mathbf{l} \in \mathbb{N}^m \mid |\mathbf{l}|_{\ell_1} = i - l \text{ and } Z(\mathbf{l}) = l\}| &= |\{\mathbf{l} \in (\mathbb{N} \setminus \{0\})^{m-l} \mid |\mathbf{l}|_{\ell_1} = i - l\}| \cdot \binom{m}{l} \\ &= \binom{i-l-1}{m-l-1} \binom{m}{l} \end{aligned}$$

for arbitrary $l = 0, \ldots, m-1$ and furthermore

$$|\{\mathbf{l} \in \mathbb{N}^m \mid |\mathbf{l}|_{\ell_1} = i - m \text{ and } Z(\mathbf{l}) = m\}| = \begin{cases} 1 & \text{if } i = m \\ 0 & \text{else} \end{cases} = \delta_{im}.$$

Therefore, we have

$$\begin{aligned} S(\nu_1, \ldots, \nu_{N_k}) &\leq \sum_{i=m}^{k+m-1} 2^i \cdot \left(\delta_{im} + \sum_{l=0}^{m-1} \binom{i-l-1}{m-l-1} \binom{m}{l} \left(\frac{36}{25}\right)^{m-l}\right) \\ &= 2^m \cdot \sum_{i=0}^{k-1} 2^i \cdot \left(\delta_{i0} + \sum_{l=0}^{m-1} \binom{i+m-l-1}{m-l-1} \binom{m}{l} \left(\frac{36}{25}\right)^{m-l}\right) \\ &= 2^m \cdot \left(1 + \sum_{l=0}^{m-1} \left(\frac{36}{25}\right)^{m-l} \binom{m}{l} \left(\sum_{i=0}^{k-1} 2^i \binom{i+m-l-1}{m-l-1}\right)\right) \\ &= 2^m + 2^m \sum_{l=0}^{m-1} \left(\frac{36}{25}\right)^{m-l} \binom{m}{l} |G_k^{m-l}|, \end{aligned}$$

where $|G_k^{m-l}|$ denotes the size of an $m-l$-dimensional level-k sparse grid without boundary, see lemma 3.6 of [4] for a proof. To derive a bound with respect to the number of grid points N_k in a sparse grid with boundary points of level k in dimension m, we rewrite the above inequality by

$$
\begin{aligned}
S(\nu_1, \dots, \nu_{N_k}) &\leq 2^m + \sum_{l=0}^{m-1} \left(2 \cdot \frac{36}{25}\right)^{m-l} \cdot 2^l \binom{m}{l} |G_k^{m-l}| \\
&\leq 2^m + \left(\frac{72}{25}\right)^m \cdot \sum_{l=0}^{m-1} 2^l \binom{m}{l} |G_k^{m-l}| = 2^m + \left(\frac{72}{25}\right)^m N_k,
\end{aligned}
$$

where the last equality is proven in lemma 2.1.2 of [8]. Since $2 < \frac{72}{25} = 2.88$, this completes the proof. □

Combining the statements of Theorems 2 and 4, we see that the sparse grid regression problem is well-posed with probability larger than $1 - 2n^{-\theta}$ if

$$
\left(\frac{72}{25}\right)^m (N_k + 1) \leq c \frac{\lambda_{\min}(M)}{1+\theta} \cdot \frac{n}{\log(n)}. \tag{18}
$$

Since the prewavelet basis of V_k is a Riesz frame with respect to the $L_{2,\rho}(\Omega)$ norm, the fraction $\frac{\lambda_{\max}(M)}{\lambda_{\min}(M)}$ is bounded from above independently of the level $k \in \mathbb{N}$. Therefore, the necessary scaling is essentially

$$
N_k \simeq 2^k k^{m-1} \lesssim \frac{n}{\log(n)},
$$

where the $\lesssim$ notation implies an m- and θ-dependent constant. The following corollary states our main result for sparse grids. There we deal with the (Bessel-potential) Sobolev spaces $H^s_{\rho,\mathrm{mix}}(\Omega)$ of dominating mixed smoothness with respect to the $L_{2,\rho}(\Omega)$ measure, see e.g. [2, 16].

Corollary 1 (Regression Error for Sparse Grids) *Let $g \in H^s_{\rho,mix}(\Omega)$ for some $0 < s \leq 2$ and let $V_k = \mathscr{V}_k^{sparse}$. Let, furthermore,* (18) *hold for an arbitrary $\theta > 0$. Then, the regression problem is well-posed in the sense of Theorem 2 with probability at least $1 - 2n^{-\theta}$ and the expected error fulfills*

$$
\mathbb{E}\left[\|\tau_r(f_{\mathscr{Z}_n, V_k}) - g\|^2_{L_{2,\rho}(\Omega)}\right] \leq C_{m,s,\theta,\sigma} \left(2^{-2sk} k^{m-1} + \frac{1}{n^\theta} + \frac{2^k k^{m-1}}{n}\right) \tag{19}
$$

with a constant $C_{m,s,\theta,\sigma}$, which depends on m, s, θ, σ and $\|g\|_{H^s_{\rho,mix}(\Omega)}$.

Proof To prove the expected error, we combine Theorems 3 and 4 and use that the squared best approximation error behaves like

$$\inf_{f \in \mathscr{V}_k^{\text{sparse}}} \|f - g\|^2_{L_{2,\rho}(\Omega)} \leq C_{m,s} 2^{-2sk} k^{m-1} \|g\|^2_{H^s_{\rho,\text{mix}}(\Omega)}$$

for $g \in H^s_{\rho,\text{mix}}(\Omega)$ with an m- and s-dependent constant $C_{m,s}$, see e.g. theorem 3.25 of [2]. Furthermore, $\frac{\lambda_{\max}(M)}{\lambda_{\min}(M)}$ is bounded from above independently of k since the prewavelet basis is a Riesz frame with respect to the $L_{2,\rho}(\Omega)$ norm. Together with the fact that $N_k \leq C_m 2^k k^{m-1}$ holds for an m-dependent constant C_m, see e.g. [8], the statement of the corollary follows immediately. □

Finally, we can ask for the optimal coupling between the number of samples n and the number of sparse grid basis functions N_k, which achieves the best possible convergence rate in the sense that the terms in the error estimate (19) are (approximately) balanced. The resulting coupling is stated in the following corollary.

Corollary 2 (Optimal Coupling and Convergence Rate for Sparse Grids) *Let $g \in H^s_{\rho,mix}(\Omega)$ for some $0 < s \leq 2$ and let $V_k = \mathscr{V}_k^{sparse}$. Then, the following holds:*

1. *Let $\sigma^2 > 0$ (noisy case) and let* (18) *hold for a $\theta \geq \frac{2s}{2s+1}$. Then, the asymptotically optimal coupling between n and N_k is*

$$N_k \sim n^{\frac{1}{2s+1}} \log(n)^{m-1} \tag{20}$$

and the resulting convergence rate for $n \to \infty$ is

$$\mathbb{E}\left[\|\tau_r\left(f_{\mathscr{Z}_n,V_k}\right) - g\|^2_{L_{2,\rho}(\Omega)}\right] = \mathscr{O}\left(n^{-\frac{2s}{2s+1}} \log(n)^{m-1}\right). \tag{21}$$

2. *Let $\sigma^2 = 0$ (noiseless case) and let* (18) *hold for a $\theta > 2s$. Then, the asymptotically optimal coupling between n and N_k is*

$$N_k \sim \frac{n}{\log(n)} \tag{22}$$

and the resulting convergence rate for $n \to \infty$ is

$$\mathbb{E}\left[\|\tau_r\left(f_{\mathscr{Z}_n,V_k}\right) - g\|^2_{L_{2,\rho}(\Omega)}\right] = \mathscr{O}\left(n^{-2s} \log(n)^{(2s+1)m-1}\right). \tag{23}$$

Proof Let $E := \mathbb{E}\left[\|\tau_r\left(f_{\mathscr{Z}_n,V_k}\right) - g\|^2_{L_{2,\rho}(\Omega)}\right]$. Note that $N_k \sim 2^k k^{m-1}$ in the sense that there exist two constants $c_1, c_2 > 0$ such that $c_1 2^k k^{m-1} \leq N_k \leq c_2 2^k k^{m-1}$ holds independently of k. Note, furthermore, that there exists a constants $C_1, C_2 > 0$

such that $C_1 \log(n) \leq k \leq C_2 \log(n)$ for $n \geq 2$ for each of the scalings (20) and (22). This can easily be obtained by taking the logarithm on both sides of (20) and (22).

We begin with the proof for the noisy case $\sigma^2 > 0$ and insert the coupling (20) into the error formula (19). Since we will see that this balances the first and third summands there, the coupling is also optimal. Indeed, we have

$$\begin{aligned}
E &\lesssim 2^{-2sk}k^{m-1} + \frac{1}{n^\theta} + \frac{2^k k^{m-1}}{n} \lesssim (N_k)^{-2s} k^{(m-1)(2s+1)} + n^{-\theta} + \frac{N_k}{n} \\
&\lesssim \left(n^{\frac{1}{2s+1}} \log(n)^{m-1}\right)^{-2s} \log(n)^{(m-1)(2s+1)} + n^{-\theta} + \frac{n^{\frac{1}{2s+1}} \log(n)^{m-1}}{n} \\
&\overset{\theta \geq \frac{2s}{2s+1}}{\lesssim} n^{-\frac{2s}{2s+1}} \left(\log(n)^{m-1} + 1 + \log(n)^{m-1}\right) = \mathcal{O}\left(n^{-\frac{2s}{2s+1}} \log(n)^{m-1}\right). \qquad (24)
\end{aligned}$$

As we see in (24), the first and third summand of the error estimate (19) are balanced for the coupling (20). Note that the coupling is valid in the sense that it (asymptotically) fulfills condition (18). This completes the proof for the noisy case.

In the noiseless case $\sigma^2 = 0$, the third summand in (19) vanishes, see also Theorem 3. Therefore, for $\theta = 2s + \delta$ with some arbitrary $\delta > 0$, the number of basis functions N_k needs to be chosen as large as possible (with respect to n) to achieve the fastest possible convergence of the first summand of (19). This is achieved by choosing n as the smallest integer such that (18) is still fulfilled, i.e. the corresponding scaling is (22). Therefore, we obtain

$$\begin{aligned}
&E \lesssim 2^{-2sk}k^{m-1} + \frac{1}{n^\theta} \lesssim (N_k)^{-2s} k^{(m-1)(2s+1)} + n^{-\theta} \\
&\lesssim \left(\frac{n}{\log(n)}\right)^{-2s} \log(n)^{(m-1)(2s+1)} + n^{-\theta} \overset{\theta=2s+\delta}{\lesssim} n^{-2s}\left(\log(n)^{(2s+1)m-1} + n^{-\delta}\right) \\
&= \mathcal{O}\left(n^{-2s} \log(n)^{(2s+1)m-1}\right),
\end{aligned}$$

which concludes the proof. □

For all of our proven convergence results, we see that the curse of dimensionality appears only in terms which scale logarithmically in the number of samples n. This is the well-known sparse grid effect, which we are used to when considering the spaces $\mathcal{V}_k^{\text{sparse}}$ for interpolation or approximation for instance, see [4].

As we see from Corollary 2, the optimal main rate that can be achieved in the noisy case is $n^{-\frac{2s}{2s+1}}$, which becomes $n^{-\frac{4}{5}}$ in the smoothest setting ($s = 2$) that the piecewise linear basis functions can exploit.[1] This comes at an expense of

[1]For higher order spline bases, a larger choice of s can be exploited here. However, one needs to prove an analogous result to Theorem 4 for the corresponding basis functions first.

oversampling by $n \sim N_k^{2s+1}$ if we neglect the logarithm. In the noiseless case, however, the much better main rate n^{-2s} can be achieved and there is only a logarithmic oversampling, see (22). This oversampling has to be present to fulfill the necessary condition (18) anyway.

Finally, note that our stability and error analysis for sparse grids heavily relies on the fact that we are dealing with a Riesz basis. Nevertheless, if we choose a basis for which $\frac{\lambda_{\max}(M)}{\lambda_{\min}(M)}$ is unbounded, e.g. the hierarchical hat basis built from $\phi_{l,i}$, see (4), we can still obtain well-posedness of the regression problem if an appropriate regularization term is added to (2), see also [2]. However, then it is not directly clear how to derive a variant of Theorem 3 for the regularized case.

5 Numerical Experiments

In this section, we have a look at numerical experiments, which illustrate our theoretical results from the previous section. To this end, we choose $\Omega = [0, 1]^2$, $V_k = \mathscr{V}_k^{\text{sparse}}$ and $\rho = \lambda_{[0,1]^2}$ as the two-dimensional Lebesgue measure. We use the example function $g : [0, 1]^2 \to \mathbb{R}$ given by

$$g(t_1, t_2) = \exp(-t_1^2 - t_2^2) + t_1 t_2. \tag{25}$$

Since g is infinitely smooth, we have $g \in H^2_{\rho,\text{mix}}((0, 1)^2)$ and we can expect our results from the previous section to hold with smoothness index $s = 2$. We now discern two cases: The noiseless case, in which our samples are given as $\mathscr{Z}_n = (\mathbf{t}_i, g(\mathbf{t}_i))_{i=1}^n$, and the noisy case, where we deal with $\mathscr{Z}_n = (\mathbf{t}_i, g(\mathbf{t}_i) + \varepsilon_i)_{i=1}^n$ and the ε_i are independent instances of a normally distributed random variable $\varepsilon \sim \mathscr{N}(0, 0.01)$.

Since $\|g\|_{L_{\infty,\rho}([0,1]^2)} < 2$ and $\mathbb{P}[|\varepsilon| > 1] < 10^{-2000}$, we can safely assume that $r = 3$ is large enough to assure that (with probability almost 1) $|g(\mathbf{t}_i) + \varepsilon_i| < r$ holds for each $i =, 1 \ldots, n$. Therefore, $\tau_r\left(f_{\mathscr{Z}_n,V_k}\right) = f_{\mathscr{Z}_n,V_k}$ since $f_{\mathscr{Z}_n,V_k}$ is the optimal piecewise linear regression function in V_k and, thus, cannot be larger than $\max_{i=1,\ldots,n} |g(\mathbf{t}_i) + \varepsilon_i|$ anywhere. Therefore, we can apply Theorem 3 and Corollaries 1 and 2 for $f_{\mathscr{Z}_n,V_k}$ instead of $\tau_r(f_{\mathscr{Z}_n,V_k})$ in our setting.

Since the prewavelet basis is a Riesz frame, we know that $\frac{\lambda_{\max}(M)}{\lambda_{\min}(M)}$ is bounded independently of k. To see that this quotient is not severely large, we exemplarily calculated it for $k = 1, \ldots, 8$ and observed that it does not exceed 5 in the two-dimensional case.

5.1 Error Decay

First, we compute the error for different pairs of grid levels k and numbers of data points n. Since our result on the regression error in Corollary 1 is only given in expectation, we compute the average AvErr of the error $\|f_{\mathscr{Z}_n,V_k} - g\|^2_{L_{2,\rho}(\Omega)}$ over 10 independent runs with different input data sets for each parameter pair (k,n). To compute the error values, we interpolated both $f_{\mathscr{Z}_n,V_k}$ and g on a full tensor-product grid of level 11, i.e. we interpolated in $\mathscr{V}_{11}^{\text{full}}$, and computed the norm of the difference there. The results can be found in Fig. 4.

We directly observe the expected error decay rates, i.e. $2^{-4k} \cdot k$ for fixed n and n^{-1} for fixed k in the noisy setting (if we tacitly assume $\theta \geq 1$), see also Corollary 1. For fixed k, we would expect the error to behave like $n^{-\theta}$ in the noiseless setting. However, since θ grows when the quotient $\frac{n}{k}$ grows, we cannot expect the error behavior to be of type n^{-p} for some p. For both, the noisy and the noiseless case, we observe that if the varying parameter (e.g. n) is too large, the error is saturated and the other parameter (e.g. k) has to be increased to guarantee a further error reduction. Note that the error for fixed n in the noisy regression setting even increases for large k. This is an overfitting effect, i.e. the basis size N_k is too large for the corresponding number of data n. Since there is no regularization in our approach, the error thus grows for large k and small n.

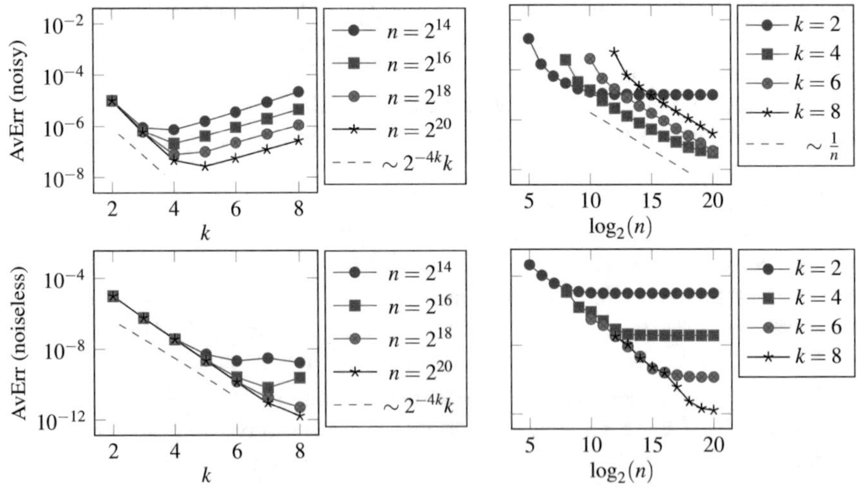

Fig. 4 The average of $\|f_{\mathscr{Z}_n,V_k} - g\|^2_{L_{2,\rho}(\Omega)}$ over 10 runs for several parameter pairs (k,n) for which $N_k \leq n$ holds. Top: Noisy data, Bottom: Noiseless data. Left: Each line represents a fixed n, Right: Each line represents a fixed k

5.2 Balancing the Error

In a next step, we balance the error terms according to Corollary 2 and inspect the resulting convergence rates. For the noisy setting, we have for $\theta \geq \frac{4}{5}$ that the optimal coupling is given by

$$N_k \sim n^{\frac{1}{5}} \log(n).$$

We, therefore, (approximately) solve $N_k^5 = n\log(n)^5$ for n and determine the optimal number of data points for $k = 1, \ldots, 6$. For $k = 6$, the amount n of data points already exceeds 2^{25}. In the noiseless setting, the picture is quite different. Here, the optimal coupling is given by

$$N_k \sim \frac{n}{\log(n)}$$

if $\theta > 4$. More accurately, we look for the smallest n such that (18) is fulfilled with $\theta > 4$. Therefore, we equate both sides of (18) and (approximately) solve for n. Here, we set $\theta = 4$ and $\lambda_{\min}(M) = 1$ and obtain that sampling by $\frac{n}{\log(n)} = 384 \cdot (N_k + 1)$ suffices to fulfill (18). The average errors (over 10 runs) for the optimal coupling in the noisy and in the noiseless setting can be found in Fig. 5.

We directly see that the convergence rate in the experimental results asymptotically matches the proven rates from Corollary 2, i.e. $n^{-\frac{2s}{2s+1}} \log(n)^{m-1} = n^{-\frac{4}{5}} \log(n)$ in the noisy case and $n^{-2s} \log(n)^{(2s+1)m-1} = n^{-4} \log(n)^9$ in the noiseless case. Furthermore, we observe that the initial error decay for noisy data

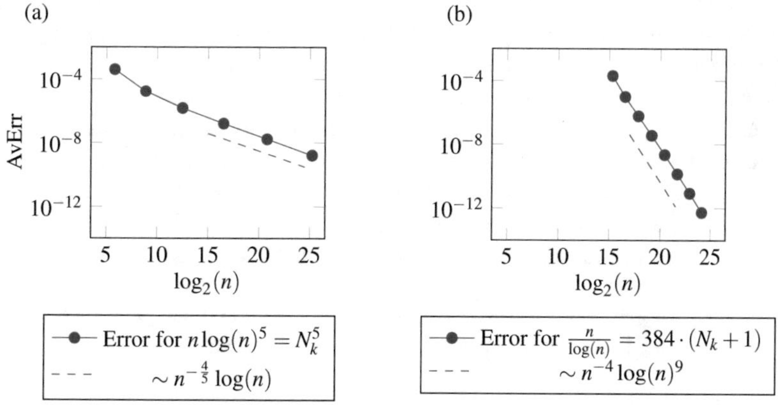

Fig. 5 The average of $\|f_{\mathscr{X}_n, V_k} - g\|^2_{L_{2,\rho}(\Omega)}$ over 10 runs for the optimal coupling between k and n. Left: Noisy data with coupling $n\log(n)^5 = N_k^5$, Right: Noiseless data with coupling $\frac{n}{\log(n)} = 384 \cdot (N_k + 1)$, which resembles (18) for our example. (**a**) Noisy data. (**b**) Noiseless data

is better than the convergence rate suggests. This is due to the fact that the noise effects the convergence behavior only if the overall error is already smaller than a certain (noise) level. Note also that the oversampling factor 384 is the reason why we already have more than 2^{15} data points for the smallest level $k = 1$ in the noiseless case. However, since our sampling resembles only a sufficient condition to ensure well-posedness of the regression problem with high probability, a much smaller oversampling constant might also do the job for practical applications.

6 Conclusion

In this article we presented error bounds, stability results and optimal parameter couplings for the least-squares regression problem and applied them to the sparse grid setting. To this end, we extended the results of [7] to arbitrary bases and provided an upper bound for the crucial quantity $S(\nu_1, \ldots, \nu_{N_k})$ from the stability and convergence estimates. Our results showed that the sparse grid prewavelet basis behaves (up to constants) like an orthonormal basis in the regression estimates because of its Riesz property. Therefore, it is a good choice for regression problems on sparse grid spaces since it employs both beneficial convergence behavior and small support of the corresponding basis functions, which is directly connected to the availability of cost-efficient linear equation system solvers, see e.g. [3, 5]. Finally, we presented a numerical example to illustrate that our results are not only of theoretical interest but resemble the true convergence behavior of actual sparse grid regression algorithms.

An interesting question which still has to be answered is if the general behavior of the growth of $S(\nu_1, \ldots, \nu_{N_k})$, see Theorem 4, carries over also to higher-order spline bases on sparse grids. This is not directly clear from the proof techniques used in this paper as they rely on the piecewise linear structure of the regression function. Furthermore, it remains open how our results generalize to the regularized case, where a penalty term is added in the minimization problem. A first step into this direction regarding the stability estimate can be found in [2]. However, the rate of error decay and the optimal parameter coupling are still unknown in this case. Finally, a thorough comparison of our derived convergence rates for the sparse grid method with the error decay behavior of other regression algorithms such as support vector machines or multilayer neural networks still has to be done.

Acknowledgements The author was supported by the Sonderforschungsbereich 1060 *The Mathematics of Emergent Effects* funded by the Deutsche Forschungsgemeinschaft.

References

1. A. Banerjee, S. Merugu, I.S. Dhillon, J. Ghosh, Clustering with Bregman divergences. J. Mach. Learn. Res. **6**, 1705–1749 (2005)
2. B. Bohn, Error analysis of regularized and unregularized least-squares regression on discretized function spaces. PhD thesis, Institute for Numerical Simulation, University of Bonn, 2017
3. B. Bohn, M. Griebel, An adaptive sparse grid approach for time series predictions, in *Sparse Grids and Applications*, ed. by J. Garcke, M. Griebel. Lecture Notes in Computational Science and Engineering, vol. 88 (Springer, Berlin, 2012), pp. 1–30
4. H.-J. Bungartz, M. Griebel, Sparse grids. Acta Numer. **13**, 147–269 (2004)
5. H.-J. Bungartz, D. Pflüger, S. Zimmer, Adaptive sparse grid techniques for data mining, in *Modelling, Simulation and Optimization of Complex Processes 2006, Proceedings of International Conference on HPSC, Hanoi*, ed. by H. Bock, E. Kostina, X. Hoang, R. Rannacher (Springer, Berlin, 2008), pp. 121–130
6. A. Chkifa, A. Cohen, G. Migliorati, F. Nobile, R. Tempone, Discrete least squares polynomial approximation with random evaluations - application to parametric and stochastic elliptic PDEs. ESAIM: Math. Modell. Numer. Anal. **49**(3), 815–837 (2015)
7. A. Cohen, M. Davenport, D. Leviatan, On the stability and accuracy of least squares approximations. Found. Comput. Math. **13**, 819–834 (2013)
8. C. Feuersänger, Sparse grid methods for higher dimensional approximation. PhD thesis, Institute for Numerical Simulation, University of Bonn, 2010
9. J. Garcke, Maschinelles Lernen durch Funktionsrekonstruktion mit verallgemeinerten dünnen Gittern. PhD thesis, Institute for Numerical Simulation, University of Bonn, 2004
10. J. Garcke, M. Griebel, M. Thess, Data mining with sparse grids. Computing **67**(3), 225–253 (2001)
11. M. Griebel, P. Oswald, Tensor product type subspace splitting and multilevel iterative methods for anisotropic problems. Adv. Comput. Math. **4**, 171–206 (1995)
12. M. Griebel, C. Rieger, B. Zwicknagl, Multiscale approximation and reproducing kernel Hilbert space methods. SIAM J. Numer. Anal. **53**(2), 852–873 (2015)
13. M. Griebel, C. Rieger, B. Zwicknagl, Regularized kernel based reconstruction in generalized Besov spaces. Found. Comput. Math. **18**(2), 459–508 (2018)
14. T. Hastie, R. Tibshirani, J. Friedman, *The Elements of Statistical Learning* (Springer, Berlin, 2001)
15. M. Hegland, Data mining techniques. Acta Numer. **10**, 313–355 (2001)
16. S. Knapek, Approximation und Kompression mit Tensorprodukt-Multiskalenräumen. PhD thesis, Institute for Numerical Simulation, University of Bonn, 2000
17. G. Migliorati, F. Nobile, E. von Schwerin, R. Tempone, Analysis of discrete L^2 projection on polynomial spaces with random evaluations. Found. Comput. Math. **14**, 419–456 (2014)
18. G. Migliorati, F. Nobile, R. Tempone, Convergence estimates in probability and in expectation for discrete least squares with noisy evaluations at random points. J. Multivar. Anal. **142**, 167–182 (2015)
19. D. Pflüger, B. Peherstorfer, H.-J. Bungartz, Spatially adaptive sparse grids for high-dimensional data-driven problems. J. Complexity **26**(5), 508–522 (2010)
20. B. Schölkopf, A. Smola, *Learning with Kernels – Support Vector Machines, Regularization, Optimization, and Beyond.* (The MIT Press, Cambridge, 2002)
21. J. Tropp, User-friendly tail bounds for sums of random matrices. Found. Comput. Math. **12**(4), 389–434 (2011)

Multilevel Adaptive Stochastic Collocation with Dimensionality Reduction

Ionuţ-Gabriel Farcaş, Paul Cristian Sârbu, Hans-Joachim Bungartz, Tobias Neckel, and Benjamin Uekermann

Abstract We present a multilevel stochastic collocation (MLSC) with a dimensionality reduction approach to quantify the uncertainty in computationally intensive applications. Standard MLSC typically employs grids with predetermined resolutions. Even more, stochastic dimensionality reduction has not been considered in previous MLSC formulations. In this paper, we design an MLSC approach in terms of adaptive sparse grids for stochastic discretization and compare two sparse grid variants, one with spatial and the other with dimension adaptivity. In addition, while performing the uncertainty propagation, we analyze, based on sensitivity information, whether the stochastic dimensionality can be reduced. We test our approach in two problems. The first one is a linear oscillator with five or six stochastic inputs. The dimensionality is reduced from five to two and from six to three. Furthermore, the dimension-adaptive interpolants proved superior in terms of accuracy and required computational cost. The second test case is a fluid-structure interaction problem with five stochastic inputs, in which we quantify the uncertainty at two instances in the time domain. The dimensionality is reduced from five to two and from five to four.

1 Introduction

One of the major goals in computational science is to obtain reliable simulation results from which useful predictions can be made. However, whether stemming from measurement errors, incomplete physics, or the variation of physical parameters, uncertainty is intrinsic to most applications. Therefore, uncertainty needs to be taken into account ab initio. The process of assessing the impact of input

I.-G. Farcaş · P. C. Sârbu · H.-J. Bungartz (✉) · T. Neckel · B. Uekermann
Technical University of Munich, Garching, Germany
e-mail: farcasi@in.tum.de; sarbu@in.tum.de; bungartz@in.tum.de; neckel@in.tum.de; uekerman@in.tum.de

J. Garcke et al. (eds.), *Sparse Grids and Applications – Miami 2016*,
Lecture Notes in Computational Science and Engineering 123,
https://doi.org/10.1007/978-3-319-75426-0_3

uncertainties in a model's output is called uncertainty propagation or analysis and it is part of the broader field of uncertainty quantification (UQ).

A prominent approach for uncertainty analysis is generalized polynomial chaos (gPC) [36]. gPC allows to approximate random functionals of second order, i.e. with finite variance, in terms of orthogonal polynomials. Furthermore, to achieve optimal convergence, the polynomials are chosen with respect to the input probability measure. One popular approach to determine the gPC coefficients is stochastic collocation (SC) (see e.g. [22]). With this method, the underlying equation needs to be satisfied at a finite set of prescribed collocation points. Being non-intrusive, SC is embarrassingly parallel, thus suitable for parallel computations, and allows the reuse of existing simulation codes. Once the gPC coefficients are available, quantities such as expectation, variance, or total Sobol' indices for global sensitivity analysis can be analytically computed. However, traditionally, the computational cost of SC scales exponentially with the dimension, introducing the "curse of dimensionality".

It is well established that exploiting the anisotropic coupling of a model's input parameters via sparse grids [37] is a suitable approach to delay the curse of dimensionality. To this end, sparse grids were extensively used in UQ. A non-exhaustive list includes [10], where adaptive sparse grid collocation was used in peridynamics problems. Adaptive sparse grid quadrature with Leja points was employed in [20]. Therein, the authors showed that, at least in interpolatory metrics, Leja sequences are often superior to more standard sparse grid constructions. In [6, 7], sparse grids and gPC were combined in the so-called sparse pseudo-spectral approximation (SPAM). SPAM aims to overcome the aliasing errors that may occur when classical sparse grid quadrature is used to evaluate the gPC coefficients. Inspired from the multigrid approach (see e.g. [32]), multilevel methods reduce the overall cost of single-level approaches, while maintaining the single-level accuracy. Multilevel SC (MLSC) was proposed in [31] for problems governed by elliptic partial differential equations (PDEs) with random input data. Therein, standard sparse grids were employed for stochastic discretization, the focus being on convergence and computational cost analysis. Furthermore, in [5] dimension-adaptive sparse grids were employed in combination with an adaptive and weighted reduced basis method to reduce the computational cost. Finally, in [26], Bayesian compressive sensing was used to reduce the stochastic dimensionality in models approximated via gPC. In recent years, the UQ research focus was generally tailored to methodologies that keep the dimensionality constant, but exploit other features of the problem at hand. A complementary approach, which we follow in this work, is to reduce the stochastic dimensionality.

To this end, we formulate a novel non-intrusive computational methodology for the uncertainty analysis in complex, high-dimensional stochastic problems. We employ gPC approximations and, to compute the gPC coefficients, we combine a non-intrusive multilevel decomposition of both deterministic and stochastic domains with stochastic dimensionality reduction. We note that even though our method is designed to be model-agnostic, the target application is fluid-structure interaction (FSI). Our methodology builds on [9]. Therein, spatially-adaptive sparse grids were used to compute the gPC coefficients in FSI simulations. However, the approach was formulated in a single-level fashion, without stochastic dimensionality reduction.

The formulation of our methodology is driven by two goals. Because one collocation point means one evaluation of the underlying solver, our first goal is to keep the number of collocation points small. Therefore, we employ adaptive sparse grids for stochastic discretization. In addition, we compare spatially- and dimension-adaptive sparse grids. For spatially-adaptive grids, we consider Newton-Cotes nodes and modified polynomial basis functions of second degree. On the other hand, for dimension-adaptive grids, we use Leja sequences and Lagrangian basis functions. Even more, for each adaptive strategy, we discuss two refinement criteria. We compare the two adaptive strategies in terms of number of grid points and accuracy with respect to reference results computed on a tensor grid. Our second goal is to reduce the stochastic dimensionality while performing the uncertainty propagation. Since MLSC relies on a sequence of problems with different resolutions, we select a subsequence of problems and asses the corresponding gPC coefficients. Afterwards, we compute total Sobol' indices for global sensitivity analysis and compare them to a user defined threshold; if a Sobol' index is less than the threshold, that uncertain input is ignored. If no uncertain input is ignored, we simulate the remaining subsequence of problems using the original stochastic grid. When stochastic dimensionality reduction is possible, we construct the corresponding lower dimensional grid. In the next step, we project the previously computed results and simulate the remaining subsequence of problems on the lower dimensional grid. Note that dimensionality reduction does not mean that the number of stochastic parameters changes; the uncertain parameters that are "ignored" are simply replaced with a corresponding deterministic value, e.g. their expectations. In addition, when dimensionality reduction is not possible, we still profit from using adaptive sparse grids and multilevel decompositions.

In Sect. 2, we introduce our notation and describe spatially- and dimension-adaptive interpolation, needed to formulate our MLSC approach. Furthermore, we discuss two refinement criteria for each adaptive strategy. Section 3 focuses on the proposed MLSC with our dimensionality reduction approach. In Sect. 4, we describe numerical results. In Sect. 4.1, we consider a linear oscillator with five or six uncertain inputs. Furthermore, we test and compare all four adaptive refinement criteria. In Sect. 4.2, we employ one spatially- and one dimension-adaptive criteria in a simple FSI scenario with five uncertain inputs. Moreover, we quantify the uncertainty at two instances in the time domain. We conclude this work in Sect. 5.

2 Adaptivity with Sparse Grids

Sparse grids were introduced in [37] for the discretization of second-order elliptic PDEs. Thereafter, sparse grids have been employed in a broad spectrum of applications, including quadrature [2, 11, 12], clustering and data mining [24, 25], or UQ [6, 7, 9, 10, 20, 22, 31]. In this section, we summarize two classes of refinement strategies for sparse grid interpolation. In Sect. 2.1, we outline sparse grids with local or spatial refinement. In Sect. 2.2, we focus on subspace- or dimension-adaptive sparse grids. Furthermore, for each strategy, we consider two different refinement criteria. For a more comprehensive overview of sparse grids, please refer to [3].

2.1 Interpolation on Spatially-Adaptive Sparse Grids

Let $l, i \in \mathbb{N}$ denote the level and spatial position, respectively. The starting point is a grid of Newton-Cotes nodes $u_{l,i} = ih_l \in [0, 1], h_l := 2^{-l}, i \geq 1$, and standard linear hat basis functions $\varphi_{l,i}(u)$ centered at $u_{l,i}$, with support $[u_{l,i} - h_l, u_{l,i} + h_l]$. $\varphi_{l,i}(u) = \varphi(2^l u - i)$, where $\varphi(u) = \max(1 - |u|, 0)$. Note that this construction leads to no boundary points. Furthermore, if $i = 1$, the number of grid points is $N_l = 1$, whereas if $i > 1$, $N_l = 2^{i-1} + 1$. The Newton-Cotes nodes are nested, i.e. the points at level $l - 1$ are a subset of the points at level l. The extension to d-dimensions is done via a tensor product construction

$$\varphi_{\mathbf{l},\mathbf{i}}(\mathbf{u}) = \prod_{j=1}^{d} \varphi_{l_j,i_j}(u_j),$$

where $\mathbf{l} = (l_1, \ldots, l_d) \in \mathbb{N}^d$ and $\mathbf{i} = (i_1, \ldots, i_d) \in \mathbb{N}^d$.

Let $W_{\mathbf{l}} = \text{span}\{\varphi_{\mathbf{l},\mathbf{i}} | \mathbf{i} \in \mathscr{I}_{\mathbf{l}}\}$ denote a so-called hierarchical increment space, where $\mathscr{I}_{\mathbf{l}} = \{\mathbf{i} \in \mathbb{N}^d : 1 \leq i_k \leq 2^{l_k} - 1, i_k \text{ odd }, k = 1 \ldots d\}$. Given a level l, the sparse grid space V_l^1 is defined as

$$V_l^1 = \bigotimes_{\mathbf{l} \in \mathscr{L}} W_{\mathbf{l}}.$$

For the standard sparse grid construction, the multi-index set $\mathscr{L}$ is

$$\mathscr{L} = \{\mathbf{l} \in \mathbb{N}^d : |\mathbf{l}|_1 \leq l + d - 1\}, \tag{1}$$

where $|\mathbf{l}|_1 := \sum_{i=1}^{d} l_i$. We depict 2 and 3D sparse grids of level three in Fig. 1.

The sparse grid interpolant $f_{\mathscr{I}_{\mathbf{l}}}$ of f reads

$$f_{\mathscr{I}_{\mathbf{l}}}(\mathbf{u}) = \sum_{\mathbf{l} \in \mathscr{L}, \mathbf{i} \in \mathscr{I}_{\mathbf{l}}} \alpha_{\mathbf{l},\mathbf{i}} \varphi_{\mathbf{l},\mathbf{i}}(\mathbf{u}), \tag{2}$$

where $\alpha_{\mathbf{l},\mathbf{i}}$ are the so-called hierarchical surpluses, computed as

$$\alpha_{\mathbf{l},\mathbf{i}} = f(\mathbf{u}_{\mathbf{l},\mathbf{i}}) - \frac{f(\mathbf{u}_{\mathbf{l},\mathbf{i}} - \mathbf{h}_{\mathbf{l}}) + f(\mathbf{u}_{\mathbf{l},\mathbf{i}} + \mathbf{h}_{\mathbf{l}})}{2}. \tag{3}$$

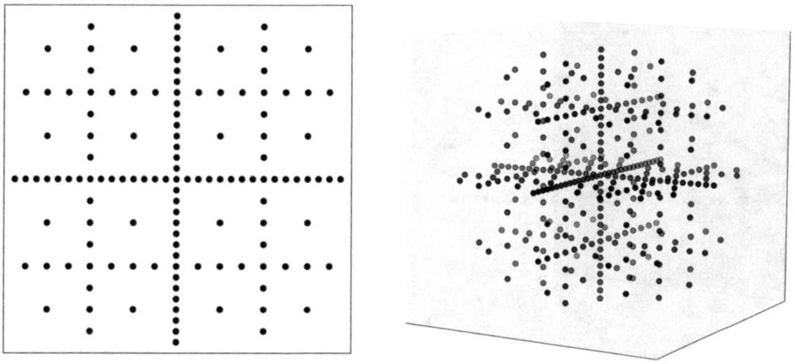

Fig. 1 Two-dimensional (left) and three-dimensional (right) standard sparse grid for $l = 5$

If $f \in H_2^{\text{mix}}([0,1]^d) = \{f : [0,1]^d \to \mathbb{R} : D^{\mathbf{l}} f \in L^2([0,1]^d), |\mathbf{l}|_\infty \leq 2\}$, where $D^{\mathbf{l}} f = \partial^{|\mathbf{l}|_1} f / \partial x_1^{l_1} \ldots \partial x_d^{l_d}$ and $|\mathbf{l}|_\infty := \max_i l_i$, then $||f(\mathbf{u}) - f_{\mathscr{I}_l}(\mathbf{u})||_{L^2} \in \mathscr{O}(h_l^2 l^{d-1})$, where $||f(\mathbf{u})||_{L^2}^2 := \int_{[0,1]^d} |f(\mathbf{u})|^2 d\mathbf{u}$. The cost is $\mathscr{O}(h_l^{-1} |\log_2 h_l|^{d-1})$ degrees of freedom (dof) (cf. [3]). When using a full grid interpolant $\tilde{f}$, $||f(\mathbf{u}) - \tilde{f}(\mathbf{u})||_{L^2} \in \mathscr{O}(h_l^2)$, at a cost of $\mathscr{O}(h_l^{-d})$ dof. Therefore, with standard sparse grids, the number of dof is significantly reduced while the accuracy is only slightly deteriorated.

The sparse grid approach can be generalized to more sophisticated basis functions, such as higher order polynomials [1]. Provided that $D^{\mathbf{p}+\mathbf{1}} f$ is bounded for some $p \geq 2$, then $||f(\mathbf{u}) - f_{\mathscr{I}_l}(\mathbf{u})||_{L^2} \in \mathscr{O}(h_l^{p+1} l^{d-1})$. In this paper, we assume that the underlying model has bounded higher order mixed derivatives and employ polynomial basis functions of degree two, depicted in Fig. 2, left, for $l \leq 3$.

As mentioned in the beginning of this section, the employed sparse grid construction leads to no points on the boundary of $[0,1]^d$. However, when the underlying function does not vanish at the boundary, we need to extend the previously described construction to incorporate boundary information, too. Our solution is to modify the standard basis functions so as to linearly extrapolate the boundary information (see [25]). In Fig. 2, right, we depict 1D modified polynomial basis functions for $l \leq 3$.

To address interpolation of computationally expensive functions, we employ spatially-adaptive or local refinement (see [17, 25]), an intrinsic property of sparse grids, due to their hierarchical construction. As hierarchical surpluses (cf. Eq. (3)) are an indicator of local interpolaton error, we employ refinement criteria based on surpluses values. Let $\bar{\mathscr{L}} = \{(\mathbf{l}, \mathbf{i}) : \mathbf{u}_{\mathbf{l},\mathbf{i}} \text{ is refinable}\}$ be the set containing all levels and indices corresponding to grid points that can be refined. We consider two refinement criteria. The first one is the maximum absolute surplus refinement [17] (MAS)

$$\max_{(\mathbf{l},\mathbf{i}) \in \bar{\mathscr{L}}} |\alpha_{\mathbf{l},\mathbf{i}}|. \tag{4}$$

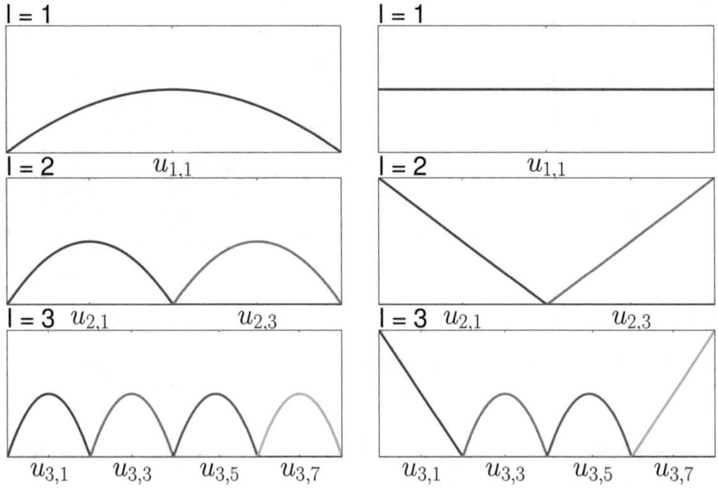

Fig. 2 Standard polynomial basis functions of second degree (left) and their modified version (right) for $l = 1, 2, 3$

The second one is the expectation value refinement (EVR), similar to [10]. It reads

$$\max_{(\mathbf{l},\mathbf{i}) \in \bar{\mathscr{L}}} |\alpha_{\mathbf{l},\mathbf{i}} \mathbb{E}[\varphi_{\mathbf{l},\mathbf{i}}(\boldsymbol{\theta})]|, \tag{5}$$

where $\mathbb{E}[\varphi_{\mathbf{l},\mathbf{i}}(\boldsymbol{\theta})] := \int_{[0,1]^d} \varphi_{\mathbf{l},\mathbf{i}}(\boldsymbol{\theta}) d\boldsymbol{\theta}$. For example, if $\varphi_{\mathbf{l},\mathbf{i}}$ are the hat functions, $\mathbb{E}[\varphi_{\mathbf{l},\mathbf{i}}(\boldsymbol{\theta})] = 2^{-|\mathbf{l}|_1}$. Note that if not all hierarchical parents exist in the refined grid, we ensure that they are added as well (see e.g. [25]).

2.2 Interpolation with Dimension-Adaptive Sparse Grids

Let $f_l^{(i)} = \sum_{j=0}^{N_l-1} f(u_{l_j}) \varphi_{l_j}(u)$ denote a sequence of one-dimensional interpolation formulae, where $i = 1, 2, \ldots, d$, $l \in \mathbb{N}$ represents the interpolation level, $u \in [0, 1]$, and $(u_{l_j})_{j=0}^{N_l-1}$ is a sequence of N_l grid points at level l. Furthermore, $\varphi_{l_j}(u)$ are Lagrange polynomials, i.e. $\varphi_{l_j}(u_{l_i}) = \delta_{i,j}$.

We construct the d-variate interpolant using the so-called Smolyak algorithm [27]. This algorithm decreases the computational effort by weakening the assumed coupling between input dimensions (cf. [6]). Let $\Delta_{l_0}^{(i)} := f_{l_0}^{(i)}$, $\Delta_{l_i}^{(i)} := f_{l_i}^{(i)} - f_{l_{i-1}}^{(i)}$, and $\Delta_{\mathbf{l}} := \Delta_{l_1}^{(1)} \otimes \ldots \otimes \Delta_{l_d}^{(d)}$. The $d-$variate sparse interpolation formula reads

$$\mathscr{A}(\mathscr{L}, d) = \sum_{\mathbf{l} \in \mathscr{L}} \Delta_{\mathbf{l}} = \sum_{\mathbf{l} \in \mathscr{L}} \Delta_{l_1}^{(1)} \otimes \ldots \otimes \Delta_{l_d}^{(d)}, \tag{6}$$

where $\mathscr{L}$ is a multi-index set. For a standard, isotropic grid of level l, $\mathscr{L}$ is defined as in Eq. (1). We remark that in this work, $\mathscr{A}(\mathscr{L}, d)$ is available only algorithmically, as a black box. Hence, we do not have direct access to $(\Delta_{l_i}^{(i)})_{i=1}^{d}$.

We construct Eq. (6) in terms of Leja points [16]. The Leja point distribution is a sequence $(u_n)_{n\in\mathbb{N}}$ in a compact domain $\mathbb{D} \subset \mathbb{C}$ that maximizes $\Psi_n(u) = \prod_{i=1}^{n}(u - u_i)$. In this work, we compute $(u_n)_{n\in\mathbb{N}}$ as

$$\begin{aligned} u_0 &= 0.5 \\ u_{n+1} &= \underset{u\in\mathbb{D}}{\operatorname{argmax}}\, |\Psi_n(u)|, \quad n = 1, 2, \ldots, \end{aligned} \tag{7}$$

where $\mathbb{D} = [0, 1]$. Leja points are nested, i.e. $(u_n)_{n=0}^{N_l-1} \subset (u_n)_{n=0}^{N_{l+1}-1}$ and they grow linearly with the level as $N_l = g_f(l-1) + 1$, where $g_f \in \mathbb{N}$ is the growth factor. Therefore, Leja points are suitable to address high-dimensional computational problems. Note that Eq. (7) may not posses a unique solution, making the Leja sequence generally not unique. In Fig. 3, we depict two- and three-dimensional Leja grids for $l = 5$ and $g_f = 2$. For more details about Leja sequences, see e.g. [18, 20].

Because the underlying basis functions are global polynomials, we can no longer perform local refinement as in Sect. 2.1. We can instead perform subspace- or dimension-adaptivity (cf. [12]). We first note that Eq. (6) assumes that $\mathscr{L}$ is admissible, i.e. it has no "holes". This means that $\forall \mathbf{l} \in \mathscr{L} : \mathbf{l} - \mathbf{e}_j \in \mathscr{L}$, where $\mathbf{e}_j$ is the jth canonical unit vector of $\mathbb{N}^d$ (see [13]). Therefore, in the following, we assume that we have a sparse grid characterized by an initial admissible multi-index set $\mathscr{L}$.

Let $\mathbf{l}', \mathbf{l}$ be two multi-indices in an admissible multi-index set $\mathscr{L}$. $\mathbf{l}'$ is called the parent of $\mathbf{l}$ if $|\mathbf{l} - \mathbf{l}'|_1 = 1$. Additionally, let $\bar{\mathscr{L}} = \{\mathbf{k} \in \mathbb{N}^d \setminus \mathscr{L} : |\mathbf{k} - \mathbf{l}|_1 = 1, \mathbf{l} \in \mathscr{L}\}$ denote the set of multi-indices whose parents are in $\mathscr{L}$. Our adaptive refinement criteria appends to $\mathscr{L}$ the multi-index $\mathbf{k} \in \bar{\mathscr{L}}$ having the highest priority. The priority of $\mathbf{k}$ is given by a function $q_{\mathbf{k}}(\Delta_{\mathbf{l}^{(1)}}, \ldots, \Delta_{\mathbf{l}^{(m)}})$, where $\mathbf{l}^{(1)} \ldots \mathbf{l}^{(m)}$ are the parents of $\mathbf{k}$. The first refinement criterion uses an averaging priority function

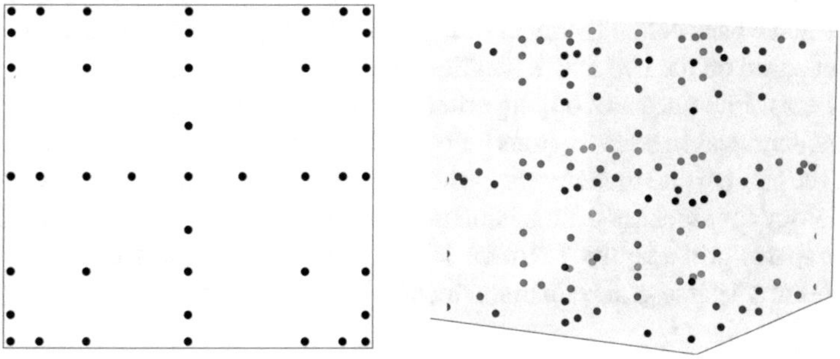

Fig. 3 Two-dimensional (left) and three-dimensional (right) sparse grid of level $l = 5$ constructed with Leja points with $g_f = 2$

(AVG), cf. [15],

$$q_{\mathbf{k}}(\Delta_{\mathbf{l}^{(1)}}, \ldots, \Delta_{\mathbf{l}^{(m)}}) = \frac{\sum_{i=1}^{m} ||\Delta_{\mathbf{l}^{(i)}}||_{L^2}}{N_{\mathbf{k}}}, \tag{8}$$

where $N_{\mathbf{k}}$ is the cost (number of grid points) needed to add $\mathbf{k}$ to $\mathscr{L}$. The second refinement criterion uses a maximum priority function (MAX)

$$q_{\mathbf{k}}(\Delta_{\mathbf{l}^{(1)}}, \ldots, \Delta_{\mathbf{l}^{(m)}}) = \max(\{q_i\}_{i=1}^{m}), \quad q_i = \max\left(r \frac{||\Delta_{\mathbf{l}^{(i)}}||_{L^2}}{||\Delta_{\mathbf{1}}||_{L^2}}, (1-r)\frac{1}{N_{\mathbf{l}}^{(i)}}\right), \tag{9}$$

where $\mathbf{1} = (1, \ldots, 1)$ is the initial level, $N_{\mathbf{l}}^{(i)}$ represents the number of grid points corresponding to $\mathbf{l}^{(i)}$, and $r \in [0, 1]$ controls the balance between error and cost. If $r \approx 0$, the refinement depends dominantly on the cost of adding a new subspace. On the other hand, when $r \approx 1$, the error dominates the refinement procedure; the choice of r is heuristic and dependents on the desired trade-off between reducing the error and controlling the computational cost. Note that the MAX refinement criterion (9) is similar to [12]. However, we add just one level at a time, thus, having a fine control over the grid points. We remark that the two aforementioned refinement strategies are heuristic. Nevertheless, since our goal is to construct a computational UQ approach suitable for computationally expensive applications, they suffice for our purposes.

We end this section summarizing our strategy for adaptive refinement. In general, adaptive refinement enriches the sparse grid based on a prescribed stopping criteria, e.g. error tolerance (emphasis on accuracy), maximum number of grid nodes (emphasis on computational cost), or combination thereof. As we aim to address computationally expensive UQ problems, we perform the refinement such that we keep the number of grid points, hence, the computational cost as small as possible while having a reasonable approximation accuracy. To this end, when employing spatially-adaptive refinement, we refine a percentage of the current grid points in one refinement step. On the other hand, in one refinement step in the dimension-adaptive case, we add a user-defined number of multi-indices to the current sparse grid multi-index set based on the two above mentioned priority functions. Note that employing only the error tolerance as stopping criteria might lead to a too early stopping of the refinement process in some regions before achieving the desired accuracy (for more details, see e.g. [5] and the references therein). Finally, we remark that our proposed methodology for stochastic dimensionality reduction is independent of the strategy used for sparse grid adaptive refinement. Therefore, our approach can be employed using arbitrary sparse grid refinement techniques.

3 Multilevel Stochastic Collocation with Dimensionality Reduction

In Sect. 3.1, we give a brief overview of gPC approximation. In Sect. 3.2, we describe our multilevel approach for computing the gPC coefficients. Finally, in Sect. 3.3, we formulate the proposed MLSC with our dimensionality reduction approach employing total Sobol' indices for variance-based global sensitivity analysis.

3.1 Generalized Polynomial Chaos

In what follows, we assume that the underlying numerical solver is given as a black box $\mathscr{M}(\mathbf{x}, \boldsymbol{\theta})$, where $\mathbf{x}$ denotes the deterministic inputs (e.g. time, boundary conditions), whereas $\boldsymbol{\theta}$ denotes the uncertain inputs. We adopt a probabilistic framework and model the uncertain inputs as a d-variate random vector $\boldsymbol{\theta} = (\theta_1, \ldots, \theta_d)$ with independent and identically distributed (i.i.d.) components from a probabilistic space $(\Omega, \mathscr{F}, \mathscr{P})$. Ω represents the sample (event) space, $\mathscr{F}$ is a σ-algebra, and $\mathscr{P}$ is a probability measure. If $\rho_i : \Gamma_i \to \mathbb{R}$ is the probability density function (PDF) of θ_i, the joint PDF $\boldsymbol{\rho}$ of $\boldsymbol{\theta}$ reads $\boldsymbol{\rho} = \prod_{i=1}^{d} \rho_i(\theta_i)$, with support $\Gamma := \prod_{i=1}^{d} \Gamma_i$.

Let $P \in \mathbb{N}$ and $\mathscr{P} = \{\mathbf{p} = (p_1, \ldots, p_d) \in \mathbb{N}^d \colon |\mathbf{p}|_1 < P\}$ a total-degree index set. Furthermore, let $\{\Phi_{p_i}(y_i)\}_{i=1}^{d}$ be univariate orthogonal polynomials of degree p_i, i.e.

$$\mathbb{E}[\Phi_{p_i}(y)\Phi_{p_j}(y)] := \int_{\Gamma_k} \Phi_{p_i}(y)\Phi_{p_j}(y)\rho_k(y)dy = c_i\delta_{ij},$$

where $\Phi_0 \equiv 1$, $c_i \in \mathbb{R}$ and δ_{ij} is the Kronecker delta. A d-variate orthogonal polynomial $\Phi_{\mathbf{p}}(\mathbf{y})$ is constructed via a tensor product of univariate polynomials, i.e.

$$\Phi_{\mathbf{p}}(\mathbf{y}) := \Phi_{p_1}(y_1) \ldots \Phi_{p_d}(y_d).$$

For a given $P \in \mathbb{N}$, the number of d-variate orthogonal polynomials is $N := \binom{d+P}{d}$. To simplify notation, in what follows, we assume that the polynomials are orthonormal, i.e. $c_i \equiv 1$. In addition, instead of using the multi-index subscript $\mathbf{p}$, we employ a scalar index $n = 0 \ldots N-1$.

The Nth order gPC approximation of $\mathscr{M}(\mathbf{x}, \boldsymbol{\theta})$ reads (see [36])

$$\mathscr{M}(\mathbf{x}, \boldsymbol{\theta}) \approx \mathscr{M}_N(\mathbf{x}, \boldsymbol{\theta}) := \sum_{n=0}^{N-1} m_n(\mathbf{x})\Phi_n(\boldsymbol{\theta}),$$

where the coefficients $m_n(\mathbf{x})$ are obtained via projection

$$m_n(\mathbf{x}) = \int_{\Gamma} \mathcal{M}(\mathbf{x}, \boldsymbol{\theta})\Phi_n(\boldsymbol{\theta})\rho(\boldsymbol{\theta})d\boldsymbol{\theta} = \mathbb{E}[\mathcal{M}(\mathbf{x}, \boldsymbol{\theta})\Phi_n(\boldsymbol{\theta})], \quad n = 0, \ldots, N-1. \tag{10}$$

We address the gPC coefficients computation in Sect. 3.2.

Because of the orthogonality of the gPC polynomial basis, quantities such as expectation, variance, or total Sobol' indices for global sensitivity analysis [28] can be computed analytically from the gPC coefficients. The expectation and the variance are obtained as (see [35] for more details)

$$\mathbb{E}[\mathcal{M}_N(\mathbf{x}, \boldsymbol{\theta})] = m_0(\mathbf{x}), \tag{11}$$

$$\mathrm{Var}[\mathcal{M}_N(\mathbf{x}, \boldsymbol{\theta})] = \sum_{n=1}^{N-1} m_n^2(\mathbf{x}). \tag{12}$$

A total Sobol' index S_i^T accounts for the total contribution of an uncertain input to the resulted variance. This comprises the contribution of the input taken individually and contributions due to its interactions with other uncertain inputs. The connection between total Sobol' indices and gPC coefficients was shown in [30]. It reads

$$S_i^T(x) = \frac{\sum_{k \in A_n} m_k^2(\mathbf{x})}{\mathrm{Var}[\mathcal{M}_N(\mathbf{x}, \boldsymbol{\theta})]}, \tag{13}$$

where $i = 1, \ldots, d$. $A_n = \{n : \mathrm{ind}(\mathbf{p}) = n \wedge \mathbf{p} \in \mathcal{P}, p_i \in \mathbf{p}, p_i \neq 0\}$, where $\mathrm{ind}(\mathbf{p})$ denotes the index of $\mathbf{p}$ in $\mathcal{P}$. For more details about gPC, please refer to [35].

3.2 Multilevel Approaches for Generalized Polynomial Chaos

We model the uncertain inputs $\{\theta_i\}_{i=1}^d$ as uniform random variables in a compact domain $[a_i, b_i]$. Therefore, the associated gPC basis consists in Legendre polynomials (cf. [36]). Before going further, note that sparse grids are defined in $[0, 1]^d$, whereas the uncertain inputs belong to $\Gamma = [a_1, b_1] \times \ldots \times [a_d, b_d]$. To this end, let $\mathbf{T} : [0, 1]^d \to \Gamma$ be a linear transformation. Then, $\mathcal{M}(\mathbf{x}, \mathbf{T}(\boldsymbol{\omega})) := \mathcal{M}(\mathbf{x}, \boldsymbol{\theta}) \circ \mathbf{T}(\boldsymbol{\omega})$ denotes the model whose stochastic space is $[0, 1]^d$. In addition, the proposed approach can be generalized to arbitrary domains, employing suitable, possibly non-linear transformations, such as the inverse cumulative distribution function (see e.g. [29]).

As previously mentioned, we assume that $\mathcal{M}(\mathbf{x}, \mathbf{T}(\boldsymbol{\omega}))$ is available as a black box. Therefore, we assess the gPC coefficients in Eq. (10) numerically. Note that as the model is assumed to be resource-intensive, even for small dimensionality, standard tensor-product quadrature is computationally too expensive. An alternative is (adaptive) sparse grid quadrature (see e.g. [2, 11]). However, in [6, 7], it was shown that significant aliasing errors may occur unless a possibly high-order scheme is employed. Furthermore, the quadrature weights can become negative (see e.g. [14]).

To evaluate the gPC coefficients in Eq. (10), we combine adaptive sparse grid interpolation with quadrature, as follows. We discretize $\boldsymbol{\omega}$ on a d-dimensional sparse grid and construct an interpolant $\mathscr{M}_d^{sg}(\mathbf{x}, \mathbf{T}(\boldsymbol{\omega}))$ of $\mathscr{M}(\mathbf{x}, \mathbf{T}(\boldsymbol{\omega})$ using either Eq. (2) or (6). Furthermore, we adaptively refine $\mathscr{M}_d^{sg}(\mathbf{x}, \mathbf{T}(\boldsymbol{\omega}))$ based on Eq. (4), (5), (8), or (9). Afterwards, we plug the interpolant into Eq. (10) and obtain

$$m_n(\mathbf{x}) = \mathbb{E}[\mathscr{M}(\mathbf{x}, \mathbf{T}(\boldsymbol{\omega}))\Phi_n(\boldsymbol{\omega})] = \mathbb{E}[\mathscr{M}_d^{sg}(\mathbf{x}, \mathbf{T}(\boldsymbol{\omega}))\Phi_n(\boldsymbol{\omega})]. \tag{14}$$

Finally, we compute the resulting integral via Gauss-Legendre quadrature. We note that when using spatially-adaptive sparse grids, we can exploit the tensor structure of both sparse grid and gPC basis to break the integral in Eq. (14) down into a weighted sum of products of one-dimensional integrals (see [9] for more details).

MLSC was initially formulated in [31] in terms of a hierarchy of multivariate sparse grid interpolants computed on standard sparse grids. In this work, we formulate an MLSC approach based on adaptive sparse grid interpolation to compute gPC approximation coefficients. Our aim is to have a computational method suited for resource-intensive applications. Let $h, q \in \mathbb{N}$ and let M_h denote the number of dof employed to discretize the problem domain at level h. Typically, if M_0 corresponds to the coarsest level, $M_h = g(h)M_0$, for some function $g(h)$. Additionally, let L_q denote the sparse grid resolution level. We note that when using adaptive sparse grids, L_q refers to L_0 after q refinement steps. Finally, let $m_n^{M_h, L_q}(\mathbf{x})$ denote the nth gPC coefficient computed using a problem domain resolution M_h and sparse grid resolution L_q. Given a $K \in \mathbb{N}$, we approximate $m_n^{M_K, L_K}(\mathbf{x})$ as

$$\begin{aligned}
m_n^{M_K, L_K}(\mathbf{x}) \approx m_n^{M_0, L_K}(\mathbf{x}) \quad &+ \\
(m_n^{M_1, L_{K-1}}(\mathbf{x}) - m_n^{M_0, L_{K-1}}(\mathbf{x})) \quad &+ \\
\dots \quad &+ \\
(m_n^{M_K, L_0}(\mathbf{x}) - m_n^{M_{K-1}, L_0}(\mathbf{x}))& \\
\approx \sum_{k=0}^{K} (m_n^{M_k, L_{K-k}}(\mathbf{x}) - m_n^{M_{k-1}, L_{K-k}}(\mathbf{x})),&
\end{aligned} \tag{15}$$

where $m_n^{M_{-1}, L_K}(\mathbf{x}) := 0$. Therefore, in Eq. (15), we begin with the coarse grid solution $m_n^{M_0, L_K}(\mathbf{x})$ and, for $k \geq 1$, add fine grid corrections $(m_n^{M_k, L_{K-k}}(\mathbf{x}) - m_n^{M_{k-1}, L_{K-k}}(\mathbf{x}))$. Note that this procedure is similar to the multigrid algorithm (see e.g. [32]).

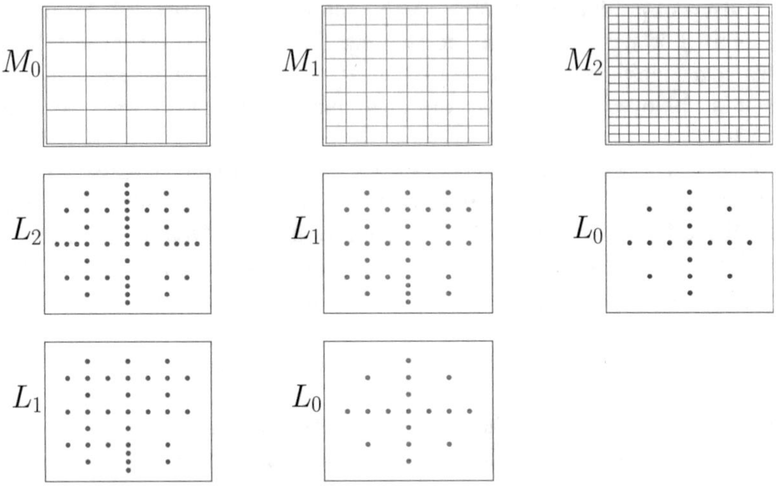

Fig. 4 Three-level decomposition of model and stochastic domains. M_0, M_1, and M_2 denote the deterministic domain resolution, whereas L_0, L_1, and L_2 denote the stochastic domain resolution. The three-level decomposition leads to five combinations: M_0 with L_2 and L_1 (left part), M_1 with L_1 and L_0 (middle part), and M_2 with L_0 (right part). Note that because we use nested sparse grids, $L_0 \subset L_1 \subset L_2$

Equation (15) requires $2K + 1$ gPC coefficients, one corresponding to M_K and two to each $M_{K-k}, k = 1, \ldots, K$. However, $\{m_n^{M_{K-k}, L_{k-1}}(\mathbf{x})\} \subset \{m_n^{M_{K-k}, L_k}(\mathbf{x})\}$, since we employ nested sparse grids. Thus, we need only solve a total of $K + 1$ problems, reducing the computational cost even further. We depict a three-level approach with adaptive sparse grids for stochastic discretization in Fig. 4.

3.3 Stochastic Dimensionality Reduction

To our knowledge, stochastic dimensionality reduction has not been considered in previous MLSC formulations. Let $\mathscr{K} = \{m_n^{M_0, L_K}(\mathbf{x}), m_n^{M_1, L_{K-1}}(\mathbf{x}), \ldots, m_n^{M_K, L_0}(\mathbf{x})\}$ denote the set of $K+1$ coefficients in Eq. (15). Additionally, consider an integer $J < K$ and a subset $\mathscr{J} = \{m_n^{M_j, L_{K-j}}(\mathbf{x}), m_n^{M_{j+1}, L_{K-j-1}}(\mathbf{x}), \ldots, m_n^{M_{j+J}, L_{K-j-J}}(\mathbf{x})\} \subset \mathscr{K}$, for some $j \geq 0$. Therefore, $\mathscr{J}$ contains $J + 1$ coefficients from the total of $K + 1$. Note that the coefficients from $\mathscr{K}$ and, implicitly from $\mathscr{J}$, too, can be computed independently from each other. Because j and J determine the subset $\mathscr{J}$, we recommend to start with a $\mathscr{J}$ of low cardinality, e.g. one, and afterwards, depending on the available computational budged and desired accuracy, to increase the cardinality of $\mathscr{J}$, if necessary. Finally, let $\tau \in (0, 1]$ denote a user defined threshold.

We formulate our dimensionality reduction approach as follows. Initially, the stochastic dimension is the original one, i.e. d. In the first step, we compute the coefficients from $\mathscr{J}$ employing a d-dimensional stochastic grid. In the second step, we obtain a multilevel approximation of $m_n^{M_J,L_J}(\mathbf{x})$ (cf. Eq. (15)),

$$m_n^{M_J,L_J}(\mathbf{x}) = \sum_{k=j}^{j+J} (m_n^{M_k,L_{K-k}}(\mathbf{x}) - m_n^{M_{k-1},L_{K-k}}(\mathbf{x})). \tag{16}$$

Next, using Eq. (13), we compute the total Sobol' indices corresponding to each uncertain input and obtain a set $\mathscr{S} = \{S_i^T(x), i = 1, \ldots, d\}$. Note that up to this point, the stochastic dimensionality is still d. In the fourth step, we analyze whether we can reduce the stochastic dimensionality. We compare each $S_i^T(x) \in \mathscr{S}$ to the user defined threshold τ; the choice of τ is heuristic and can depend on the number of uncertain inputs and desired results accuracy. If $S_i^T(x) < \tau$, we consider the ith input's contribution as unimportant and we ignore it. By ignoring an uncertain input, we mean that we no longer consider it uncertain, but deterministic or known. In this paper, the deterministic value associated to an uncertain input is its expectation. If $S_i^T(x) \geq \tau$, the ith uncertain input remains stochastic. On the one hand, if no uncertain input is ignored, we calculate the coefficients from $\mathscr{K} \setminus \mathscr{J}$ using the original stochastic grid and, afterwards, compute the multilevel gPC coefficients using Eq. (15). On the other hand, if $s < d$ uncertain inputs are ignored, we compute the coefficients from $\mathscr{K} \setminus \mathscr{J}$ using a stochastic grid of dimension $d - s$. Because the coefficients from $\mathscr{J}$ and $\mathscr{K} \setminus \mathscr{J}$ are computed using different stochastic dimensionality, we need to project the results from the original grid onto the reduced grid. Finally, we compute the multilevel gPC coefficients (cf. Eq. (15)) using the grid of dimensionality $d - s$.

To project the results from the stochastic grid with dimensionality d onto a grid with dimensionality $d - s$, we simply evaluate the d-dimensional interpolant at the s deterministic inputs. Let $m_n^{M_k,L_{K-k}}(\mathbf{x}) \in \mathscr{J}$ and $\mathbf{T}(\boldsymbol{\omega}) = (T_1(\omega_1), T_2(\omega_2), \ldots, T_d(\omega_d))$. Assuming, without loss of generality, that the inputs $d - s + 1, \ldots, d$ are ignored, the projection on a $d - s$ dimensional grid reads (cf. Eq. (14))

$$m_n^{M_k,L_{K-k}}(\mathbf{x}) = \mathbb{E}[\mathscr{M}_d^{sg}(\mathbf{x}, T_1(\omega_1), \ldots, T_{d-s}(\omega_{d-s}), \bar{u}_{d-s+1}, \ldots, \bar{u}_d)\Phi_n(\omega_1, \ldots, \omega_{d-s})], \tag{17}$$

where $\bar{u}_{d-s+1}, \ldots, \bar{u}_d$ are the deterministic values corresponding to the ignored stochastic inputs, mapped to $[0, 1]$. On the one hand, when employing dimension-adaptive interpolation, we simply substitute Eq. (6) into Eq. (17) and then evaluate $\mathscr{M}_d^{sg}$. On the other hand, when using spatially-adaptive grids, we transform the

d-dimensional interpolant from Eq. (2) into a $d-s$-dimensional version, as

$$\begin{aligned} f_{\mathscr{I}_1}(x, \mathbf{T}(\omega)) &= \sum_{\mathbf{l}\in\mathscr{L},\mathbf{i}\in\mathscr{I}_1} \alpha_{\mathbf{l},\mathbf{i}}\varphi_{\mathbf{l},\mathbf{i}}(\mathbf{u}) \\ &= \sum_{\mathbf{l}\in\mathscr{L},\mathbf{i}\in\mathscr{I}_1} \Big(\sum_k \alpha_{(l_1,\ldots,l_k,\ldots,l_d),(i_1,\ldots,i_k,\ldots,i_d)} \prod_{k=d-s+1}^{d} \varphi_{l_k,i_k}(\bar{u}_k)\Big) \prod_{j=1}^{d-s} \varphi_{l_j,i_j}(u_j) \\ &= \sum_{\mathbf{l}\in\mathscr{L}',\mathbf{i}\in\mathscr{I}_1'} \alpha'_{\mathbf{l},\mathbf{i}}\varphi'_{\mathbf{l},\mathbf{i}}(\mathbf{u}'), \end{aligned}$$

and afterwards, we plug it into Eq. (17). In the above formula, $\alpha'_{\mathbf{l},\mathbf{i}}$ is obtained from $\alpha_{\mathbf{l},\mathbf{i}}$ by "absorbing" $\prod_{k=d-s+1}^{d} \varphi_{l_k,i_k}(\bar{u}_k)$, i.e. the sparse grid basis functions evaluated at the deterministic inputs. Furthermore, $\varphi'_{\mathbf{l},\mathbf{i}}$ are basis functions of the new $d-s$ dimensional space and $\mathbf{u}'$ represents $\mathbf{u}$ without the components $d-s+1,\ldots,d$. We summarize our proposed MLSC with dimensionality reduction in Algorithm 1.

Before ending this section, we note that although in our formulation we analyze only once whether the stochastic dimensionality can be reduced, our approach can be easily generalized. Therefore, instead of considering just one $\mathscr{J} \subset \mathscr{K}$, we could consider a sequence $\mathscr{J}_1, \mathscr{J}_2, \ldots \subset \mathscr{K}$, possibly such that $\mathscr{J}_m \subseteq \mathscr{J}_{m+1}, m = 1, 2, \ldots$. After we compute the gPC coefficients from $\mathscr{J}_m$, we analyze whether the stochastic dimensionality could be reduced. In our view, the generalized approach is suitable when the initial number of uncertain parameters is large or when some total Sobol' indices are close to the prescribed threshold.

Algorithm 1: The proposed MLSC with stochastic dimensionality reduction

Data: initial stochastic dimensionality d, $\mathscr{J}$, $\mathscr{K}$, τ
Result: number of ignored dimensions s, gPC coefficients
compute the gPC coefficients from $\mathscr{J}$ using Eq. (14);
compute the partial multilevel gPC coefficients using Eq. (16);
$s \leftarrow 0$;
for $i \leftarrow 1$ **to** d **do**
    compute $S_i^T(x)$ based on the gPC coefficients from $\mathscr{J}$ using Eq. (13);
    if $S_i^T(x) < \tau$ **then**
        $s \leftarrow s+1$;
if $s > 0$ **then**
    project coefficients from $\mathscr{J}$ onto a stochastic grid of dimensionality $d-s$ using Eq. (17);
compute the gPC coefficients from $\mathscr{K} \setminus \mathscr{J}$ on a grid of dimensionality $d-s$ using Eq. (14);
compute the final multilevel gPC coefficients on a grid of dimensionality $d-s$ using Eq. (15) ;

4 Numerical Results

In Sect. 4.1, we consider a linear oscillator model with two stochastic setups, one with five and the other with six uncertain inputs. Furthermore, we compare the four considered adaptive refinement criteria with respect to reference results from a full tensor grid. In Sect. 4.2, we employ one spatially- and one dimension-adaptive criterion in an FSI scenario with five uncertain inputs. Moreover, we quantify the uncertainty at two instances in the time domain. Throughout this work, the entire sparse grid functionality was provided by the open-source library SG++[1] [25].

4.1 Second-Order Linear Oscillator with External Forcing

The first considered test case is a damped linear oscillator subject to external forces, modeled as a second-order ordinary differential equation

$$\begin{cases} \frac{d^2y}{dt^2}(t) + c\frac{dy}{dt}(t) + ky(t) = f\cos(\omega t) \\ y(0) = y_0 \\ \frac{dy}{dt}(0) = y_1, \end{cases}$$

where c is the damping coefficient, k the spring constant, f the forcing amplitude, and ω the frequency. Furthermore, y_0 represents the initial position, whereas y_1 is the initial velocity. Throughout our simulations, $t \in [0, 20]$.

To gain a broad overview of our approach, we consider two stochastic setups. In the first one, the deterministic inputs are $\mathbf{x}_1 = (t, \omega)$, $\omega = 1.05$, whereas the vector of stochastic inputs is $\boldsymbol{\theta}_1 = (c, k, f, y_0, y_1)$. In the second setup, $\mathbf{x}_2 = t$ and $\boldsymbol{\theta}_2 = (c, k, f, \omega, y_0, y_1)$. $\boldsymbol{\theta}_1$ and $\boldsymbol{\theta}_2$ have i.i.d. uniform components. The corresponding a_i and b_i are listed in Table 1. These values are taken from [20, Sect. 5.1.] and yield an underdamped regime. In addition, the deterministic values associated to the stochastic inputs are $\bar{u}_i = (a_i + b_i)/2$, i.e. their expectations. Note that all $\bar{u}_i \in [0, 1]$, the standard sparse grid domain; otherwise, we would need to scale them (cf. Sect. 3.3).

Table 1 Second-order linear oscillator test case: uniform bounds for $\boldsymbol{\theta}_1, \boldsymbol{\theta}_2$ and associated deterministic values

	c [N s/m]	k [N/m]	f [N]	ω [rad/s]	y_0 [m]	y_1 [m/s]
a_i	8.00e−02	3.00e−02	8.00e−02	8.00e−01	4.50e−01	−5.00e−02
b_i	1.20e−01	4.00e−02	1.20e−01	1.20e+00	5.50e−01	5.00e−02
$\bar{u}_i$	1.00e−01	3.50e−02	1.00e−01	1.00e+00	5.00e−01	0.00e+00

[1] http://sgpp.sparsegrids.org/.

Table 2 Three-level MLSC for the linear oscillator test case: 5D stochastic setup, no dimensionality reduction

L_0	L_1	L_2	g_f	rc	nrp/l	r	Error $\mathbb{E}$	Error Var
71	351	1471	–	–	–	–	1.40e−05	2.04e−05
71	187	413	–	MAS	20%	–	1.34e−04	2.64e−05
71	187	407	–	EVR	20%	–	1.63e−04	3.10e−05
71	251	799	–	MAS	40%	–	1.42e−05	2.04e−05
71	251	799	–	EVR	40%	–	1.42e−04	2.04e−05
61	231	681	2	–	–	–	5.63e−09	3.64e−08
61	125	177	2	AVG	10	–	2.78e−09	8.56e−08
61	97	133	2	MAX	10	0.05	2.36e−08	2.10e−05
61	121	171	2	MAX	10	0.50	5.19e−09	1.88e−07
61	121	181	2	MAX	10	0.95	3.17e−09	9.71e−08

The quantities of interest (QoI) are the expectation and variance of the displacement $y(t)$ measured at $t_0 = 10$, computed using Eqs. (11) and (12). We employ a gPC expansion with index set $\mathscr{P} = \{\mathbf{p} = (p_1, \ldots, p_d) \in \mathbb{N}^d \colon |\mathbf{p}|_1 < 5\}$ (cf. Sect. 3.1).

For the hierarchical discretization of the problem and the stochastic domains, we consider three levels. Therefore, $K = 2$ and $\mathscr{K} = \{m_n^{M_0,L_2}(t_0), m_n^{M_1,L_1}(t_0), m_n^{M_2,L_0}(t_0)\}$ (cf. Sect. 3.2). We discretize the model using Adams predictor-corrector methods from the `scicpy.integrate`[2] package and consider $M_0 = 500$, $M_1 = 2000$, and $M_2 = 8000$ time steps in the discretization scheme. Furthermore, for the stochastic domain discretization, we consider both standard and adaptive sparse grids. To analyze the possibility for dimensionality reduction, we consider two choices for $\mathscr{J} \subset \mathscr{K}$. In the first choice, $\mathscr{J}_1 = \{m_n^{M_1,L_1}(t_0)\}$; we choose the "middle" configuration (cf. Fig. 4) because it comprises grids of intermediate accuracy, thus avoiding potentially inaccurate results. In the second choice, $\mathscr{J}_2 = \{m_n^{M_0,L_2}(t_0), m_n^{M_1,L_1}(t_0)\}$. Finally, we compute reference results using M_2 time steps for the time domain discretization and a full tensor grid with 32768 Gauss-Legendre nodes for the stochastic space discretization. We assess the difference between reference and our MLSC results via their relative error, i.e. $|(QoI_{ref} - QoI)/QoI_{ref}|$. The reference results are $\mathbb{E}_{ref}[y(t_0)] = 1.95e - 02$ and $\text{Var}_{ref}[y(t_0)] = 1.03e - 02$.

The results of the corresponding simulations are listed in Tables 2, 3, 4, 5, 6 and 7. In the first three columns, we have the number of grid points corresponding to $L_q, q = 0, 1, 2$. For both spatially- and dimension-adaptive sparse grids, $L_0 = 3$. In the following three columns, g_f denotes the growth factor for Leja points, rc stands for refinement criterion, and nrp/l stands for the number of refinement points/levels. For spatially-adaptive sparse grids, nrp/l denotes the percentage of the current grid points to be refined. In the dimension-adaptive case, nrp/l stands for the number of levels added to the current index set (cf. Sect. 2.2). In the fifth

[2]https://docs.scipy.org/doc/scipy-0.18.1/reference/integrate.html.

Table 3 Three-level MLSC for the linear oscillator test case: 5D stochastic setup, dimensionality reduction for $\mathscr{I}_1 = \{m_n^{M_1,L_1}(t_0)|t_0 = 10\}$

L_0	L_1	L_2	g_f	rc	nrp/l	r	Error $\mathbb{E}$	$S_1^T(t_0)$	$S_2^T(t_0)$	$S_3^T(t_0)$	$S_4^T(t_0)$	$S_5^T(t_0)$
17	351	129	–	–	–	–	1.69e−02	0.021	0.106	7.3e−05	3.9e−04	0.879
17	187	41	–	MAS	20%	–	1.64e−02	0.021	0.106	7.3e−05	3.9e−04	0.879
17	187	41	–	EVR	20%	–	1.64e−02	0.021	0.106	7.3e−05	3.9e−04	0.879
17	251	81	–	MAS	40%	–	1.60e−02	0.021	0.106	7.3e−05	3.9e−04	0.879
17	251	80	–	EVR	40%	–	1.60e−02	0.021	0.106	7.3e−05	3.9e−04	0.879
13	231	41	2	–	–	–	1.69e−02	0.021	0.106	7.3e−05	3.9e−04	0.879
13	125	81	2	AVG	10	–	1.69e−02	0.021	0.106	7.3e−05	3.9e−04	0.879
13	97	73	2	MAX	10	0.05	1.69e−02	0.021	0.106	7.3e−05	3.9e−04	0.879
13	121	73	2	MAX	10	0.50	1.69e−02	0.021	0.106	7.3e−05	3.9e−04	0.879
13	121	73	2	MAX	10	0.95	1.69e−02	0.021	0.106	7.3e−05	3.9e−04	0.879

Table 4 Three-level MLSC for the linear oscillator test case: 5D stochastic setup, dimensionality reduction for $\mathscr{I}_2 = \{m_n^{M_0,L_2}(t_0), m_n^{M_1,L_1}(t_0)|t_0 = 10\}$

L_0	L_1	L_2	g_f	rc	nrp/l	r	Error $\mathbb{E}$	$S_1^T(t_0)$	$S_2^T(t_0)$	$S_3^T(t_0)$	$S_4^T(t_0)$	$S_5^T(t_0)$
17	351	1471	–	–	–	–	1.72e−02	0.021	0.106	7.3e−05	3.9e−04	0.879
17	187	413	–	MAS	20%	–	1.59e−02	0.021	0.106	7.3e−05	3.9e−04	0.879
17	187	407	–	EVR	20%	–	1.58e−02	0.021	0.106	7.3e−05	3.9e−04	0.879
17	251	799	–	MAS	40%	–	1.63e−02	0.021	0.106	7.3e−05	3.9e−04	0.879
17	251	854	–	EVR	40%	–	1.59e−02	0.021	0.106	7.3e−05	3.9e−04	0.879
13	231	681	2	–	–	–	9.40e−02	0.021	0.106	7.3e−05	3.9e−04	0.879
13	125	177	2	AVG	10	–	1.69e−02	0.021	0.106	7.3e−05	3.9e−04	0.879
13	97	133	2	MAX	10	0.05	1.69e−02	0.021	0.106	7.3e−05	3.9e−04	0.879
13	121	171	2	MAX	10	0.50	1.69e−02	0.021	0.106	7.3e−05	3.9e−04	0.879
13	121	181	2	MAX	10	0.95	1.69e−02	0.021	0.106	7.3e−05	3.9e−04	0.879

Table 5 Three-level MLSC for the linear oscillator test case: 6D stochastic setup, no dimensionality reduction

L_0	L_1	L_2	g_f	rc	nrp/l	r	Error $\mathbb{E}$	Error Var
545	2561	10,625	–	–	–	–	2.37e−05	3.81e−05
545	1452	3709	–	MAS	20%	–	2.19e−05	3.74e−05
545	1452	3691	–	EVR	20%	–	2.20e−05	3.75e−05
545	1970	7142	–	MAS	40%	–	2.36e−05	3.81e−05
545	1970	7084	–	EVR	40%	–	2.36e−05	3.81e−05
377	1289	3653	2	–	–	–	1.07e−07	4.34e−05
377	489	601	2	AVG	10	–	1.09e−04	1.81e−04
377	415	451	2	MAX	10	0.05	3.11e−06	4.37e−05
377	415	451	2	MAX	10	0.50	3.11e−07	4.37e−05
377	497	609	2	MAX	10	0.95	1.09e−04	1.81e−04

Table 6 Three-level MLSC for the linear oscillator test case: 6D stochastic setup, dimensionality reduction for $\mathscr{J}_1 = \{m_n^{M_1,L_1}(t_0) | t_0 = 10\}$

L_0	L_1	L_2	g_f	rc	nrp/l	r	Error $\mathbb{E}$	$S_1^T(t_0)$	$S_2^T(t_0)$	$S_3^T(t_0)$	$S_4^T(t_0)$	$S_5^T(t_0)$	$S_6^T(t_0)$
111	2561	1023	–	–	–	–	1.16e−02	0.0181	0.0824	0.0034	0.2359	0.0003	0.6675
111	1452	417	–	MAS	20%	–	1.67e−02	0.0181	0.0824	0.0034	0.2359	0.0003	0.6675
111	1452	417	–	EVR	20%	–	1.67e−02	0.0181	0.0824	0.0034	0.2359	0.0003	0.6675
111	1970	717	–	MAS	40%	–	1.67e−02	0.0181	0.0824	0.0034	0.2359	0.0003	0.6675
111	1970	717	–	EVR	40%	–	1.67e−02	0.0181	0.0824	0.0034	0.2359	0.0003	0.6675
63	1289	231	2	–	–	–	1.33e−02	0.0181	0.0824	0.0033	0.2358	0.0003	0.6676
63	489	149	2	AVG	10	–	1.22e−02	0.0181	0.0824	0.0033	0.2357	0.0003	0.6676
63	415	131	2	MAX	10	0.05	1.12e−02	0.0181	0.0825	0.0033	0.2357	0.0003	0.6677
63	415	131	2	MAX	10	0.50	1.12e−02	0.0181	0.0825	0.0033	0.2357	0.0003	0.6677
63	497	131	2	MAX	10	0.95	1.12e−02	0.0181	0.0825	0.0033	0.2357	0.0003	0.6677

Table 7 Three-level MLSC for the linear oscillator test case: 6D stochastic setup, dimensionality reduction for $\mathscr{J}_2 = \{m_n^{M_0,L_2}(t_0), m_n^{M_1,L_1}(t_0) | t_0 = 10\}$

L_0	L_1	L_2	g_f	rc	nrp/l	r	Error $\mathbb{E}$	$S_1^T(t_0)$	$S_2^T(t_0)$	$S_3^T(t_0)$	$S_4^T(t_0)$	$S_5^T(t_0)$	$S_6^T(t_0)$
111	2561	10,625	–	–	–	–	1.21e−02	0.0182	0.0825	0.0034	0.2358	0.0003	0.6675
111	1452	3709	–	MAS	20%	–	1.72e−02	0.0181	0.0825	0.0034	0.2359	0.0003	0.6675
111	1452	3961	–	EVR	20%	–	1.72e−02	0.0181	0.0825	0.0034	0.2359	0.0003	0.6675
111	1970	7142	–	MAS	40%	–	1.72e−02	0.0181	0.0825	0.0034	0.2358	0.0003	0.6675
111	1970	7084	–	EVR	40%	–	1.72e−02	0.0181	0.0825	0.0034	0.2358	0.0003	0.6675
63	1289	3653	2	–	–	–	1.33e−02	0.0181	0.0825	0.0033	0.2358	0.0003	0.6775
63	489	601	2	AVG	10	–	1.22e−02	0.0181	0.0825	0.0033	0.2357	0.0003	0.6677
63	415	451	2	MAX	10	0.05	1.27e−02	0.0181	0.0825	0.0033	0.2358	0.0003	0.6676
63	415	451	2	MAX	10	0.50	1.27e−02	0.0181	0.0825	0.0033	0.2358	0.0003	0.6676
63	497	609	2	MAX	10	0.95	1.22e−02	0.0181	0.0825	0.0033	0.2357	0.0003	0.6676

column, r is the weight between cost and error in Eq. (9). Finally, when performing dimensionality reduction, $\{S_i^T(t_0)\}_{i=1}^d$ denotes the total Sobol' index computed for the gPC coefficients in $\mathscr{J}$.

In the first test, we perform standard MLSC. The results are outlined in Table 2. First, we observe that although we begin with a similar number of grid points for L_0 (71 for spatially-adaptive, 61 for dimension-adaptive sparse grids), for L_1 and L_2 the linear growth of Leja leads to considerably fewer points for the dimension-adaptive interpolation. Moreover, we observe that interpolation on dimension-adaptive grids yields smaller expectation errors by at least a factor of $\mathscr{O}(10^{-3})$, whereas, the variance errors are similar or smaller. Therefore, for this setup, dimension-adaptive interpolation is superior in terms of accuracy and required computational cost.

Next, we analyze the possibility to reduce the stochastic dimensionality using a threshold $\tau = 0.05$ and $\mathscr{J}_1 = \{m_n^{M_1,L_1}(t_0)\}$ or $\mathscr{J}_2 = \{m_n^{M_0,L_2}(t_0), m_n^{M_1,L_1}(t_0)\}$. The results are listed in Tables 3 and 4. We observe that for both $\mathscr{J}_1$ and $\mathscr{J}_2$ the total Sobol' indices corresponding to c, f, and y_0 are at least two times smaller than τ. Therefore, we can reduce the dimensionality from five to two, significantly reducing the number of grid points for L_0, and, when employing J_1, for L_2, too. Note that in both Tables 3 and 4, $\{S_i^T(t_0)\}_{i=1}^5$ are identical rounded up to at least three digits of precision; this is due to very similar corresponding gPC coefficients. Furthermore, in our tests, the total Sobol' indices corresponding to the ignored uncertain inputs are significantly smaller than τ. In situations when some of the Sobol' indices are close to τ, to avoid repeating potentially expensive computations, we recommend enlarging the initial J (e.g. in the current setup, J_2 is a viable extension to J_1).

In the 6D stochastic scenario, the setup is the same as in the 5D case, except that $L_0 = 4$ for both spatially- and dimension-adaptive sparse grids. The reference results $\mathbb{E}_{ref}[y(t_0)] = 2.85e-02$ and $\mathrm{Var}_{ref}[y(t_0)] = 1.35e-02$ are computed using M_2 time steps and a full stochastic grid comprising 262144 Gauss-Legendre nodes.

In Table 5, we outline the results for the standard MLSC. Similar to the 5D case, the accuracy/cost ratio of dimension-adaptive grids is larger compared to spatially-adaptive grids. Therefore, for this setup, too, dimension-adaptive interpolation is superior in terms of accuracy and required computational cost. Tables 6 and 7 contain the results for the two dimensionality reduction strategies. Similar to the 5D setup, the total Sobol' indices for c ($S_1^T(t_0)$), f ($S_3^T(t_0)$), and y_0 ($S_5^T(t_0)$) are smaller than τ by more than a factor of two; the difference is that the major contributions are due to ω and y_1. Therefore, we can reduce the dimensionality from six to three, thus, significantly lowering the number of grid points and the overall computational cost.

We end this section with an important remark referring to the total Sobol' indices in our results. Since the total Sobol' indices comprise both the first and higher order contributions (cf. Sect. 3.1), they generally sum up to more than 100%. However, in both 5D and 6D scenarios, the sum of the total Sobol' indices is only a little more than 100%. Thus, the interactions between the uncertain inputs are negligible and the underlying stochastic model could be well approximated by a multi-linear functional.

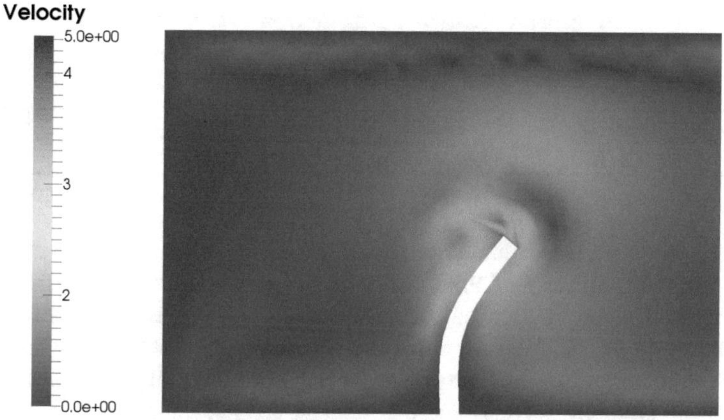

Fig. 5 Channel flow with an elastic vertical flap. We have a parabolic inflow at the left and an outflow at the right boundary. Furthermore, at the top and bottom boundaries, we prescribe no-slip conditions. The inflow excites a periodic bending movement of the vertical structure

4.2 A simple Fluid-Structure Interaction Example

The second test case, a simple FSI example, is depicted in Fig. 5. The fluid flow is governed by the incompressible Navier-Stokes equations, formulated in the arbitrary Lagrangian-Eulerian approach [8] to allow for moving geometries. A non-linear Saint-Venant-Kirchhoff model governs the elastic structure. The fluid and structure domains are discretized via finite elements (FEM). The FEM meshes match at the boundary, where we enforce balancing between stresses and displacements. To couple the fluid flow and structure solvers, we use a Dirichlet-Neumann implicit partitioned approach and sub-iterate until convergence in every time step. A quasi-Newton scheme [19] is used to stabilize and accelerate the sub-iterations. In this way, both solvers can be simultaneously executed. Therefore, taking advantage of the non-intrusiveness of MLSC, we end up with a total of three layers of parallelism: both solvers run on distributed data, they run in parallel to each other, and simulations corresponding to different points in the stochastic domain can be performed in parallel, too (cf. [9]).

To simulate the fluid flow and structure solvers, we employ the high-performance multi-physics code `Alya` [34] as a black box. For coupling, we use the open-source library `preCICE`[3] [4]. For more details about the Alya-preCICE coupling, see [33]. For each simulation, the initial data in the fluid domain is obtained via precomputing a full period of the time domain with a rigid structure to ensure a stabilized initial fluid field and only then start the coupled simulation; the structure is initially at rest.

[3] http://www.precice.org/.

Table 8 FSI test case: uniform bounds for $\boldsymbol{\theta}$ and associated deterministic values

	ρ_f [kg/m^3]	μ [kg/m s]	ρ_s [kg/m^3]	E [kg/m s^2]	ν
a_i	7.40e−01	7.40e−02	7.40e−01	3.70e+05	2.22e−01
b_i	1.26e+00	1.26e−01	1.26e+00	6.30e+05	3.78e−01
$\bar{u}_i$	1.00e+00	1.00e−01	1.00e+00	5.00e+05	3.00e−01

In this test case, $\mathbf{x} = t$ and $\boldsymbol{\theta} = (\rho_f, \mu, \rho_s, E, \nu)$, where ρ_f is the density and μ the dynamic viscosity of the fluid, whereas ρ_s is the density, E the Young's modulus, and ν the Poisson ratio corresponding to the elastic structure. We use a time step $\delta t = 10^{-3}$ to discretize the time domain. For both solvers, we use implicit time integrators (implicit Euler for the flow solver and a Newmark scheme [21] for the structure solver) such that we can use the same timestep size for all meshes. We do, hence, not expect stability issues and have also not observed any in our experiments. We simulate 500 time steps, therefore, $t \in [0, 0.5]$. Moreover, $\boldsymbol{\theta}$ is a multivariate random vector with uniform i.i.d. components, whose a_i, b_i, and associated deterministic values (i.e. their expectations), are listed in Table 8.

We employ a multi-variate gPC expansion such that $|\mathbf{p}|_1 < 5$. The QoI are the expectation and variance of the x axis displacement (x disp.) measured at the upper right corner of the structure. For a broader overview of our approach, we measure the displacement at two time instances, one close to center of the time domain, i.e. $t_0 = 0.235$, and another one at the right boundary, i.e. $t_1 = 0.500$. We employ a three-level discretization hierarchy in both the problem and the stochastic domain. $M_0 = \{\mathrm{f} : 1568,\ \mathrm{s} : 40\}$ triangular elements for the fluid (f) and structure (s) domains, $M_1 = \{\mathrm{f} : 6272,\ \mathrm{s} : 160\}$, and $M_2 = \{\mathrm{f} : 25088,\ \mathrm{s} : 640\}$. On a 16 core Intel Sandy Bridge processor on the CoolMAC cluster,[4] with 12 cores for the fluid flow and four cores for the structure solver, one run with resolution M_0 takes about 10 min, about 15 min for M_1, and about 60 min for M_2. To discretize the stochastic domain, we use spatially-adaptive grids with the MAS criterion and dimension-adaptive grids with the AVG priority function. For both spatially- and dimension-adaptive grids, $L_0 = 3$. Moreover, L_1 and L_2 refer to L_0 after 1 and 2 refinement steps, respectively. Finally, to analyze the possibility for dimensionality reduction, we consider $\tau = 0.05$ and $\mathscr{I} = \{m_n^{M_1,L_1}(t_0)\}$.

In Table 9, we list the results when $t_0 = 0.235$. We observe that the Sobol' indices associated to the fluid's dynamic viscosity and structure's density and Poisson ratio are significantly smaller than τ. Therefore, we can reduce the dimensionality from five to two, hence, the sparse grids of resolution L_0 and L_2 are now two-dimensional. In terms of computational savings, this translates into reducing the number of grid points from 71 to 17—for spatially-adaptive—and from 61 to 13—for dimension-adaptive grids for the combination $M_2 - L_0$. Therefore, we save approximately 54 and 48 h of computing time, respectively. In Table 10, we present our results when $t_1 = 0.500$. In this case, we observe that the Sobol' index assciated

[4]http://www.mac.tum.de/wiki/index.php/MAC$_$Cluster.

Table 9 Three-level MLSC for the FSI test case: 5D stochastic setup, dimensionality reduction for $\mathscr{J} = \{m_n^{M_1,L_1}(t_0)|t_0 = 0.235\}$

L_0	L_1	L_2	g_f	rc	nrp/l	$\mathbb{E}$[x disp.(t_0)]	Var[x disp.(t_0)]	$S_1^T(t_0)$	$S_2^T(t_0)$	$S_3^T(t_0)$	$S_4^T(t_0)$	$S_5^T(t_0)$
17	184	43	–	MAS	20%	3.23e−01	8.14e−04	0.423	0.007	0.009	0.598	0.002
13	117	75	2	AVG	10	3.23e−01	8.12e−04	0.423	0.007	0.008	0.596	0.002

Table 10 Three-level MLSC for the FSI test case: 5D stochastic setup, dimensionality reduction for $\mathscr{J} = \{m_n^{M_1,L_1}(t_1)|t_1 = 0.500\}$

L_0	L_1	L_2	g_f	rc	nrp/l	$\mathbb{E}$[x disp.(t_1)]	Var[x disp.(t_1)]	$S_1^T(t_1)$	$S_2^T(t_1)$	$S_3^T(t_1)$	$S_4^T(t_1)$	$S_5^T(t_1)$
49	179	214	–	MAS	20%	2.88e−01	4.60e−04	0.297	0.354	0.163	0.572	0.007
41	121	161	2	AVG	10	2.88e−01	5.22e−04	0.284	0.370	0.152	0.572	0.006

to the Poisson ratio is less that τ by more than a factor of 10. Thus, we can reduce the dimensionality from five to four, saving approximately 22—for spatially adaptive—and 20—for dimension-adaptive grids—hours of computing time at level L_0. In addition, in both dimensionality reduction cases, the computational cost is further decreased due to the significant reduction of number of grid points on level L_2.

Note that in Table 9, on the one hand, the total Sobol' indices sum up to only a little more than 100%, indicating that the interactions between the uncertain inputs are very week (cf. Sect. 4.1). On the other hand, in Table 10, the total Sobol' indices add up to approximately 140%. Therefore, the interactions between the uncertain inputs are stronger than in the previous case. We assume that this difference is due to both the setup of the underlying scenario and nonlinearity of FSI. However, as our proposed approach relies on the individual values of the total Sobol' indices, the result of their summation does not impose a limitation to our method.

Finally, we remark that the single level variant corresponding—with respect to accuracy—to our multi-level formulation would be computationally more expensive, since it requires the evaluation of the time discretization with the highest fidelity (M_2) at each point from the sparse grid having the highest resolution (L_2). Even more, no dimensionality reduction based on total Sobol' indices could be performed.

5 Conclusions and Outlook

We have presented a novel multilevel stochastic collocation with a dimensionality reduction approach to compute the coefficients of generalized polynomial chaos approximations. Our focus was on formulating a computational method suitable to quantify the uncertainty in resource-intensive problems. To this end, our first contribution was to employ both spatially- and dimension-adaptive sparse grids for the stochastic domain discretization. Our second and main contribution was stochastic dimensionality reduction based on total Sobol' indices for global sensitivity analysis. Therefore, our method exploits the advantages of adaptive sparse grids, multi-level decompositions, and, when feasible, stochastic dimensionality reduction, to lower the overall computational cost of the uncertainty analysis. We remark that when dimensionality reduction is not possible, we still profit from using adaptive sparse grids and multilevel decompositions.

The results in two test cases—a second-order linear oscillator and a fluid-structure interaction problem—showed that the stochastic dimensionality could be reduced. Furthermore, the dimension-adaptive interpolants proved superior.

We analyzed once whether the stochastic dimensionality can be reduced. Nevertheless, this approach can be easily generalized. The generalized approach can be suitable for high-dimensional UQ problems or problems in which some total Sobol' indices are close to the prescribed threshold. Furthermore, our approach can be also formulated in terms of other sensitivity measures, such as the Shapley value [23].

Acknowledgements We thank David Holzmueller for developing the dimension-adaptive interpolation module in SG++, employed in this paper. Moreover, we thankfully acknowledge the financial support of the German Academic Exchange Service (http://daad.de/), of the German Research Foundation through the TUM International Graduate School of Science and Engineering (IGSSE) within the project 10.02 BAYES (http://igsse.tum.de/), and the financial support of the priority program 1648 - Software for Exascale Computing of the German Research Foundation (http://www.sppexa.de).

References

1. H.-J. Bungartz, Finite elements of higher order on sparse grids. Habilitationsschrift, Fakultät für Informatik, Technische Universität München, Shaker Verlag, Aachen, 1998
2. H.-J. Bungartz, S. Dirnstorfer, Multivariate quadrature on adaptive sparse grids. Computing **71**(1), 89–114 (2003)
3. H.-J. Bungartz, M. Griebel, Sparse grids. Acta Numer. **13**, 147–269 (2004)
4. H.-J. Bungartz, F. Lindner, B. Gatzhammer, M. Mehl, K. Scheufele, A. Shukaev, B. Uekermann, preCICE – a fully parallel library for multi-physics surface coupling. Comput. Fluids (2016). https://doi.org/10.1016/j.compfluid.2016.04.003
5. P. Chen, A. Quarteroni, A new algorithm for high-dimensional uncertainty quantification based on dimension-adaptive sparse grid approximation and reduced basis methods. J. Comput. Phys. **298**(Supplement C), 176–193 (2015)
6. P.R. Conrad, Y.M. Marzouk, Adaptive smolyak pseudospectral approximations. SIAM J. Sci. Comput. **35**(6), A2643–A2670 (2013)
7. P.G. Constantine, M.S. Eldred, E.T. Phipps, Sparse pseudospectral approximation method. Comput. Methods Appl. Mech. Eng. **229–232**, 1–12 (2012)
8. J. Donea, A. Huerta, J.-P. Ponthot, A. Rodriguez-Ferran, Arbitrary lagrangian-eulerian methods. Encycl. Comput. Mech. **1**, 413–437 (2004)
9. I.-G. Farcas, B. Uekermann, T. Neckel, H.-J. Bungartz, Nonintrusive uncertainty analysis of fluid-structure interaction with spatially adaptive sparse grids and polynomial chaos expansion. SIAM J. Sci. Comput. **40**(2), B457–B482. https://doi.org/10.1137/16M1093975
10. F. Franzelin, P. Diehl, D. Pflüger, Non-intrusive uncertainty quantification with sparse grids for multivariate peridynamic simulations, in *Meshfree Methods for Partial Differential Equations VII*, ed. by M. Griebel, M.A. Schweitzer. Lecture Notes in Computational Science and Engineering, vol. 100 (Springer International Publishing, Cham, 2015), pp. 115–143

11. T. Gerstner, M. Griebel, Numerical integration using sparse grids. Numer. Algorithms **18**(3), 209–232 (1998)
12. T. Gerstner, M. Griebel, Dimension–adaptive tensor–product quadrature. Computing **71**(1), 65–87 (2003)
13. M. Griebel, M. Schneider, C. Zenger, A combination technique for the solution of sparse grid problems, in *Iterative Methods in Linear Algebra*, ed. by P. de Groen, R. Beauwens (IMACS, Elsevier, North Holland, 1992), pp. 263–281
14. F. Heiss, V. Winschel, Likelihood approximation by numerical integration on sparse grids. J. Econometrics **144**(1), 62–80 (2008)
15. A. Klimke, Uncertainty modeling using fuzzy arithmetic and sparse grids. PhD thesis, Universität Stuttgart, Shaker Verlag, Aachen, 2006
16. F. Leja, Sur certaines suites liées aux ensemble plan et leur application à la representation conforme. Ann. Polon. Math. **5**, 8–13 (1957)
17. X. Ma, N. Zabaras, An adaptive hierarchical sparse grid collocation algorithm for the solution of stochastic differential equations. J. Comput. Phys. **228**(8), 3084–3113 (2009)
18. S.D. Marchi, On leja sequences: some results and applications. Appl. Math. Comput. **152**(3), 621–647 (2004)
19. M. Mehl, B. Uekermann, H. Bijl, D. Blom, B. Gatzhammer, A. van Zuijlen, Parallel coupling numerics for partitioned fluid-structure interaction simulations. Comput. Math. Appl. **71**(4), 869–891 (2016)
20. A. Narayan, J.D. Jakeman, Adaptive leja sparse grid constructions for stochastic collocation and high-dimensional approximation. SIAM J. Sci. Comput. **36**(6), A2952–A2983 (2014)
21. N.M. Newmark, A method of computation for structural dynamics. J .Eng. Mech. Div. **85**(3), 67–94 (1959)
22. F. Nobile, R. Tempone, C.G. Webster, A sparse grid stochastic collocation method for partial differential equations with random input data. SIAM J. Numer. Anal. **46**(5), 2309–2345 (2008)
23. A.B. Owen, Sobol' indices and shapley value. SIAM/ASA J. Uncertain. Quantif. **2**(1), 245–251 (2014)
24. B. Peherstorfer, Model order reduction of parametrized systems with sparse grid learning techniques. Dissertation, Department of Informatics, Technische Universität München, 2013
25. D. Pflüger, *Spatially Adaptive Sparse Grids for High-Dimensional Problems* (Verlag Dr. Hut, München, 2010)
26. K. Sargsyan, C. Safta, H.N. Najm, B.J. Debusschere, D. Ricciuto, P. Thornton, Dimensionality reduction for complex models via bayesian compressive sensing. Int. J. Uncertain. Quantif. **4**(1), 63–93 (2014)
27. S. Smolyak, Quadrature and interpolation formulas for tensor products of certain classes of functions. Sov. Math. Dokl. **4**, 240–243 (1963)
28. I. Sobol, Global sensitivity indices for nonlinear mathematical models and their monte carlo estimates. Math. Comput. Simul. **55**(1–3), 271–280 (2001)
29. D. Stirzaker, *Elementary Probability* (Cambridge University Press, Cambridge, 2003)
30. B. Sudret, Global sensitivity analysis using polynomial chaos expansions. Reliab. Eng. Syst. Saf. **93**(7), 964–979 (2008)
31. A.L. Teckentrup, P. Jantsch, C.G. Webster, M. Gunzburger, A multilevel stochastic collocation method for partial differential equations with random input data.SIAM/ASA J. Uncertain. Quantif. **3**(1), 1046–1074 (2015)
32. U. Trottenberg, A. Schuller, *Multigrid* (Academic, Orlando, 2001)
33. B. Uekermann, J.C. Cajas, B. Gatzhammer, G. Houzeaux, M. Mehl, M. Vazquez, Towards partitioned fluid-structure interaction on massively parallel systems, in *11th World Congress on Computational Mechanics (WCCM XI)*, ed. by E. Oñate, J. Oliver, A. Huerta (2014), pp. 1–12
34. M. Vázquez, G. Houzeaux, S. Koric, A. Artigues, J. Aguado-Sierra, R. Arís, D. Mira, H. Calmet, F. Cucchietti, H. Owen, A. Taha, E.D. Burness, J.M. Cela, M. Valero, Alya: multiphysics engineering simulation toward exascale. J. Comput. Sci. **14**, 15–27 (2016)

35. D. Xiu, *Numerical Methods for Stochastic Computations: A Spectral Method Approach* (Princeton University Press, Princeton, 2010)
36. D. Xiu, G.E. Karniadakis, The Wiener-Askey polynomial chaos for stochastic differential equations. SIAM J. Sci. Comput. **24**, 619–644 (2002)
37. C. Zenger, Sparse grids, in *Parallel Algorithms for Partial Differential Equations, Proceedings of the Sixth GAMM-Seminar, Kiel, 1990*, ed. by W. Hackbusch. Notes on Numerical Fluid Mechanics, vol. 31 (Vieweg Verlag, Braunschweig, 1991), pp. 241–251

Limiting Ranges of Function Values of Sparse Grid Surrogates

Fabian Franzelin and Dirk Pflüger

Abstract Sparse grid interpolants of high-dimensional functions do not maintain the range of function values. This is a core problem when one is dealing with probability density functions, for example. We present a novel approach to limit range of function values of sparse grid surrogates. It is based on computing minimal sets of sparse grid indices that extend the original sparse grid with properly chosen coefficients such that the function value range of the resulting surrogate function is limited to a certain interval. We provide the prerequisites for the existence of minimal extension sets and formally derive the intersection search algorithm that computes them efficiently. The main advantage of this approach is that the surrogate remains a linear combination of basis functions and, therefore, any problem specific post-processing operation such as evaluation, quadrature, differentiation, regression, density estimation, etc. can remain unchanged. Our sparse grid approach is applicable to arbitrarily refined sparse grids.

1 Introduction

A core problem of the representation of high-dimensional functions on sparse grids (interpolation, approximation) is that they do not maintain the range of function values. For example, the interpolation of a strictly positive function in more than one variable can lead to negative function values of the interpolant. This can be a severe limitation to a wide range of problems. Think about physical problems where values outside of a certain interval do not make sense and a surrogate leading to such values is useless. In the context of environmental engineering, for example, one is interested in finding storage sites for carbon dioxide with a small leakage, which is non-negative. An early example for the problem that the sparse grid function is not necessarily positive everywhere, even if the function values at

F. Franzelin · D. Pflüger (✉)
Institute for Parallel and Distributed Systems, University of Stuttgart, Stuttgart, Germany
e-mail: fabian.franzelin@ipvs.uni-stuttgart.de; dirk.pflueger@ipvs.uni-stuttgart.de

J. Garcke et al. (eds.), *Sparse Grids and Applications – Miami 2016*,
Lecture Notes in Computational Science and Engineering 123,
https://doi.org/10.1007/978-3-319-75426-0_4

all interpolation points are positive, stems from the interpolation of multivariate Gaussian densities [5, 12], which will serve as an extreme scenario in this paper.

This problem translates to approximation problems such as regression or density estimation. There, the probability density function of some unknown distribution is recovered based on samples drawn from that distribution. In [8], the authors published the theoretical fundamentals for a density estimation method based on spline smoothing. Their idea became very popular in the following years in the context of sparse grids in various research areas [6, 9–11], the question of how to conserve fundamental properties of probability density functions such as the non-negativity, has not yet been investigated so far. Furthermore, the sparse grid function is computed via the solution of a linear system: The approximation process does not even ensure that the sparse grid function is positive at the grid points themselves.

The general problem of limiting the range of sparse grid function values to a certain interval can be transformed into the problem of assuring non-negative function values. A proper transformation of the original approximation problem is an elegant way to solve this problem. Such an approach has been introduced in [5] in the context of likelihood estimation and in [7] for density estimation. It is based on the idea of approximating the logarithm of the probability density instead of the density itself. They characterize the density by the maximum a posteriori method with Gaussian process priors. By definition, this approach solves the non-negativity problem but makes it difficult to assert the normalization condition since the computation of a high-dimensional integral of an exponential function with an exponent that is a sparse grid function is required. Simpler approaches, such as a truncation of the range of function values, are not an option here either. They usually raise other problems, for example efficient integration or differentiation of the resulting function [4], which we want to avoid.

In this paper we introduce a new method that restricts the function value range of a sparse grid surrogate to the one of the original function while the original structure of the surrogate as a linear combination of basis functions is preserved. The method adds a minimal extension set with properly chosen hierarchical coefficients to the sparse grid surrogate to achieve that goal. We formally derive various methods to efficiently compute minimal extension sets for arbitrarily refined sparse grids.

This paper is structured as follows: First, we introduce briefly sparse grids and describe the properties that are required for our new method. In Sect. 3 we derive the method that limits ranges of function values of arbitrarily refined sparse grid surrogates by adding an extension set with properly chosen coefficients to the original sparse grid. To evaluate the method, we provide results for the approximation of multivariate Gaussians with extended sparse grids. In Sect. 5 we summarize the paper and give an outlook to future work.

2 Sparse Grids

Let $\mathbf{l} := \{l_1, \ldots, l_D\}$ and $\mathbf{i} := \{i_1, \ldots, i_D\}$ be multi-indices with dimensionality $0 < D \in \mathbb{N}$ and level $d = 1, \ldots, D$: $l_d > 0$. We define a nested level-index set as

$$\mathscr{I}_{\mathbf{l}}^{\mathrm{SG}} := \{\mathbf{i} \in \mathbb{N}^D : 1 \leq i_d < 2^{l_d}, i_d \text{ odd}, d = 1, \ldots, D\} . \quad (1)$$

The subsets for various levels and $D = 2$ are shown in Fig. 1 (center), the joint sets in Fig. 1 (right). With the piecewise linear one-dimensional reference basis function $\varphi(x) := \max\{1 - |x|, 0\}$ we obtain the higher-dimensional piecewise D-linear basis function of an arbitrary level-index pair $(\mathbf{l}, \mathbf{i})$ via tensor product and by translation and scaling as

$$\varphi_{\mathbf{l},\mathbf{i}}(\mathbf{x}) := \prod_{d=1}^{D} \varphi(2^{l_d} x_d - i_d) . \quad (2)$$

The level-index sets $\mathscr{I}_{\mathbf{l}}^{\mathrm{SG}}$ define a unique set of hierarchical increment spaces

$$\mathscr{W}_{\mathbf{l}} := \operatorname{span}\{\varphi_{\mathbf{l},\mathbf{i}}(\mathbf{x}) : \mathbf{i} \in \mathscr{I}_{\mathbf{l}}^{\mathrm{SG}}\} , \quad (3)$$

which are shown in Fig. 1 (left) for $D = 1$. The space $\mathscr{V}_\ell$ of piecewise D-linear functions for some level $\ell \in \mathbb{N}$ is now formed by all spaces with $|\mathbf{l}|_\infty \leq \ell$ where each increment space adds the difference of the corresponding nodes with respect to the nodal space. $\mathscr{V}_\ell$ and the corresponding full grid index set $\mathscr{I}_\ell^{\mathrm{FG}}$ are defined as

$$\mathscr{V}_\ell := \bigoplus_{\mathbf{l} \in \mathbb{N}^D : |\mathbf{l}|_\infty \leq \ell} \mathscr{W}_{\mathbf{l}} , \qquad \mathscr{I}_\ell^{\mathrm{FG}} := \bigcup_{\mathbf{l} \in \mathbb{N}^D : |\mathbf{l}|_\infty \leq \ell} \{(\mathbf{l}, \mathbf{i}) : \mathbf{i} \in \mathscr{I}_{\mathbf{l}}^{\mathrm{SG}}\} . \quad (4)$$

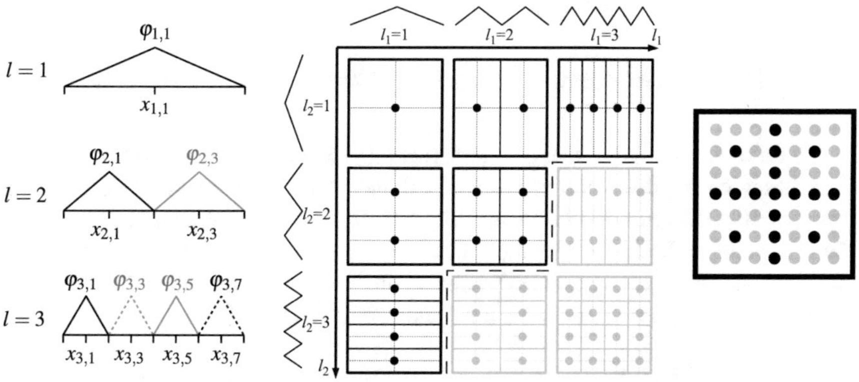

Fig. 1 One-dimensional piecewise linear basis functions up to level 3 with increment spaces $\mathscr{W}_1, \mathscr{W}_2, \mathscr{W}_3$ in one dimension (left), the tableau of hierarchical increments $\mathscr{W}_{\mathbf{l}}$ up to level 3 in two dimensions (center) and the resulting sparse grid $\mathscr{I}_3^{\mathrm{SG}}$ (black) and full grid $\mathscr{I}_3^{\mathrm{FG}}$ (black and gray) (right) [12]

With this, we can write a full grid function based on the hierarchical increment spaces as

$$u_{\mathscr{I}_\ell^{\mathrm{FG}}}(\mathbf{x}) := \sum_{(\mathbf{l},\mathbf{i})\in\mathscr{I}_\ell^{\mathrm{FG}}} v_{\mathbf{l},\mathbf{i}}\varphi_{\mathbf{l},\mathbf{i}}(\mathbf{x}) = \sum_{\mathbf{l}\in\mathbb{N}^D:\,|\mathbf{l}|_\infty<\ell}\ \sum_{\mathbf{i}\in\mathscr{I}_\mathbf{l}^{\mathrm{SG}}} v_{\mathbf{l},\mathbf{i}}\varphi_{\mathbf{l},\mathbf{i}}(\mathbf{x})\ . \tag{5}$$

We take now advantage of the hierarchical definition of the basis and reduce the number of increment spaces by measuring the contribution of each one to the approximation divided by the work we need to spend on it, i.e. the number of grid points. Based on this measure, we solve a continuous knapsack problem [1] and obtain the sparse grid space $\mathscr{V}_\ell^{(1)}$, which is optimal in the L_2 and L_∞-norms, and the corresponding sparse grid index set as

$$\mathscr{V}_\ell^{(1)} := \bigoplus_{|\mathbf{l}|_1\le\ell+D-1} \mathscr{W}_\mathbf{l}\,, \qquad \mathscr{I}_\ell^{\mathrm{SG}} := \bigcup_{\mathbf{l}\in\mathbb{N}^D:\,|\mathbf{l}|_1<\ell+D-1} \{(\mathbf{l},\mathbf{i}) : \mathbf{i}\in\mathscr{I}_\mathbf{l}^{\mathrm{SG}}\}\,, \tag{6}$$

see the left upper triangle in Fig. 1 (center). We write a sparse grid function $u_{\mathscr{I}_\ell^{\mathrm{SG}}} \in \mathscr{V}_\ell^{(1)}$ as

$$u_{\mathscr{I}_\ell^{\mathrm{SG}}}(\mathbf{x}) := \sum_{(\mathbf{l},\mathbf{i})\in\mathscr{I}_\ell^{\mathrm{SG}}} v_{\mathbf{l},\mathbf{i}}\varphi_{\mathbf{l},\mathbf{i}}(\mathbf{x}) = \sum_{\mathbf{l}\in\mathbb{N}^D:\,|\mathbf{l}|_1\le\ell+D-1}\ \sum_{\mathbf{i}\in\mathscr{I}_\mathbf{l}^{\mathrm{SG}}} v_{\mathbf{l},\mathbf{i}}\varphi_{\mathbf{l},\mathbf{i}}(\mathbf{x})\,, \tag{7}$$

where we call $v_{\mathbf{l},\mathbf{i}} \in \mathbb{R}$ hierarchical coefficients. Note that we omit the index ℓ and the superscript SG when it is obvious that we are referring to a sparse grid index set and the level is unimportant, e.g. for an adaptively refined sparse grid. Furthermore, for simplicity of notation we refer to a level-index pair $(\mathbf{l}, \mathbf{i}) \in \mathscr{I}$ via a simple index $i \in \mathscr{I}$ whenever the actual level and index are not important.

2.1 Hierarchical Ancestors and the Fundamental Property

In the following we introduce the fundamental property of sparse grid basis functions based on the definition of hierarchical ancestors. To this end it is convenient to interpret a one-dimensional sparse grid as a binary tree. As shown in Fig. 1 (left), the grid point on level 1 forms the root node; the grid points at level 2 are the left and right successor of the root node, and so on. With this idea, we define the set of hierarchical ancestors $\mathscr{A}$ for some level-index pair $(\mathbf{l}, \mathbf{i}) \in \mathscr{I}$ as

$$\mathscr{A}(\mathbf{l},\mathbf{i}) := \{(\mathbf{k},\mathbf{j}) \in \mathscr{I}_{|\mathbf{l}|_\infty}^{\mathrm{FG}} : k_d < l_d \wedge j_d = 2\lfloor i_d/2^{l_d-k_d+1}\rfloor + 1, d = 1,\dots,D\}\,, \tag{8}$$

and the set of ancestors that are part of an arbitrary level-index set $\mathscr{I}$ as

$$\mathscr{A}_{\mathscr{I}}(\mathbf{l}, \mathbf{i}) := \mathscr{A}(\mathbf{l}, \mathbf{i}) \cap \mathscr{I} . \tag{9}$$

We define two indices $(\mathbf{l}, \mathbf{i}), (\mathbf{k}, \mathbf{j}) \in \mathscr{I}$ to be hierarchically independent if

$$(\mathbf{k}, \mathbf{j}) \notin \mathscr{A}(\mathbf{l}, \mathbf{i}) \wedge (\mathbf{l}, \mathbf{i}) \notin \mathscr{A}(\mathbf{k}, \mathbf{j}) . \tag{10}$$

For all basis functions of a sparse grid basis that fulfill the *fundamental property* it holds that

$$\forall (\mathbf{l}, \mathbf{i}) \in \mathscr{I} : \begin{cases} \varphi_{\mathbf{k},\mathbf{j}}(\mathbf{x}_{\mathbf{l},\mathbf{i}}) = 1 & \text{for } \mathbf{l} = \mathbf{k} \wedge \mathbf{i} = \mathbf{j} \\ \varphi_{\mathbf{k},\mathbf{j}}(\mathbf{x}_{\mathbf{l},\mathbf{i}}) > 0 & \text{for } (\mathbf{k}, \mathbf{j}) \in \mathscr{A}(\mathbf{l}, \mathbf{i}) \\ \varphi_{\mathbf{k},\mathbf{j}}(\mathbf{x}_{\mathbf{l},\mathbf{i}}) = 0 & \text{else} . \end{cases} \tag{11}$$

This means that a basis function $\varphi_{\mathbf{k},\mathbf{j}}$ evaluates to zero at some grid point $\mathbf{x}_{\mathbf{l},\mathbf{i}}$ if $(\mathbf{k}, \mathbf{j})$ is not a hierarchical ancestor of $(\mathbf{l}, \mathbf{i})$. This is trivial to see for the one-dimensional case, due to the limitation of the support of the basis functions, see Fig. 1 (left). This property is preserved by the tensor product: If an index $(\mathbf{k}, \mathbf{j})$ is not a hierarchical ancestor of $(\mathbf{l}, \mathbf{i})$, then there exists at least one term in the tensor product which is equal to zero making the whole product zero.

2.2 *Interpolation on Sparse Grids*

Interpolation on sparse grids can be interpreted as a basis transformation from the nodal basis to the hierarchical basis. Assume that the sparse grid index set $\mathscr{I}$ is suitably sorted. Then the corresponding system of linear equations that solves the interpolation problem reads as

$$\mathbf{V}^{\mathrm{SG}}\mathbf{v} = \mathbf{u} \tag{12}$$

where $V^{\mathrm{SG}}_{(\mathbf{l},\mathbf{i}),(\mathbf{k},\mathbf{j})} = \varphi_{\mathbf{l},\mathbf{i}}(\mathbf{x}_{\mathbf{k},\mathbf{j}})$ and $u_{\mathbf{l},\mathbf{i}} = u(\mathbf{x}_{\mathbf{l},\mathbf{i}})$. Due to the fundamental property, this system of linear equations can be solved by forward substitution and the basis transformation can be written as

$$v_{\mathbf{l},\mathbf{i}} = u_{\mathbf{l},\mathbf{i}} - \sum_{(\mathbf{k},\mathbf{j}) \in \mathscr{A}_{\mathscr{I}}(\mathbf{l},\mathbf{i})} v_{\mathbf{k},\mathbf{j}} \varphi_{\mathbf{k},\mathbf{j}}(\mathbf{x}_{\mathbf{l},\mathbf{i}}) . \tag{13}$$

We see here directly that we have to solve all equations for the ancestors of $(\mathbf{l}, \mathbf{i})$ before we can solve the equation for $(\mathbf{l}, \mathbf{i})$. This is possible since the ancestor set just includes grid points which lie on subspaces with smaller level. Hence, a top-down approach solves the interpolation: One starts with the smallest available level, solves all the equations for grid points that lie on a subspace with the same levelsum, and proceeds with the next one until the maximum level is reached.

3 Limiting Ranges of Sparse Grid Function Values

In the following we derive a method that preserves the range of function values of some model function by a sparse grid approximation. Our approach is based on proper extension of the sparse grid level-index set such that the structure of the sparse grid function as a linear combination of basis functions is preserved. Let $u\colon [0,1]^D \to [y^{\mathrm{l}}, y^{\mathrm{u}}] \subset \mathbb{R}$ be some function with $y^{\mathrm{l}} < y^{\mathrm{u}}$ and $u_{\mathscr{I}^{\mathrm{SG}}}\colon [0,1]^D \to \mathbb{R}$ be a sparse grid function on a piecewise D-linear basis that approximates u. In the following we assume that $u_{\mathscr{I}^{\mathrm{SG}}}$ interpolates u at the sparse grid points. Note that the approach of extending the sparse grid index set is not directly applicable to other sparse grid approximations that stem from density estimation [11] or regression [12] if the function values at the grid points lie outside of $[y^{\mathrm{l}}, y^{\mathrm{u}}]$.

The problem of preserving ranges of function values can be reduced to the problem of enforcing non-negative function value ranges

$$u(\mathbf{x}) \in [y^{\mathrm{l}}, y^{\mathrm{u}}] :\Rightarrow u_{\mathscr{I}^{\mathrm{SG}}}(\mathbf{x}) \in [y^{\mathrm{l}}, y^{\mathrm{u}}] \Leftrightarrow u_{\mathscr{I}^{\mathrm{SG}}}(\mathbf{x}) - y^{\mathrm{l}} \geq 0 \wedge y^{\mathrm{u}} - u_{\mathscr{I}^{\mathrm{SG}}}(\mathbf{x}) \geq 0\,, \tag{14}$$

which allows us to generalize the problem to

$$u(\mathbf{x}) \geq 0 :\Rightarrow u_{\mathscr{I}^{\mathrm{SG}}}(\mathbf{x}) = \sum_{i \in \mathscr{I}^{\mathrm{SG}}} v_i \varphi_i(\mathbf{x}) \geq 0\,. \tag{15}$$

Note that adding constants to a sparse grid function can be done easily in the nodal space. Our approach to fulfill Eq. (15) is based on the minimum principle that holds for full grids with a nodal basis.

Theorem 1 (Minimum Principle for Full Grids with Nodal Basis) *Let* $u(\mathbf{x}) \geq 0$ *and* $u_{\mathscr{I}^{FG}}\colon [0,1]^D \to \mathbb{R}\,, \mathbf{x} \mapsto \sum_{i \in \mathscr{I}^{FG}} u_i \varphi_i(\mathbf{x})$ *be a full grid approximation of* u *on a piecewise D-linear nodal basis* φ. *If* $\forall i \in \mathscr{I}^{FG}\colon u_i \geq 0$ *and* $u_{\mathscr{I}^{FG}}$ *is* 0 *at the boundary of the input domain then it holds that*

$$\min_{\mathbf{x} \in [0,1]^D} u_{\mathscr{I}^{FG}}(\mathbf{x}) = \min_{i \in \mathscr{I}^{FG}} \{0, u_i\} \geq 0\,. \tag{16}$$

Proof We can interpret the full grid discretization as a set of hypercubes that have in each corner a grid point or a boundary point. Within each hypercube the minimum principle holds since $u_{\mathscr{I}^{\mathrm{FG}}}$ is locally a harmonic function due to the φ_i being piecewise D-linear functions and due to the linearity of the Laplace operator

$$\Delta u_{\mathscr{I}^{\mathrm{FG}}}(\mathbf{x}) = \Delta \sum_{i \in \mathscr{I}^{\mathrm{FG}}} u_i \varphi_i(\mathbf{x}) = \sum_{i \in \mathscr{I}^{\mathrm{FG}}} u_i \underbrace{\Delta \varphi_i(\mathbf{x})}_{=0} = 0\,. \tag{17}$$

Therefore, the minimum of $u_{\mathscr{I}^{\mathrm{FG}}}$ must lie at a full grid point or a boundary point. If we extend this argument to the global domain $[0,1]^D$, we come to the conclusion that the global minimum must be either at the boundary of the input domain or at a full grid point. □

Both, the nodal and the hierarchical basis with piecewise D-linear basis functions, span the same function space. Therefore, a basis transformation from the nodal basis to the piecewise D-linear hierarchical basis preserves the non-negativity property even though some hierarchical coefficients can be negative. However, once the full grid is truncated to overcome the curse of dimensionality this property can not be guaranteed anymore. Nevertheless, there must always exist a set of level-index pairs $\mathscr{I}^{\text{ext}} \subseteq \mathscr{I}^{\text{FG}} \setminus \mathscr{I}^{\text{SG}}$ with $\mathscr{I}^{\text{SG}} \subset \mathscr{I}^{\text{FG}}$ that extends the actual sparse grid index set $\mathscr{I}^{\text{SG}}$ such that it holds

$$\forall \mathbf{x} \in [0, 1]^D \colon u_{\mathscr{I}^{\text{SG}}}(\mathbf{x}) + u_{\mathscr{I}^{\text{ext}}}(\mathbf{x}) = \sum_{i \in \mathscr{I}^{\text{SG}}} v_i \varphi_i(\mathbf{x}) + \sum_{i \in \mathscr{I}^{\text{ext}}} w_i \varphi_i(\mathbf{x}) \geq 0 , \tag{18}$$

for properly chosen coefficients $\mathbf{w}$ of $u_{\mathscr{I}^{\text{ext}}}$. We call $\mathscr{I}^{\text{ext}}$ extension set. With Eq. (18) we reformulate Eq. (15) as follows: Assume that $u(\mathbf{x}) \geq 0$ and an approximation $u_{\mathscr{I}^{\text{SG}}}(\mathbf{x}) \approx u(\mathbf{x})$ for which it holds that $\forall i \in \mathscr{I}^{\text{SG}} \colon u_{\mathscr{I}^{\text{SG}}}(\mathbf{x}_i) \geq 0$ are given. For

$$\ell_{\max} := \max_{(\mathbf{l},\mathbf{i}) \in \mathscr{I}^{\text{SG}}} |\mathbf{l}|_\infty , \tag{19}$$

there always exists an extension set $\mathscr{I}^{\text{ext}}$ ensuring positivity according to Eq. (18), being a subset of the full grid points $\mathscr{I}^{\text{FG}}_{\ell_{\max}}$, which is minimal,

$$\forall \tilde{\mathscr{I}}^{\text{ext}} \subseteq \mathscr{I}^{\text{ext}}, \exists i \in \mathscr{I}^{\text{FG}}_{\ell_{\max}} \colon u_{\mathscr{I}^{\text{SG}}}(\mathbf{x}_i) + u_{\mathscr{I}^{\text{ext}} \setminus \tilde{\mathscr{I}}^{\text{ext}}}(\mathbf{x}_i) < 0 . \tag{20}$$

The coefficients $\mathbf{w}$ must fulfill according to Eq. (13)

$$\forall i \in \mathscr{I}^{\text{ext}} \colon w_i = u_i - \left(\sum_{j \in \mathscr{A}_{\mathscr{I}^{\text{SG}}}(i)} v_j \varphi_j(\mathbf{x}_i) + \sum_{j \in \mathscr{A}_{\mathscr{I}^{\text{ext}}}(i)} w_j \varphi_j(\mathbf{x}_i) \right) . \tag{21}$$

3.1 Limitation from Above and Below

We can iteratively solve Eq. (14) with Eq. (18) for a known extension set. First we limit the range of sparse grid function values from below, then we limit it from above. In both operations we perform the following steps: (1) Transform the range of function values such that the function evaluates to zero at the upper or the lower limit, (2) search for an extension set that assures a non-negative surrogate, (3) transform the range of function values back to the original space including the new grid points. It is required to repeat this process until the grid is unchanged since the

Algorithm 1: Iterative limitation of the range of function values of a sparse grid approximation to a certain interval defined by $y^{\mathrm{l}}, y^{\mathrm{u}} \in \mathbb{R}$, $y^{\mathrm{l}} < y^{\mathrm{u}}$

Data: index set $\mathscr{I}^{\mathrm{SG}}$, coefficients $\mathbf{v}$ and interval defined by $y^{\mathrm{l}}, y^{\mathrm{u}} \in \mathbb{R}$
Result: extended index set $\mathscr{I}^{\mathrm{SG}}$ and corresponding coefficients $\mathbf{v}$

$N \leftarrow 0$
while $N < |\mathscr{I}^{SG}|$ **do**
  $N \leftarrow |\mathscr{I}^{\mathrm{SG}}|$
  `// ------------------------------------`
  `// Limit the range of function values from below`
  $\mathbf{v}' \leftarrow \text{interpolate}(\mathscr{I}^{\mathrm{SG}}, \{(\mathbf{x}_i, \sum_{j \in \mathscr{I}^{\mathrm{SG}}} v_j \varphi_j(\mathbf{x}_i) - y^{\mathrm{l}})\}_{i \in \mathscr{I}^{\mathrm{SG}}})$
  $\mathscr{I}^{\mathrm{ext}}, \mathbf{w}' \leftarrow \text{makePositive}(\mathscr{I}^{\mathrm{SG}}, \mathbf{v}')$
  $\mathscr{U}^{\mathrm{l}} \leftarrow \{(\mathbf{x}_i, \sum_{j \in \mathscr{I}^{\mathrm{SG}}} v'_j \varphi_j(\mathbf{x}_i) + \sum_{j \in \mathscr{I}^{\mathrm{ext}}} w'_j \varphi_j(\mathbf{x}_i) + y^{\mathrm{l}})\}_{i \in \mathscr{I}^{\mathrm{SG}} \cup \mathscr{I}^{\mathrm{ext}}}$
  $\mathbf{v}'' \leftarrow \text{interpolate}(\mathscr{I}^{\mathrm{SG}} \cup \mathscr{I}^{\mathrm{ext}}, \mathscr{U}^{\mathrm{l}})$
  $\mathscr{I}^{\mathrm{SG}}, \mathbf{v} \leftarrow \mathscr{I}^{\mathrm{SG}} \cup \mathscr{I}^{\mathrm{ext}}, \mathbf{v}''$
  `// ------------------------------------`
  `// Limit the range function values from above`
  $\mathbf{v}' \leftarrow \text{interpolate}(\mathscr{I}^{\mathrm{SG}}, \{(\mathbf{x}_i, y^{\mathrm{u}} - \sum_{j \in \mathscr{I}^{\mathrm{SG}}} v_j \varphi_j(\mathbf{x}_i))\}_{i \in \mathscr{I}^{\mathrm{SG}}})$
  $\mathscr{I}^{\mathrm{ext}}, \mathbf{w}' \leftarrow \text{makePositive}(\mathscr{I}^{\mathrm{SG}}, \mathbf{v}')$
  $\mathscr{U}^{\mathrm{u}} \leftarrow \{(\mathbf{x}_i, y^{\mathrm{u}} + \sum_{j \in \mathscr{I}^{\mathrm{SG}}} v'_j \varphi_j(\mathbf{x}_i) + \sum_{j \in \mathscr{I}^{\mathrm{ext}}} w'_j \varphi_j(\mathbf{x}_i))\}_{i \in \mathscr{I}^{\mathrm{SG}} \cup \mathscr{I}^{\mathrm{ext}}}$
  $\mathbf{v}'' \leftarrow \text{interpolate}(\mathscr{I}^{\mathrm{SG}} \cup \mathscr{I}^{\mathrm{ext}}, \mathscr{U}^{\mathrm{u}})$
  $\mathscr{I}^{\mathrm{SG}}, \mathbf{v} \leftarrow \mathscr{I}^{\mathrm{SG}} \cup \mathscr{I}^{\mathrm{ext}}, \mathbf{v}''$
return $\mathscr{I}^{\mathrm{SG}}, \mathbf{v}$

limitation from below can cancel the limitation from above and vice versa. We will discuss this issue in Sect. 3.3. The complete procedure is shown in Algorithm 1. In the following we address the questions on how to compute the minimal extension set and how to compute the corresponding coefficients $\mathbf{w}$ in order to guarantee non-negative surrogates.

3.2 *Minimal Extension Set*

Due to Theorem 1, we know that the minimum of a sparse grid function is at a full grid point or a boundary point. Together with the fundamental property for the piecewise D-linear basis functions we can define a greedy optimization approach that constructs the minimal extension set $\mathscr{I}^{\mathrm{ext}}$ as follows: Let $\ell_{\max} := \max_{(\mathbf{l},\mathbf{i}) \in \mathscr{I}^{\mathrm{SG}}} |\mathbf{l}|_\infty$ be the maximum level of an arbitrarily refined sparse grid index set $\mathscr{I}^{\mathrm{SG}}$ and $\mathscr{I}^{\mathrm{cand}} := \mathscr{I}^{\mathrm{FG}}_{\ell_{\max}} \setminus \mathscr{I}^{\mathrm{SG}}$ be the candidate set that includes all level-index pairs of the corresponding full grid $\mathscr{I}^{\mathrm{FG}}_{\ell_{\max}}$ which are not yet part of $\mathscr{I}^{\mathrm{SG}}$. We split the candidate set according to the levelsum of the grid points and obtain $\mathscr{I}^{\mathrm{cand}}_k := \{(\mathbf{l}, \mathbf{i}) \in \mathscr{I}^{\mathrm{cand}} \colon |\mathbf{l}|_1 = k\}$. Assume $\exists i \in \mathscr{I}^{\mathrm{cand}}_{k_{\min}} \colon u_{\mathscr{I}^{\mathrm{SG}}}(\mathbf{x}_i) < 0$ where $k_{\min}$ is the smallest levelsum available in $\mathscr{I}^{\mathrm{cand}}$. We know from Eq. (11)

Algorithm 2: Computation of the minimal extension set

Data: index set $\mathscr{I}^{\mathrm{SG}}$, coefficients $\mathbf{v}$, dimensionality D
Result: extending index set $\mathscr{I}^{\mathrm{ext}}$ and corresponding coefficients $\mathbf{w}$
// compute the maximum grid level to be considered
$\ell_{\max} \leftarrow \max_{(\mathbf{l},\mathbf{i}) \in \mathscr{I}^{\mathrm{SG}}} |\mathbf{l}|_{\infty}$
// compute candidate grid points
$\mathscr{I}^{\mathrm{cand}} \leftarrow$ computeCandidates($\mathscr{I}^{\mathrm{SG}}$, $\mathbf{v}$)
// run over the candidate set starting with a levelsum of 1
$\mathscr{I}^{\mathrm{ext}} \leftarrow \{\}$
$\mathbf{w}_0 \leftarrow ()$
$k \leftarrow 1$
while $k \leq D\ell_{max}$ **do**
  // extract those grid points with negative function value
  $\mathscr{I}_k^{\mathrm{ext}} \leftarrow \{(\mathbf{l},\mathbf{i}) \in \mathscr{I}^{\mathrm{cand}} \colon |\mathbf{l}|_1 = k \wedge u_{\mathscr{I}^{\mathrm{SG}}}(\mathbf{x}_{\mathbf{l},\mathbf{i}}) + u_{\mathscr{I}^{\mathrm{ext}}}(\mathbf{x}_{\mathbf{l},\mathbf{i}}) < 0\}$
  // compute coefficients $\mathbf{w}_k$ for all indices in $\mathscr{I}^{\mathrm{ext}} \cup \mathscr{I}_k^{\mathrm{ext}}$
  $\mathbf{w}_k \leftarrow$ computeCoefficients($\mathscr{I}^{\mathrm{SG}}$, $\mathbf{v}$, $\mathscr{I}^{\mathrm{ext}}$, $\mathbf{w}_{k-1}$, $\mathscr{I}_k^{\mathrm{ext}}$)
  // update the extension set and continue
  $\mathscr{I}^{\mathrm{ext}} \leftarrow \mathscr{I}^{\mathrm{ext}} \cup \mathscr{I}_k^{\mathrm{ext}}$
  $k \leftarrow k + 1$
return $\mathscr{I}^{\mathrm{ext}}$, $\mathbf{w}_{D\ell_{\max}}$

that $\forall j \in \mathscr{I}^{\mathrm{cand}}, j \neq i \colon \varphi_j(\mathbf{x}_i) = 0$ since $\mathscr{I}^{\mathrm{SG}}$ contains all the hierarchical ancestors of i. Consequently, i must be part of the extension set in order to achieve $u_{\mathscr{I}^{\mathrm{SG}} \cup \mathscr{I}^{\mathrm{ext}}}(\mathbf{x}_i) > 0$. We obtain, therefore, the minimal extension set $\mathscr{I}_{k_{\min}}^{\mathrm{ext}} = \{i \in \mathscr{I}_{k_{\min}}^{\mathrm{cand}} \colon u_{\mathscr{I}^{\mathrm{SG}}}(\mathbf{x}_i) < 0\}$ for the current levelsum $k_{\min}$. Once we have considered all indices of $\mathscr{I}_{k_{\min}}^{\mathrm{cand}}$ we compute their coefficients according to some method from the following section and continue with $k := k_{\min} + 1$. We repeat the procedure and collect all minimum extension sets $\mathscr{I}_k^{\mathrm{ext}}$ up to the maximum levelsum $D\ell_{\max}$. You find the complete algorithm in Algorithm 2.

There are two important aspects to mention on this iterative algorithm: First, the size of the extended sparse grid index set is in the worst case equal to the size of the corresponding full grid. Second, the size of the extension set depends on how the coefficients of the new grid points are computed. As a consequence, the minimal extension set is uniquely defined by the sparse grid surrogate $u_{\mathscr{I}^{\mathrm{SG}}}$ and the method used to estimate the function values for the indices in the extension set. To minimize the work, good methods for computing the candidate set and for computing the corresponding coefficients are required. In the following sections we present such methods.

3.3 Computing Coefficients of the Extension Set

Assume that we have found a minimal extension set $\mathscr{I}^{\mathrm{ext}}$ for a sparse grid $\mathscr{I}^{\mathrm{SG}}$ and need to compute the coefficients for $\mathscr{I}_k^{\mathrm{ext}} \subseteq \mathscr{I}^{\mathrm{ext}}$ for iteration k. From Eq. (18) it follows that $\forall i \in \mathscr{I}_k^{\text{ext}}\colon u_i \geq 0$. Due to the minimality of the extension set from Eq. (20) it holds that

$$u_{\mathscr{I}^{\text{SG}}}(\mathbf{x}_i) + u_{\mathscr{A}_{\mathscr{I}^{\text{ext}}}(i)}(\mathbf{x}_i) < 0 \,, \tag{22}$$

which we insert in Eq. (21) and obtain

$$w_i = \underbrace{u_i}_{\geq 0} - \underbrace{(u_{\mathscr{I}^{\text{SG}}}(\mathbf{x}_i) + u_{\mathscr{A}_{\mathscr{I}^{\text{ext}}}(i)}(\mathbf{x}_i))}_{<0} > 0 \,, \tag{23}$$

thus that all new hierarchical coefficients must be larger than zero. This is the reason why we need an iterative approach to limit the range of function values from above and below: The limitation from below leads to additional positive coefficients in the linear combination. To limit the function from above we flip the sign of the function in order to apply the limitation algorithm from below. The positive coefficients turn into negative ones after this flip and could potentially cause that function values become negative which were positive before. This increases the costs of the algorithm, since we need to find a minimal extension set and the corresponding coefficients several times. We will show how one can compute a minimal extension set efficiently in the next section.

Here, we want to list three methods to estimate function values at the new grid points:

set-to-zero: The simplest approach is to set all the function values to zero, i.e.

$$\forall (\mathbf{l}, \mathbf{i}) \in \mathscr{I}^{\text{ext}}\colon u_{\mathbf{l},\mathbf{i}} := 0 \,. \tag{24}$$

This method leads to the largest possible extension sets and is, therefore, the baseline for all more sophisticated methods that use the information about the function at hand.

interpolate-function: This is the optimal case, where one interpolates the true function values

$$\forall (\mathbf{l}, \mathbf{i}) \in \mathscr{I}^{\text{ext}}\colon u_{\mathbf{l},\mathbf{i}} := u(\mathbf{x}_{\mathbf{l},\mathbf{i}}) \,. \tag{25}$$

This is, however, unfeasible if evaluating u is computationally expensive.

interpolate-boundaries: This approach incorporates the function values that are located in the neighborhood of the new level-index pair $(\mathbf{l}, \mathbf{i})$ into the estimation of the function value at the new point $\mathbf{x}_{\mathbf{l},\mathbf{i}}$. To estimate $u_{\mathbf{l},\mathbf{i}}$ we interpolate linearly between the function values which lie at the boundary of $\text{supp}(\varphi_{\mathbf{l},\mathbf{i}})$ in each direction d and take the minimum of it, i.e.

$$\forall (\mathbf{l}, \mathbf{i}) \in \mathscr{I}^{\text{ext}}\colon u_{\mathbf{l},\mathbf{i}} := \max \left\{ 0, \min_{d \in \{1,\ldots,D\}} \frac{u_{\mathscr{I}^{\text{SG}}}(\mathbf{x}_{l_d, i_d - 1}) + u_{\mathscr{I}^{\text{SG}}}(\mathbf{x}_{l_d, i_d + 1})}{2} \right\} . \tag{26}$$

The minimum operation ensures monotonicity for the probability density function we will consider as an example. Alternatively, operations such as the average can be considered as well.

3.4 Intersection Search

In the following we present a method that reduces the size of the candidate set by searching for intersections of sparse grid indices. As we have seen before, grid points for which the sparse grid function becomes negative form part of the extension set. We exploit this knowledge and reduce the number of candidates by just considering those grid points for which it is theoretically possible that the function value is negative. According to the definition of a sparse grid function in Eq. (7) this can just be the case for grid points that have ancestors with negative coefficients.

We define the intersection $(\mathbf{l}^*, \mathbf{i}^*)$ of two level-index pairs $(\mathbf{l}, \mathbf{i})$, $(\mathbf{k}, \mathbf{j})$ as

$$\forall d \in \{1, \ldots, D\} \colon i_d^* = \begin{cases} i_d & \text{for } l_d \geq k_d \\ j_d & \text{else} \end{cases} , \qquad l_d^* = \begin{cases} l_d & \text{for } l_d \geq k_d \\ k_d & \text{else} \end{cases} , \tag{27}$$

and show that the extension set is a subset of the set that contains all intersections of at least two hierarchically independent sparse grid points with negative coefficient and overlapping support.

Theorem 2 (Intersections of Grid Points with Negative Coefficients) *Let ℓ_{max} be the maximum level in an arbitrarily refined sparse grid index set $\mathcal{I}^{SG}$ of dimensionality $0 < D \in \mathbb{N}$. Let $i \in \mathcal{I}_{\ell_{max}}^{FG} \setminus \mathcal{I}^{SG}$ with $u_{\mathcal{I}^{SG}}(\mathbf{x}_i) < 0$. Then $\exists i' \in \mathcal{A}(i) \cup \{i\}$ with $u_{\mathcal{I}^{SG}}(\mathbf{x}_{i'}) < 0$ and i' is the intersection of at most D hierarchically independent indices with negative coefficients.*

Proof Let $\mathcal{A}^-_{\mathcal{I}^{SG}}(i) := \{j \in \mathcal{A}_{\mathcal{I}^{SG}}(i) \colon v_j < 0\}$ and let k^* be the intersection of all indices in $\mathcal{A}^-_{\mathcal{I}^{SG}}(i)$. Then k^* is as well the intersection of at most D hierarchically independent indices in $\mathcal{A}^-_{\mathcal{I}^{SG}}(i)$: The level of k^* has D components, which are defined according to Eq. (27) as the componentwise maximum of the levels in $\mathcal{A}^-_{\mathcal{I}^{SG}}(i)$. One can successively construct k^* by taking an index of $\mathcal{A}^-_{\mathcal{I}^{SG}}(i)$ that defines a maximal number of level components of k^* until the collected indices define all level components. Each of the first D indices must define at least one component and k^* is defined uniquely after at most D indices. All of them are hierarchically independent since at least one level component of each index is larger than the corresponding level component of the other indices.

We define

$$u^*(\mathbf{x}) := \sum_{j \in \mathcal{A}_{\mathcal{I}^{SG}}(k^*) \cup \{k^*\}} v_j \varphi_j(\mathbf{x}) , \tag{28}$$

to be a local approximation of $u_{\mathcal{I}^{SG}}$ around $\mathbf{x}_{k^*}$. Since k^* is the intersection of all indices in $\mathcal{A}^-_{\mathcal{I}^{SG}}(i)$ it holds that $\mathcal{A}^-_{\mathcal{I}^{SG}}(i) \subset \mathcal{A}_{\mathcal{I}^{SG}}(k^*) \cup \{k^*\}$ and, therefore, $u^*(\mathbf{x}_i) \leq u_{\mathcal{I}^{SG}}(\mathbf{x}_i) < 0$. Let $\mathcal{K}^*$ be the set of all up to $3^D - 1$ ancestors that lie at the boundary of $\text{supp}(\varphi_{k^*})$. Then if follows from $\forall j \in \mathcal{K}^* \colon \mathcal{A}_{\mathcal{I}^{SG}}(j) \subset \mathcal{A}_{\mathcal{I}^{SG}}(k^*) \subset \mathcal{A}(i)$ that $u^*(\mathbf{x}_j) = u_{\mathcal{I}^{SG}}(\mathbf{x}_j)$. The minimum of u^* must lie at a grid point at the boundary of $\text{supp}(\varphi_{k^*})$ or at its center due to the minimum principle. Let $\hat{k}$ be the index where u^* is minimal and $u_{\mathscr{G}\text{SG}}(\mathbf{x}_{\hat{k}}) = u^*(\mathbf{x}_{\hat{k}}) < 0$ then we can distinguish two cases:

1. $\hat{k} = k^*$: By construction, k^* satisfies the condition on i'.
2. $\hat{k} \in \mathscr{K}^*$: We set $i := \hat{k}$ and continue. The number of iterations is limited by the level. Hence, we must find an index where $\hat{k} = k^*$ after a finite number of steps.

□

The remaining task is to enumerate these intersections. We call this method *intersection search*. Figure 2 shows the intersections of grid points with negative coefficients (red circles) for a sparse grid function of level 2.

A naive enumeration of all combinations that neglects hierarchical dependencies and overlapping supports becomes quickly too expensive. Let $M \geq 2$ be the number of negative coefficients of a sparse grid surrogate and $D \geq 2$ the dimensionality of the input space, then an upper bound for the number of intersections to be checked is

$$\#\text{intersections}(M, D) = \sum_{d=2}^{D} \binom{M}{d} \tag{29}$$

We consider this number as a baseline for more sophisticated algorithms.

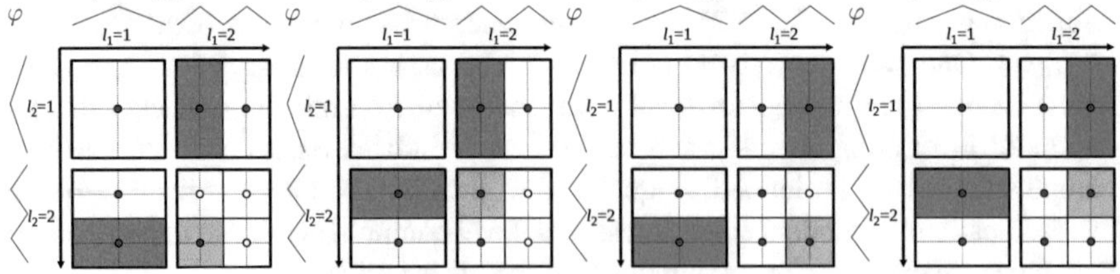

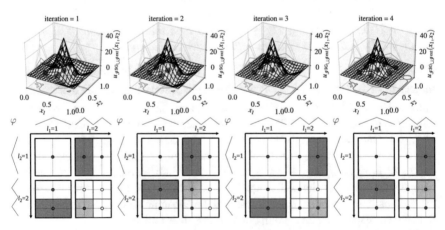

Fig. 2 Example for the interpolation of a normal distribution with $\mu = 1/2, \sigma = 1/16$ with a sparse grid of level 2 and $N = 5$, and the evolution of the extended grid using intersection search to find the extension set and the set-to-zero method to compute their coefficients. The upper plots show the extended sparse grid interpolant and the lower plots the subspace tableau that contains all the corresponding full grid points. The grid points with positive coefficients are shown as blue dots, grid points with negative coefficient as red dots. The basis function's support of independent grid points which are used to compute the intersection points as candidates for the extension set are shown in green. The joint support is shown in orange. This example presents the worst case for the extended sparse grid: Preserving non-negativity of the normal distribution leads to a full grid with $N = 9$

In the following, we present such an algorithm that considers the results from Theorem 2. The intersection operation is commutative and associative since it is based on computing maxima of levels. This allows us to compute intersections of an arbitrary set of grid points with negative coefficients as successive intersection operations of two grid points. For example, let i and j be grid points from which the intersection is i^*. Then, the intersection of i, j and a third grid point k is the same as the intersection of i^* and k. Moreover, we can describe each grid point i as an intersection of at most D other grid points, which allows us computing intersections with a bottom-up scheme from $2, 3, \dots, D$.

These deductions lead to Algorithm 3. First, we run over the index set that contains all level-index pairs with negative coefficient in some arbitrary order and store all combinations of grid points that fulfill Theorem 2. Due to the commutative property of the intersection operation, it is enough to store each valid combination just once. Second, we iterate over the number of dimensions and compute intersections based on the generated combinations: In the first iteration

Algorithm 3: Intersection search algorithm

Data: index set $\mathscr{I}^{\mathrm{SG}}$, coefficients $\mathbf{v}$
Result: candidate set $\mathscr{I}^{\mathrm{cand}}$

// load pairwise valid intersections
$\mathscr{J}_2 \leftarrow \{k \in \mathscr{I}^{\mathrm{SG}} : v_k < 0\}$
for $i \in \mathscr{J}_2$ **do**
  $I_i \leftarrow \{\}$
  for $j \in \mathscr{J}_2$ **do**
    // consider each intersection once in the first run
    if $i < j$ **then**
      // search for valid intersections
      if $\mathrm{supp}(\varphi_i) \cap \mathrm{supp}(\varphi_j) \neq \emptyset \wedge i \notin \mathscr{A}(j) \wedge j \notin \mathscr{A}(i)$ **then**
        $I_i \leftarrow I_i \cup \{j\}$

// compute intersections dimensionwise
$\mathscr{I}^{\mathrm{cand}} \leftarrow \{\}$
for $d \in \{2, 3, \dots, D\}$ **do**
  // check every index of the current candidate set
  for $i \in \mathscr{J}_d$ **do**
    // check all the indices that overlap with i
    for $j \in I_i$ **do**
      $k^* \leftarrow$ computeIntersection(i, j)
      // add intersection to candidate set
      if $k^* \notin \mathscr{I}^{SG} \cup \mathscr{I}^{cand}$ **then**
        $\mathscr{I}^{\mathrm{cand}} \leftarrow \mathscr{I}^{\mathrm{cand}} \cup \{k^*\}$
      // prepare intersections for the next iteration
      $I_{k^*} \leftarrow \{k \in I_j : \mathrm{supp}(\varphi_k) \cap \mathrm{supp}(\varphi_{k^*}) \neq \emptyset \wedge k \notin \mathscr{A}(k^*) \wedge k^* \notin \mathscr{A}(k)\}$
      if $|I_{k^*}| > 0$ **then**
        $\mathscr{J}_{d+1} \leftarrow \mathscr{J}_{d+1} \cup \{k^*\}$

return $\mathscr{I}^{\mathrm{cand}}$

$d = 2$ we compute the intersection k of two grid points i, j, add it to the candidate set and generate the set of overlapping grid points for k out of the set of overlapping grid points of j. We store the result and compute the intersections of k in the next iteration $d = 3$, and so on. Note that all the operations in the innermost loop of Algorithm 3 have a complexity of $\mathcal{O}(1)$ or $\mathcal{O}(D)$.

For $D = 2$ this algorithm is optimal in a sense that we enumerate each intersection just once. This does not hold for $D > 3$. Empirical tests show that we compute $\mathcal{O}(\ell 2^{\ell D})$ intersections of grid points for regular grids of level ℓ. The size of the candidate set, i.e. the number of unique intersections, is in general significantly smaller than a full grid, especially for adaptively refined grids. You find a detailed analysis of the presented algorithm in Sect. 4.

4 Approximation of Gaussians with Extended Sparse Grids

A very challenging function to be approximated by sparse grids are peaked Gaussian probability density functions. It was shown that interpolating such functions with a sparse grid is very costly in terms of the number of grid points [2, 12]. The reason is the exponentially growing fore-factor that is usually omitted in the asymptotic error. The large gradient around the mean of the Gaussian leads to large coefficients with alternating sign in the sparse grid function. This causes large negative function values where the Gaussian is close to zero, see Fig. 3. The convergent phase of the function is shifted towards prohibitively large grids. Therefore, the approximation of a multivariate distribution with independent Gaussian marginals is well suited to study the extended version of the sparse grid density estimation.

We use a multivariate Gaussian with the same marginals as in [12], i.e.

$$u(\mathbf{x}) := \prod_{d=1}^{D} \frac{1}{\sqrt{2\pi\sigma_d^2}} \exp\left(-\frac{(x_d - \mu_d)^2}{2\sigma_d^2}\right), \tag{30}$$

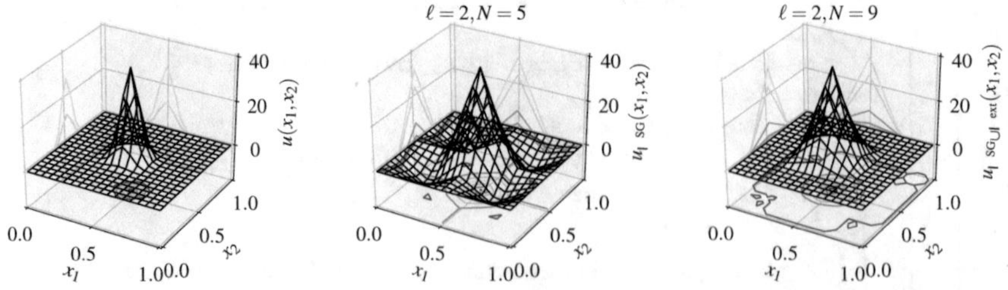

Fig. 3 On the left, a multivariate Gaussian distribution with independent marginals $\mu = 1/2$ and $\sigma = 1/16$ is shown. The plot in the center shows the sparse grid interpolant of level 2 with a piecewise bilinear basis. One can see that the function values are negative almost everywhere, due to the overlapping support of the grid points with levelsum 3, which have all large negative coefficients. The plot on the right shows the extended sparse grid function that is non-negative everywhere

with $\mu_d = 1/2, \sigma_d = 1/16$. We vary the dimensionality and investigate the properties of extended regular and adaptively refined sparse grid interpolants of various levels. The main questions we want to answer with this example are the following:

1. How many intersections do we need to compute to obtain the candidate set $\mathscr{I}^{\text{cand}}$ using the intersection search algorithm?
2. How large is the final candidate set $\mathscr{I}^{\text{cand}}$ and the extension set $\mathscr{I}^{\text{ext}}$?
3. How does the extension affect the convergence rate of the sparse grid function?

4.1 *Intersection Search and Candidate Sets for Regular Sparse Grids*

As the baseline, we use the approach where the candidate set is simply equal to a full grid. With respect to the intersection search it is important to distinguish between the number of intersections we compute during the intersection search and the number of unique intersections that result from this process. The larger the difference between these numbers, the less efficient is the intersection search.

Results for the intersection search for regular sparse grids are shown in Fig. 4. We observe that independent of the dimensionality it holds that the intersection search computes fewer intersections (purple) than the upper bound provided by Eq. (29)

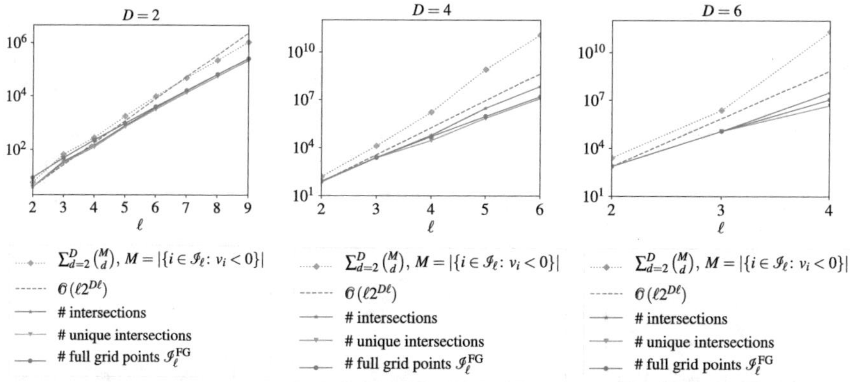

Fig. 4 Comparison of several quantities that describe the growth of the candidate set $\mathscr{I}^{\text{cand}}$ of the intersection search algorithm for a multivariate Gaussian distribution of independent marginals with $\mu = 1/2$ and $\sigma = 1/16$: The dotted green line is an upper bound for the number of intersections that form the candidate set, see Eq. (29). The blue line is the number of full grid points for the corresponding level ℓ. The purple line shows the number of intersections, for which we see clearly that they have full grid complexity for lower levels and grow slightly quicker for $D > 2$ and $\ell > 2$. The dashed purple line shows an asymptotic growth of $\mathcal{O}(\ell 2^{\ell D})$, which seems to be a good indicator for the growth of the number of intersections for regular grids in this extreme setting. Out of these intersections one obtains the final candidate set that contains just the unique intersections (orange). The size of the candidate set is never larger than the size of the corresponding full grid

(dotted green). Furthermore, the number of unique intersections (orange) is always smaller than the number of full grid points (blue). This is a necessary prerequisite for a valid implementation and in particular if the interpolate-function strategy is to be used. For $D > 2$ and $\ell > 2$, however, we compute slightly more intersections than there are full grid points. The number of intersections seem to grow as $\mathcal{O}(\ell 2^{D\ell})$ (dashed purple), which is more than there are full grid points $\mathcal{O}(2^{D\ell})$. The number of unique intersections (orange) is equal to the size of the candidate set, which seems to be almost as large as the corresponding full grid. This shows that the candidate set has full grid complexity but, nevertheless, the log-scale on the y-axis is misleading. We take the results for the highest levels in each dimension, i.e. $\ell = 9, 6, 4$ for $D = 2, 4, 6$, split up the grid points of the corresponding candidate set by their levelsum and plot the sizes of these subsets on a linear scale, see Fig. 5. The upper plots show that the efficiency of the intersection search increases with increasing dimensionality since the difference between the number of unique intersections and the corresponding full grid points grows.

The gain, however, is just linear and not enough to overcome the curse of dimensionality: While the number of full grid points grows exponentially, the size of the candidate set does not shrink with the same speed, see the lower plots of

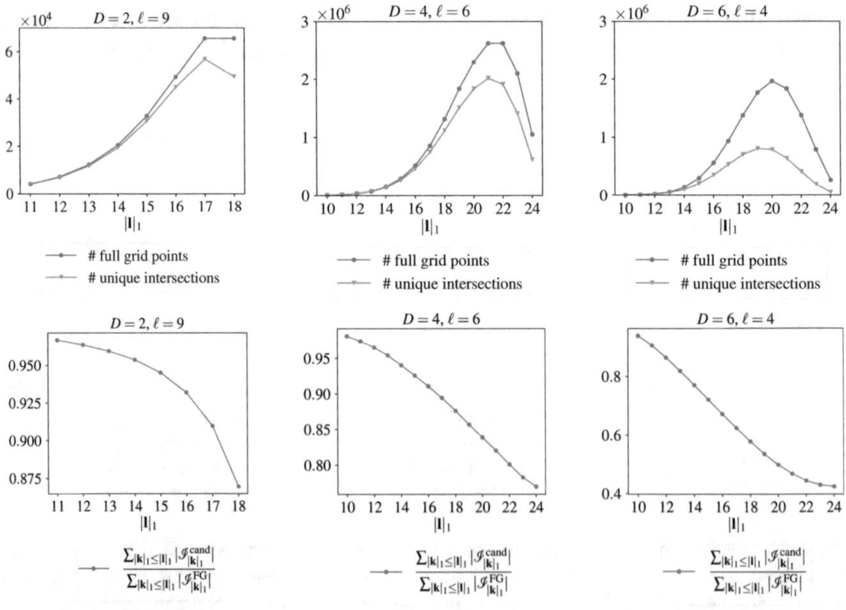

Fig. 5 The upper plots show the number of full grid points and unique intersections split up by levelsum for $D = 2, 4, 6$. The lower plots show the ratio of the accumulated number of unique intersections and the number of full grid points ordered again by levelsum. These plots show how much we save in the iterative approach of finding the minimal extension set using the intersection search method for finding candidates compared to enumerating the complete full grid. The higher the dimensionality, the more we save

Fig. 5. They show the ratio of the accumulated number of unique intersections and the full grid points with respect to the levelsum. The smaller this ratio, the higher the efficiency of the intersection search algorithm. One can see that the efficiency increases with increasing levelsum and increasing dimensionality. The final candidate set contains for $D = 2$ approximately 87%, for $D = 4$ around 75% and for $D = 6$ just 40% of the full grid points. Still, the computation of the candidate set $\mathscr{I}^{\text{cand}}$ is where the curse of dimensionality hits the extension set algorithm: In the worst case, the size of $\mathscr{I}^{\text{cand}}$ is equal to the full grid.

4.2 *Extension Sets and Convergence for Regular Grids*

The candidate set just serves for the purpose to compute the actual extension set of the original sparse grid function. The extension set is a subset of the candidate set and it is computed in a top-down approach, starting with the smallest levelsum which is not yet part of the grid.

Figure 6 shows the size of the extension set after each extension iteration. These plots show that most of the grid points in the extension set are grid points with a small levelsum for $D > 2$. For $D = 4$ the grid points with the largest levelsum is equal to 15, for $D = 6$, it is 12. Up to these levelsums, we include most of the grid points available in the candidate set, but none of the higher ones, from which the candidate set contains much more (compare Fig. 5).

If we were able to predict this behavior and could generalize it to arbitrary sparse grid functions, we could restrict the intersection search to smaller subspaces and overcome again the curse of dimensionality to some extent. However, Theorem 2 does not allow such a truncation of the search space. Further investigation is required, which goes beyond this work.

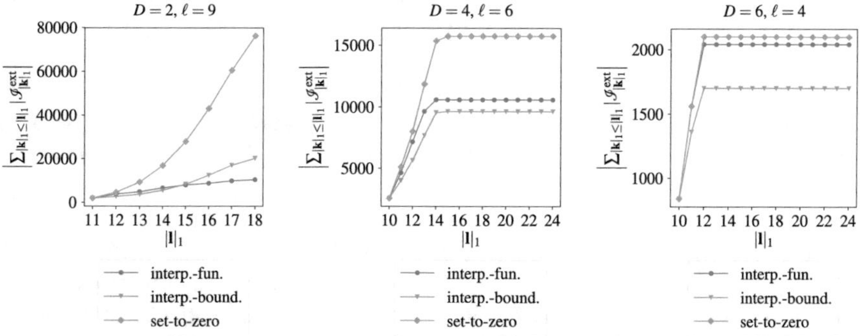

Fig. 6 The plots show the size of the extension set after iterating the grid points of the candidate set in ascending order with respect to their levelsum for $D = 2, 4, 6$ and $\ell = 9, 6, 4$. One can see that the set-to-zero method leads to the largest extension sets. The extension sets stop to grow after $|\mathbf{l}|_1 = 15$ and $|\mathbf{l}|_1 = 12$ for $D = 4, 6$. This means that we achieve non-negative sparse grid surrogates already by adding points with smaller levelsum

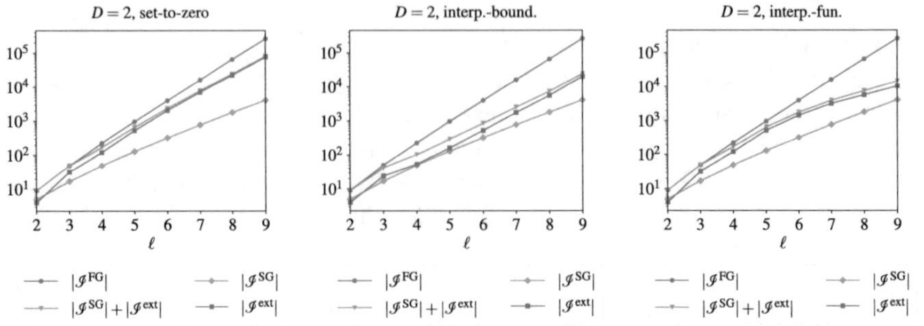

Fig. 7 These plots show the size of extended sparse grids (orange) for different coefficient estimation methods (set-to-zero, interpolate-boundaries, interpolate-function) in relation to the size of the corresponding full grids (blue) and the original sparse grids (green). In the left figure, where we use the set-to-zero method, we observe that the size of the extended sparse grid grows similar to the full grid size. This is not the case if we interpolate the original function or the interpolate-boundaries method. For these approaches (center, right), the size of the extended sparse grid grows as fast or slower as the size of the original sparse grid

However, the observation that the extended set is not growing anymore after a certain levelsum influences the size of the resulting extended sparse grids significantly: It scales as the size of the original sparse grid for a wide range of levels if the coefficients of the grid points of the extended set are chosen properly, as shown in Fig. 7. These plots show the grid size of a full grid, the original sparse grid, the extension set, and the sum of the latter two grids for $D = 2$. They show the growth of the extension set and the growth of the original sparse grid in relation to the coefficient estimation method for the new grid points. The size of the extension set scales as the size of a full grid for the set-to-zero method. This is no longer the case if we evaluate the original function to compute the coefficients. The size of the extension set grows slower for higher levels than it does for smaller ones. The third method, the linear interpolation of the function values at the boundaries of the domain of each new grid point, leads to similar results than the interpolation of the original function.

For higher-dimensional settings, we observe similar behavior, see Fig. 8. Here, we just provide the grid sizes of the extended sparse grids using the interpolation of boundaries method since the grid sizes are very similar for all the presented coefficient estimation methods.

From these results we conclude that the extension of the sparse grid is feasible in terms of the grid size. However, we need to know how the error of the extended interpolant evolves. To this end, we provide estimated interpolation errors for the extended sparse grid method in Fig. 9. We used 10,000 randomly chosen samples, evaluated the original Gaussian at these points and approximated the L_2 and the L_∞-error of the interpolants and computed the volume of the sparse grid functions $I(u_{\mathcal{G}})$, which should converge to 1, of course.

As an upper bound for the accuracy of the method we use a *truncated sparse grid function* where we set its function value simply to $\max\{0, u_{\mathcal{G}}(\mathbf{x})\}$.

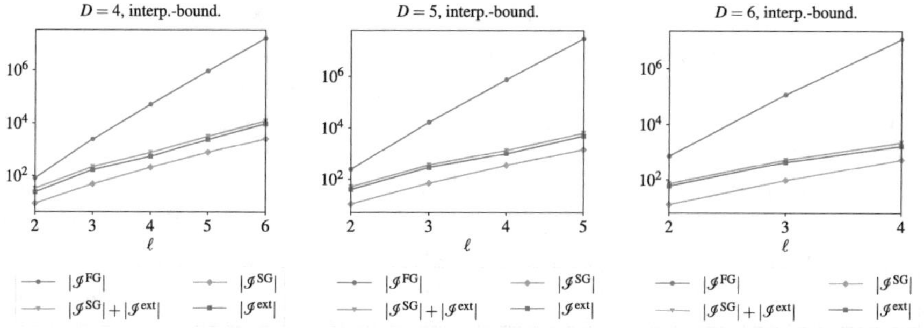

Fig. 8 Comparison of the sparse grid (green), the size of the minimal extension set if the interpolate-boundaries coefficient strategy is used (purple), the corresponding size of the extended sparse grid (orange) and the size of the corresponding full grid (blue). We observe that the size of the extension set grows similarly to the size of the sparse grid and, therefore, the extended sparse grid is still significantly smaller than a full grid and makes the approach applicable to moderate dimensional problems

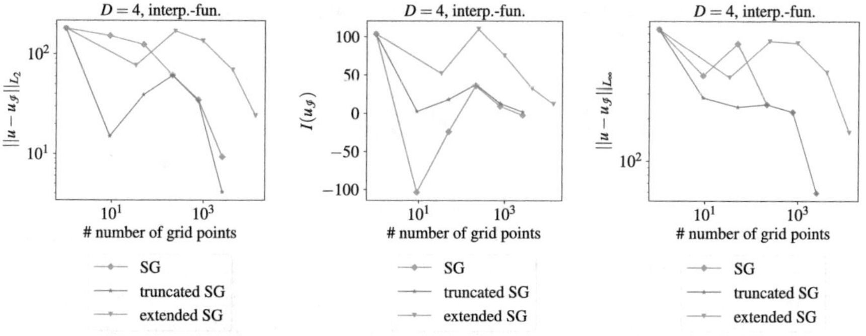

Fig. 9 The plots show the L_2 and the L_∞-error of the original sparse grid interpolant (green), a truncated version of it (purple), where we limit the range of function values by taking $\max\{0, u_{\mathcal{G}}(\mathbf{x})\}$ and the extended sparse grid interpolant (orange)

The estimated L_2-errors are illustrated in Fig. 9 (left). They show that up to $\ell = 3$ at 50 grid points we gain significantly from truncating the function value range of the sparse grid function. The truncation leads to a sharp transition of the function around the Gaussian peak. The extension, however, leads to a smooth transition and, therefore, reduces the gradient of the Gaussian peak itself due to the repeated over and underestimation of the surrounding function. As a consequence, the error of the extended sparse grid is not equal to the error of the truncated function. These over and underestimations cause the peaks in the error of the truncated sparse grid function and the extended sparse grid method. For the latter method the peaks are shifted towards the right and have a smaller height.

If we compare the error of the extended sparse grid with the original sparse grid we see that for $\ell = 2$ we gain accuracy by extending the grid. But for $\ell \geq 3$ we loose accuracy. However, once we reach the convergent phase with the original sparse grid, at $\ell = 5$ with $N = 1000$, the extended sparse grid converges as well, with a certain delay due to the extension set. Furthermore, we observe that the speed of convergence seems to be the same for the extended sparse grid and the original one. This is coherent with the previous observation that the complexity of the extension set is the same as the complexity of a sparse grid.

The grid points in the extension set are mainly located in areas where the Gaussian is close to zero. Therefore, the error is very similar for all the coefficient estimation methods of the extended sparse grid.

The integral of the approximations, which one can see in Fig. 9 (center), reveals another interesting aspect of the extension method: The estimated volume of the truncated sparse grid function (purpler line) is very close to the volume of the original sparse grid function (green line) for $\ell > 3$. The difference between these two lines indicates how much of the volume of the original function is negative. We need a large amount of additional grid points to cope with negativity even tough this difference is small. This means that the area where the original sparse grid function evaluates to negative values is large but the function values are small. The error in the maximum norm supports this deduction, see Fig. 9 (right).

4.3 Extension Sets for Adaptively Refined Grids

We extend the discussion now to adaptively refined grids. In [12] the authors have shown that adaptively refined grids cannot cope with the difficulties that the Gaussian distribution imposes on a sparse grid approximation. Therefore, we restrict ourselves to the results on the extension of the sparse grid and omit error plots.

The setting of the experiment is the following: We start with regular sparse grids of different levels to approximate the same Gaussian distribution as before. Then we refine each sparse grid function 10 times and add all the $2D$ successors of the grid point with the largest absolute coefficient. Afterwards, we make sure that the resulting sparse grid is consistent and add missing grid points.

Due to the design of the experiment and the Gaussian distribution, the adaptively refined sparse grids have the following property: By refining a small grid we obtain a non-regularly shaped grid. On the other hand, if we refine a large grid we obtain a still regularly shaped grid. We expect the intersection search to be significantly more effective for non-regular grids because the maximum level grows with every refinement step in the worst case and the corresponding full grid grows exponentially. But the number of intersections grows slower since we just add a constant number of new grid points in each refinement step. If all the associated coefficients of these grid points were positive, the complexity of the intersection search would not even increase at all.

The results in Fig. 10 show the expected behavior for the Gaussian distribution. First, we focus on Fig. 10 (left), which shows that, indeed, the number of intersections is two orders of magnitude smaller than the number of full grid points for adaptively refined grids that start with a regular grid with a level smaller than

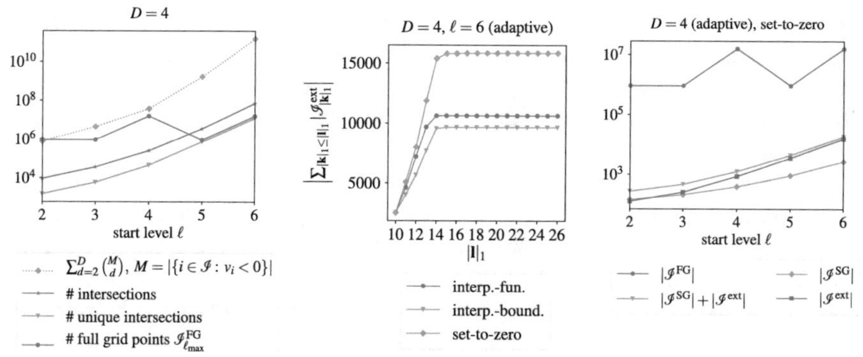

Fig. 10 The left plot shows the number of intersections (purple) we need to compute to generate the candidate set (orange). The number of intersections is several orders of magnitude smaller than the number of full grid points for $\ell \leq 4$. The plot in the center shows the size of the extension set for $\ell = 6$ that, again, stops growing at $|\mathbf{l}|_1 > 15$ for each coefficient estimation method (set-to-zero, interpolate-boundaries, interpolat-function). As you can see in the right plot, the extended sparse grid is significantly smaller than the corresponding full grid

5. The intersection search algorithm pays off significantly. For $\ell \geq 5$, where the refined grids have again a regular shape, the results are similar to the ones from the last section.

The plot in Fig. 10 (center) shows that the maximum levelsum of the adaptive sparse grid increases from 24 to 26 for $\ell = 6$ compared to the regular grid. Again, most grid points in the extension set have a small levelsum. There are none available with $|\mathbf{l}|_1 \geq 16$, independent of the coefficient estimation method. The numbers on the y-axis show that the size of the extension set is smaller than 2×10^3. This is significantly less than the size of the full grid, which is larger than 10^6, as one can see in Fig. 10 (right). This confirms the previous results and shows again that the size of the extended sparse grid grows as the original sparse grid.

To conclude this example we summarize the main observations we have made with respect to the extended sparse grid method: First, the size of the extended set has sparse grid complexity with properly chosen coefficients for the grid points of the extension set. The interpolate-boundaries method shows good results. Second, finding the candidate set is computationally demanding. It requires the enumeration of a non-negligible fraction of the corresponding full grid points. Nevertheless, the more the sparse grid is irregularly shaped the more the intersection search pays off.

5 Conclusions

In this paper we presented a new method to limit the range of function values of arbitrarily refined sparse grid surrogates with a piecewise D-linear basis. It is based on computing a minimal extension set that is added to the original sparse grid and guarantees for properly chosen hierarchical coefficients that the range of function values of the resulting surrogate function is limited to a certain interval. Furthermore, we provided a variety of methods to, first, compute efficiently the minimal extension set based on intersections of sparse grid points and, second, to compute the corresponding hierarchical coefficients the need to evaluate the original model function.

This approach has two main advantages over common methods: First, the extension can be interpreted as a second level of adaptive refinement and the surrogate remains a linear combination of basis functions. Hence, any further operation on the surrogate, such as evaluation, quadrature, interpolation, regression, density estimation, etc. can be applied without further restrictions. Second, the experimental results indicate that the size of the extension set itself grows as the original sparse grid itself, which makes it applicable to a large variety of problems. Nevertheless, the bottleneck of the extension approach is the size of the candidate set. Its size grows exponentially for the peaked Gaussian scenario that we considered in this paper. In this extreme scenario it can contain a non-negligible fraction of full grid points. Truncating the candidate set seems to be possible, but no general approach is known yet. Nevertheless, for real world examples in the context of density estimation with sparse grids (see [3], for example), we observed competitive approximation errors at small additional costs for extending the sparse grid.

The presented method is a can not only be applied to interpolation problems but to approximation problems in general if the presented prerequisites are met. The solution of the unconstrained optimization problem presented in [11], for example, describes the estimation of a sparse grid density function. The optimization could be extended by constraints that guarantee non-negative function values at all grid points, which would enable sparse grid densities that fulfill the properties of probability density functions. Probability density functions play an important role in the context of uncertainty quantification or the solution of the Fokker-Plank equations. It is, however, unclear how the approach we presented in this paper translates to these fields.

All the algorithms presented in this paper are publicly available in the sparse grid toolbox SG^{++} .

Acknowledgements The authors acknowledge the German Research Foundation (DFG) for its financial support of the project within the Cluster of Excellence in Simulation Technology at the University of Stuttgart.

References

1. H.-J. Bungartz, M. Griebel, Sparse grids. Acta Numer. **13**, 147–269 (2004)
2. C. Feuersänger, Sparse grid methods for higher dimensional approximation. Ph.D. thesis, Rheinischen Friedrich–Wilhelms–Universität Bonn, 2010
3. F. Franzelin, D. Pflüger, From data to uncertainty: an efficient integrated data-driven sparse grid approach to propagate uncertainty, in *Sparse Grids and Applications - Stuttgart 2014*, ed. by J. Garcke, D. Pflüger (Springer International Publishing, Cham, 2016), pp. 29–49

4. F. Franzelin, P. Diehl, D. Pflüger, Non-intrusive uncertainty quantification with sparse grids for multivariate peridynamic simulations, in *Meshfree Methods for Partial Differential Equations VII*, ed. by M. Griebel, M. A. Schweitzer. Lecture Notes in Computational Science and Engineering, vol. 100 (Springer International Publishing, Berlin, 2015), pp. 115–143
5. M. Frommert, D. Pflüger, T. Riller, M. Reinecke, H.-J. Bungartz, T. Enßlin, Efficient cosmological parameter sampling using sparse grids. Mon. Not. R. Astron. Soc. **406**(2), 1177–1189 (2010)
6. J. Garcke, Maschinelles Lernen durch Funktionsrekonstruktion mit verallgemeinerten dünnen Gittern. Ph.D. thesis, University of Bonn, Institute for Numerical Simulation, 2004
7. M. Griebel, M. Hegland, A finite element method for density estimation with Gaussian process priors. SIAM J. Numer. Anal. **47**(6), 4759–4792 (2010)
8. M. Hegland, G. Hooker, S. Roberts, Finite element thin plate splines in density estimation. ANZIAM J. **42**, 712–734 (2000)
9. B. Peherstorfer, D. Pflüger, H.-J. Bungartz, Clustering based on density estimation with sparse grids, in *KI 2012: Advances in Artificial Intelligence*, ed. by B. Glimm, A. Krüger. Lecture Notes in Computer Science, vol. 7526 (Springer, Berlin, 2012), pp. 131–142
10. B. Peherstorfer, F. Franzelin, D. Pflüger, H.-J. Bungartz, *Classification with Probability Density Estimation on Sparse Grids* (Springer International Publishing, Cham, 2014), pp. 255–270
11. B. Peherstorfer, D. Pflüger, H.-J. Bungartz, Density estimation with adaptive sparse grids for large data sets, in *Proceedings of the 2014 SIAM International Conference on Data Mining* (2014), pp. 443–451
12. D. Pflüger, *Spatially Adaptive Sparse Grids for High-Dimensional Problems* (Verlag Dr. Hut, München, 2010)

Scalable Algorithmic Detection of Silent Data Corruption for High-Dimensional PDEs

Alfredo Parra Hinojosa, Hans-Joachim Bungartz, and Dirk Pflüger

Abstract In this paper we show how to benefit from the numerical properties of a well-established extrapolation method—the *combination technique*—to make it tolerant to silent data corruption (SDC). The term SDC refers to errors in data not detected by the system. We use the hierarchical structure of the combination technique to detect if parts of the floating point data are corrupted. The method we present is based on robust regression and other well-known outlier detection techniques. It is a lossy approach, meaning we sacrifice some accuracy but we benefit from the small computational overhead. We test our algorithms on a d-dimensional advection-diffusion equation and inject SDC of different orders of magnitude. We show that our method has a very good detection rate: large errors are always detected, and the small errors that go undetected do not noticeably damage the solution. We also carry out scalability tests for a 5D scenario. We finally discuss how to deal with false positives and how to extend these ideas to more general quantities of interest.

1 Introduction

Existing high-performance computing systems exhibit various forms of anomalous behavior. Most commonly, hardware components are prone to fail while performing computations, which often leads to undesirable outcomes. This is simply a result of having a large amount of hardware elements in a system, each of which has a certain probability of failing after a given (possibly very long) time so that the probability of any single component failing within a given time increases proportionally.

A. P. Hinojosa · H.-J. Bungartz
Chair of Scientific Computing, Technische Universität München, München, Germany
e-mail: alfredo.parra@tum.de; bungartz@in.tum.de

D. Pflüger (✉)
Institute for Parallel and Distributed Systems, University of Stuttgart, Stuttgart, Germany
e-mail: dirk.pflueger@ipvs.uni-stuttgart.de

J. Garcke et al. (eds.), *Sparse Grids and Applications – Miami 2016*,
Lecture Notes in Computational Science and Engineering 123,
https://doi.org/10.1007/978-3-319-75426-0_5

Additionally, there are various ways in which the component can fail and some types of errors are expected to become more critical as node count increases towards exascale.

One class of errors that has caught the attention of supercomputer users over the past few years is what is known as *silent data corruption*, or SDC. In the most general sense, SDC can be any undetected system error, usually in the form of arithmetic computation errors, control errors, or wrong network transfer of data [29]. A well-known type of SDC is an undetected bit flip, which can be caused by cosmic rays interacting with the silicon die or other hardware defects [1, 7]. SDC is still poorly understood, but there is a growing consensus among computational scientists supporting the claim that these errors will affect simulations in future exascale systems [29]. Despite the uncertainty surrounding the frequency of such errors, one single occurrence of SDC can have fatal consequences in a simulation [9], as we observed in our experiments.

The effect of SDC has been studied in a wide variety of scenarios and there exists several tools and algorithms to simulate, detect and recover from SDC. Notably, one can find implementations of MPI that include redundancy to detect and recover from SDC [13]; others have applied techniques of data analysis and time series to determine whether numerical computations fall within expected bounds [2]; some groups focus on developing robust numerical algorithms that can tolerate SDC, notably GMRES [10] and other Krylov subspace iterative methods [8]; and processor designers are rethinking arithmetic circuits in CMOS technologies to make them more robust to SDC [26]. But given that a one-size-fits-all solution is unlikely to be developed for the vast amount of application codes out there, it is crucial that algorithm designers try to exploit the numerical properties of their schemes to overcome the challenge of silent faults in the foreseeable future. This is the approach we adopt in this paper, which we aim to demonstrate in the context of the solution of high-dimensional PDEs, one of the most challenging problems in HPC.

1.1 High-Dimensional PDEs in High-Performance Computing

Many interesting physical phenomena are modeled via partial differential equations (PDEs). As a motivating example, consider the problem of microturbulence arising from the confinement of hot plasma with a strong magnetic field. This is an ellusive problem in plasma physics, since the appearance of microturbulence in fusion reactors hinders the production of clean energy. Mathematically, the evolution of the plasma field in such a scenario can be described by a PDE, namely, the gyrokinetic Vlasov equation, given in its most general form by [24]

$$\frac{\partial \mathbf{u}}{\partial t} = \mathscr{L}(\mathbf{u}) + \mathscr{N}(\mathbf{u})\,. \tag{1}$$

This equation describes the time evolution of a (5+1)-dimensional plasma field $\mathbf{u} \equiv \mathbf{u}(x, y, z, v_{\parallel}, \mu; t)$, with x, y and z being the spatial coordinates and $v_{\parallel}$ and μ the velocity coordinates. The differential operators $\mathscr{L}$ and $\mathscr{N}$ describe the linear and nonlinear spatial evolution of $\mathbf{u}$, respectively. The explicit form of the operators $\mathscr{L}$ and $\mathscr{N}$ necessitate numerical methods to solve Eq. (1). The physics code GENE does this for a wide variety of scenarios [23]. It uses a Runge-Kutta scheme in time and a combination of high-order finite differences and Fourier discretization in space, resulting in a 5D cartesian grid $\Omega_{\mathbf{i}}$ with $2^{i_1} \times 2^{i_3} \times 2^{i_3} \times 2^{i_4} \times 2^{i_5}$ discretization points. A grid for a typical scenario has $128 \times 64 \times 512 \times 64 \times 16$ points (2^{32} in total), and requires roughly 2TB just to be stored [25]. Increasing the number of points in a given dimension increases the computational costs considerably, which slows down the research in plasma fusion. This is currently a major challenge in the plasma physics community.

Sparse grids are an attractive option to reduce the number of discretization points in high-dimensional problems, while keeping the error small [5]. Discretizing the computational domain using sparse grids is possible in many cases, but for a legacy code like GENE, with thousands of lines of code and highly optimized routines, it might not be realistic. The *combination technique* helps to overcome this difficulty. It is an extrapolation method that approximates the sparse grid solution, but with the advantage of not having to change the discretization scheme in the original code. In this paper we describe the advantages of using the combination technique to solve high-dimensional PDEs and the properties it has that can be exploited to deal with SDC. In previous work we have outlined two such strategies [21], but our original formulation had several drawbacks and were tested on small 2D examples in serial. The main contribution of this paper is to show an efficient and scalable implementation of those algorithms, which requires some modifications to the original formulation. We show that our algorithms can detect and recover from SDC at a very small computational cost in 2, 3 and up to 5 dimensions. We also demonstrate that one of our algorithms scales to up to 32k cores for a 5D simulation. We are not aware of any results in the literature showing scalability results of this magnitude for SDC detection and correction.

2 Theory of the Classical Combination Technique

We start by introducing some basic notations used throughout the paper. Consider the unit interval [0, 1] discretized with a grid Ω_l with $2^l - 1$ inner points and one point on each boundary ($2^l + 1$ points in total). The mesh size is $h_l := 2^{-l}$ and grid points are $x_{l,j} := j \cdot h_l$ for $0 \le j \le 2^l$, with $l \in \mathbb{N} = \{1, 2, \ldots\}$. In arbitrary dimensions, bold letters denote multi-indices: $\mathbf{l} = (l_1, \ldots, l_d) \in \mathbb{N}^d$. The d-unit cube can be discretized with a Cartesian grid $\Omega_{\mathbf{l}} := \Omega_{l_1} \times \cdots \times \Omega_{l_d}$ with mesh sizes $h_{\mathbf{l}} := (h_{l_1}, \ldots, h_{l_d}) := 2^{-\mathbf{l}} := (2^{-l_1}, \ldots, 2^{-l_d}) \in \mathbb{R}^d$ and grid points $x_{\mathbf{l},\mathbf{j}} := (x_{l_1,j_1}, \ldots, x_{l_d,j_d})$ for $\mathbf{0} \le \mathbf{j} \le 2^{\mathbf{l}}$. We compare multi-indices componentwise: $\mathbf{i} \le \mathbf{j}$ means $i_k \le j_k$ for all $k \in \{1, \ldots, d\}$. Discrete l_p-norms $|\cdot|_p$ for multi-indices will also be used. For example, $|\mathbf{l}|_1 := l_1 + \cdots + l_d$. The wedge operator $\mathbf{i} \wedge \mathbf{j}$ denotes the componentwise minimum of $\mathbf{i}$ and $\mathbf{j}$: $\mathbf{i} \wedge \mathbf{j} := (\min\{i_1, j_1\}, \ldots, \min\{i_d, j_d\})$. Also, the grids are defined according to the standard nested doubling rule, which means that the number of grid points at each level doubles with increasing level i and $\Omega_i \subset \Omega_{i+1}$.

Suppose $u(\mathbf{x}) \in V \subset C([0,1]^d)$ is the exact solution of a PDE in d dimensions. We will denote a numerical approximation of u by $u_{\mathbf{i}}(\mathbf{x}) \in V_{\mathbf{i}} \subset V$, where $V_{\mathbf{i}} = \bigotimes_{k=1}^{d} V_{i_k}$ is the space of piecewise d-linear functions defined on a grid $\Omega_{\mathbf{i}}$ [15]. The combination technique involves solving a PDE on a set of anisotropic grids and adding them together with certain weights to approximate a full grid solution $u_{\mathbf{n}}$:

$$u_n^{(c)} = \sum_{q=0}^{d-1} (-1)^q \binom{d-1}{q} \sum_{\mathbf{i} \in \mathscr{I}_q} u_{\mathbf{i}} \approx u_{\mathbf{n}}. \tag{2}$$

This is the classical formulation of the combination technique, with the index set $\mathscr{I}_q = \{\mathbf{i} : |\mathbf{i}|_1 = n + (d-1) - q\}$. It is worth mentioning that the grid on which the combined solution $u_n^{(c)}$ lives is a sparse grid. As an example, the index sets resulting from choosing $n = 7$ in 2D are

$$\mathscr{I}_0 = \{(7,1), (6,2), (5,3), (4,4), (3,5), (2,6), (1,7)\},$$

$$\mathscr{I}_1 = \{(6,1), (5,2), (4,3), (3,4), (2,5), (1,6)\}.$$

This translates into solving the PDE on 13 different grids and combining them to approximate the full grid solution $u_{7,7}$. It is worth mentioning that the grid on which the combined solution $u_n^{(c)}$ lives is a sparse grid.

In order to combine the different grids, it is common to transform each $u_{\mathbf{i}}$ into the *hierarchical basis* given by

$$u_{\mathbf{i}}(\mathbf{x}) = \sum_{\mathbf{l} \le \mathbf{i}} \sum_{\mathbf{j} \in \mathscr{I}_{\mathbf{l}}} \alpha_{\mathbf{l},\mathbf{j}}^{(\mathbf{i})} \phi_{\mathbf{l},\mathbf{j}}(\mathbf{x}). \tag{3}$$

The $\alpha_{\mathbf{l},\mathbf{j}}^{(\mathbf{i})} \in \mathbb{R}$ are called the *hierarchical coefficients* or *hierarchical surpluses*, and can be computed from the function values as follows:

$$\alpha_{\mathbf{l},\mathbf{j}}^{(\mathbf{i})} = \left(\prod_{k=1}^{d} \left[-\tfrac{1}{2} \quad 1 \quad -\tfrac{1}{2} \right]_{l_k, j_k} \right) u_{\mathbf{i}}(x_{\mathbf{l},\mathbf{j}}). \tag{4}$$

In 1D this stencil notation means

$$\alpha_{l,j}^{(i)} = \begin{bmatrix} -\frac{1}{2} & 1 & -\frac{1}{2} \end{bmatrix}_{l,j} u_i(x_{l,j}) \\ = -\frac{1}{2}u_i(x_{l,j-1}) + u_i(x_{l,j}) - \frac{1}{2}u_i(x_{l,j+1}) \tag{5}$$

The basis functions $\phi_{\mathbf{l},\mathbf{j}}$ are d-dimensional hat functions:

$$\phi_{\mathbf{l},\mathbf{j}}(\mathbf{x}) := \prod_{k=1}^{d} \phi_{l_k, j_k}(x_k), \tag{6}$$

with

$$\phi_{l,j}(x) := \max(1 - |2^l x - j|, 0). \tag{7}$$

The hierarchical index set $\mathscr{I}_\mathbf{l}$ is given by

$$\mathscr{I}_\mathbf{l} := \left\{ \mathbf{j} : 1 \leq j_k \leq 2^{l_k} - 1,\ j_k \text{ odd},\ 1 \leq k \leq d \right\}. \tag{8}$$

The classical combination technique is based on the premise that each $u_\mathbf{i}$ satisfies the *error splitting assumption* (ESA):

$$u - u_\mathbf{i} = \sum_{k=1}^{d} \sum_{\substack{\{e_1,\ldots,e_k\} \\ \subset\{1,\ldots,d\}}} C_{e_1,\ldots,e_k}(\mathbf{x}, h_{i_{e_1}}, \ldots, h_{i_{e_k}}) h_{i_{e_1}}^p \cdots h_{i_{e_k}}^p, \tag{9}$$

where $p \in \mathbb{N}$. It is also assumed that each $\{e_1, \ldots, e_k\} \subset \{1, \ldots, d\}$ is bounded by $|C_{e_1,\ldots,e_k}(\mathbf{x}, h_{i_{e_1}}, \ldots, h_{i_{e_k}})| \leq \kappa_{e_1,\ldots,e_k}(\mathbf{x})$, and that all $\kappa_{e_1,\ldots,e_k}$ are bounded by $\kappa_{e_1,\ldots,e_k}(\mathbf{x}) \leq \kappa(\mathbf{x})$. It is important to note that Eq. (9) is a *pointwise* relation, which means that it must hold for all points $\mathbf{x}$.

The ESA in one dimension reduces to

$$u - u_i = C_1(x_1, h_i) h_i^p, \quad |C_1(x_1, h_i)| \leq \kappa_1(x_1). \tag{10}$$

In two dimensions it becomes

$$u - u_\mathbf{i} = C_1(x_1, x_2, h_{i_1}) h_{i_1}^p + C_2(x_1, x_2, h_{i_2}) h_{i_2}^p + C_{1,2}(x_1, x_2, h_{i_1}, h_{i_2}) h_{i_1}^p h_{i_2}^p. \tag{11}$$

It is possible to show that, if all $u_{\mathbf{i}}$ satisfy the ESA, then the pointwise error of the combination technique is [16]

$$|u - u_n^{(c)}| = \mathcal{O}(h_n^2(\log(h_n^{-1}))^2), \tag{12}$$

which is only slightly worse than the error on a full grid, $\mathcal{O}(h_n^2)$. The ESA is satisfied for the Laplace equation solved with finite differences [6] and for the advection equation solved with both an implicit first order scheme [27] and second order centered finite differences in space, and the fourth order Runge-Kutta scheme in time [17, Section 4.5].

Strongly anisotropic combination solutions cause instabilities in the combination technique, so it is common to increase the minimum level of discretization per dimension using a truncation parameter τ, which results in the *truncated combination technique*

$$u_{n,\tau}^{(c)} = \sum_{q=0}^{d-1}(-1)^q \binom{d-1}{q} \sum_{\mathbf{i}\in\mathcal{I}_{q,\tau}} u_{\mathbf{i}}, \tag{13}$$

with the index set $\mathcal{I}_{q,\tau} = \{\mathbf{i} : |\mathbf{i}|_1 = n + (d-1) - q + \tau, \text{ and } i_j > \tau, \quad \forall j = 1, \ldots, d\}$. By setting $\tau = 3$ in the example presented earlier we would instead obtain the index sets

$$\mathcal{I}_{0,3} = \{(7,4), (6,5), (5,6), (4,7)\},$$

$$\mathcal{I}_{1,3} = \{(6,4), (5,5), (4,6)\}.$$

Throughout this paper we will use this formulation of the combination technique.

One of the main advantages of the combination technique is that computing the various $u_{\mathbf{i}}$ can be done in parallel. We now briefly explain an efficient parallelization strategy and the fault-tolerant variant of the combination technique.

3 The Combination Technique in Parallel

The fact that we solve the same PDE on multiple grids with different resolutions means there is room for parallelization. The only step requiring communication is the combination step—adding the different solutions together to obtain the combined solution $u_n^{(c)}$. This is done either once at the end of the simulation or at every certain number of time steps, depending on the application.

The authors of [19] have described a very efficient parallelization strategy for the combination technique. It is a master-worker scheme, whereby the total available processes P in a parallel system are divided into M *process groups*. Each process group is then assigned a subset of all the grids on which the PDE has to be solved, using an appropriate load balancing scheme. The groups then solve the PDE on their set of grids, one after the other, independently of the other groups. A master process signals the groups when it is time to combine the results. The authors have shown that the computational overhead of the combination technique is usually very small compared to the actual computation of the solutions on each grid, especially as the dimension and problem size increase. The authors tested this scheme on the supercomputer *Hazel Hen* with a 5D example and the scalability results with up to 180k cores were very promising. Their largest experiments consisted of 182 combination grids with a varying number of process groups (from 11 to 88, resulting in 8192 and 1024 processes per group respectively).

4 Dealing with System Faults

Another advantage of the combination technique is that its hierarchical structure can be exploited to deal with system faults. The authors in [18] have shown that if some of the component solutions $u_{\mathbf{i}}$ go missing due to system faults, one can combine the rest of the solutions with alternative weights and still obtain a good approximation of the full grid. This fault-tolerant combination technique is illustrated in Fig. 1 for a simple 2D example. Notice that some additional solutions are needed for the alternative combination to work (in this example, the solution $u_{\mathbf{i}}$ with $\mathbf{i} = (3, 1)$ had to be added). These additional solutions are computed along with the original set of solutions, but this extra effort is small. This approach also has the advantage of not requiring any checkpointing, but as one might expect, some accuracy is lost. This loss of accuracy is nevertheless very small and worth the cost. In previous work we have shown that it is possible to incorporate this fault tolerant algorithm in the parallelization strategy described above. The idea is that if some processes in a group fail during the computation phase, the whole process group is removed from the communicator before the combination step, and all solutions in that group are given a coefficient of zero. Our experiments showed that the combined solution remains accurate, and that the algorithm scales well with up to 65k cores [20].

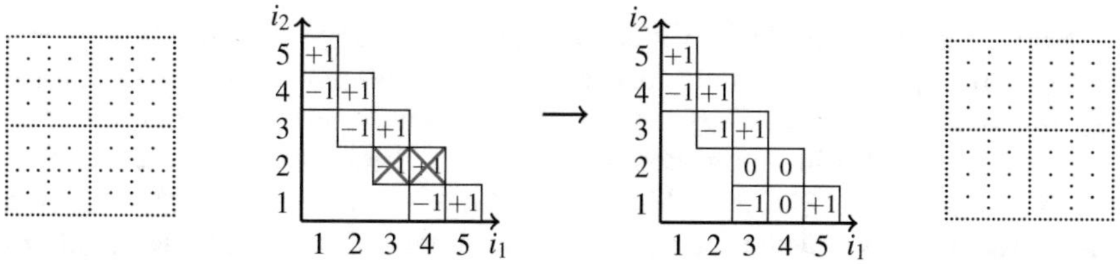

Fig. 1 Simple example of the fault-tolerant combination technique where two solutions go missing due to system faults (left) and the alternative combination coefficients (right)

5 Detecting and Recovering from SDC

In this section we address the main topic of this paper: what can be done if SDC affects one or more combination solutions while being computed? We have argued in [21] that even small arithmetic errors in the combination solutions $u_{\mathbf{i}}$ can ruin the combined solution $u_n^{(c)}$, and we have described two possible ways to detect corrupted data before the combination takes place. We now formalize and extend those ideas, and we discuss how to implement them efficiently and in parallel. The main idea is to recognize that although we do not know what the solution of the PDE will look like, if the combination technique converges, we can expect the different combination solutions look somewhat similar, so any solutions that deviate too much from the others should be inspected for SDC. Since the ESA is sufficient for convergence, we can define solutions to be "similar" to each other if they fulfil the ESA. The two methods that we will describe are based on outlier detection techniques and robust regression. Before going into the details, we emphasize that in both cases we implicitly assume that each $u_{\mathbf{i}}$ is expressed in the hierarchical basis (4). This is admissible since the hierarchical coefficients also satisfy the ESA, which is the starting point of our SDC detection algorithms.

5.1 *Method 1: Comparing Combination Solutions Pairwise via a Maximum Norm*

In Sect. 2 we argued that each combination solution $u_{\mathbf{i}}$ has to fulfill the ESA (9) pointwise. So suppose we take two arbitrary combination solutions $u_{\mathbf{t}}$ and $u_{\mathbf{s}}$ from the set of all the solutions to be combined. If the two solutions satisfy the ESA, then their difference should satisfy the relation

$$\begin{aligned} u_{\mathbf{t}}(x_{\mathbf{l},\mathbf{j}}) - u_{\mathbf{s}}(x_{\mathbf{l},\mathbf{j}}) &= C_1(x_{\mathbf{l},\mathbf{j}}, h_{t_1})h_{t_1}^p + C_2(x_{\mathbf{l},\mathbf{j}}, h_{t_2})h_{t_2}^p + C_{1,2}(x_{\mathbf{l},\mathbf{j}}, h_{t_1}, h_{t_2})h_{t_1}^p h_{t_2}^p \\ &\quad - C_1(x_{\mathbf{l},\mathbf{j}}, h_{s_1})h_{s_1}^p - C_2(x_{\mathbf{l},\mathbf{j}}, h_{s_2})h_{s_2}^p - C_{1,2}(x_{\mathbf{l},\mathbf{j}}, h_{s_1}, h_{s_2})h_{s_1}^p h_{s_2}^p. \end{aligned} \quad (14)$$

(We do the analysis in 2D for simplicity, but the idea applies to arbitrary dimensions.) This equation holds only for the grid points common to both grids $\Omega_{\mathbf{s}}$ and $\Omega_{\mathbf{t}}$, i.e. for all $x_{\mathbf{l},\mathbf{j}}$ with $(1, 1) \leq \mathbf{l} \leq \mathbf{t} \wedge \mathbf{s}$. Taking the largest value of (14) over all $(\mathbf{l}, \mathbf{j})$ can serve as an indicator of how similar or different two solutions $u_{\mathbf{t}}$ and $u_{\mathbf{s}}$ are. Now assume that SDC has affected one or more function values of either $u_{\mathbf{t}}$ or $u_{\mathbf{s}}$, causing (14) to be large for the affected grid points. This means we should be suspicious of the grid point $x_{\mathbf{l},\mathbf{j}}$ where (14) is largest, so we can measure

$$\beta_{(\mathbf{s},\mathbf{t})} := \max_{\mathbf{l} \leq \mathbf{t} \wedge \mathbf{s}} \max_{\mathbf{j} \in \mathscr{I}_{\mathbf{l}}} \left| u_{\mathbf{t}}(x_{\mathbf{l},\mathbf{j}}) - u_{\mathbf{s}}(x_{\mathbf{l},\mathbf{j}}) \right|. \quad (15)$$

We can compute $\beta_{(\mathbf{s},\mathbf{t})}$ for many pairs of grids $(\mathbf{s}, \mathbf{t})$ and try to determine whether they fit the error expansion (14). If some measurements don't fit the model well, they can be considered to be outliers. By "fitting" we mean finding the functions C_1, C_2 and $C_{1,2}$ in a least square sense. The only problem is that each of these functions depend on $x_{\mathbf{l},\mathbf{j}}$ since (14) holds pointwise. This means that if we want to fit measurements of $\beta_{(\mathbf{s},\mathbf{t})}$ to the model (14), every measurement should be made at the same grid point for all pairs $(\mathbf{s}, \mathbf{t})$. We can do this by focusing on the grid point $x^*_{\mathbf{l},\mathbf{j}}$ where $\beta_{(\mathbf{s},\mathbf{t})}$ is largest *over all pairs*:

$$(\mathbf{s},\mathbf{t})^* = \arg\max_{(\mathbf{s},\mathbf{t})\in\mathscr{V}} \beta_{(\mathbf{s},\mathbf{t})} \tag{16}$$

$$x^*_{\mathbf{l},\mathbf{j}} = \arg\max_{x_{\mathbf{l},\mathbf{j}}} \beta_{(\mathbf{s},\mathbf{t})^*} \tag{17}$$

$\mathscr{V}$ denotes the set of all pairs of multi-indices under consideration. We can then measure

$$\begin{aligned}\beta^*_{(\mathbf{s},\mathbf{t})} := u_{\mathbf{t}}(x^*_{\mathbf{l},\mathbf{j}}) - u_{\mathbf{s}}(x^*_{\mathbf{l},\mathbf{j}}) &\approx C_1(x^*_{\mathbf{l},\mathbf{j}}, h_{t_1})h^p_{t_1} + C_2(x^*_{\mathbf{l},\mathbf{j}}, h_{t_2})h^p_{t_2}\\ &\quad - C_1(x^*_{\mathbf{l},\mathbf{j}}, h_{s_1})h^p_{s_1} - C_2(x^*_{\mathbf{l},\mathbf{j}}, h_{s_2})h^p_{s_2}\\ &=: \tilde{\beta}_{(\mathbf{s},\mathbf{t})}\end{aligned} \tag{18}$$

for all pairs $(\mathbf{s}, \mathbf{t}) \in \mathscr{V}$. In the equation above, $\beta^*_{(\mathbf{s},\mathbf{t})}$ denotes the actual value we measure, while $\tilde{\beta}_{(\mathbf{s},\mathbf{t})}$ is the model for the error expansion (14) (without the higher order terms). For a given index $\mathbf{s}$ we suggest to measure $\beta^*_{(\mathbf{s},\mathbf{t})}$ with respect to the d-nearest neighbors of $\mathbf{s}$ (those for which the distance $|\mathbf{s} - \mathbf{t}|_1$ is smallest).

As an example, consider a set of 10 combination solutions in two dimensions with the following index set:

$$\overbrace{\mathscr{I}_{0,3} \cup \mathscr{I}_{1,3}}^{\text{Classical set}} \cup \overbrace{\mathscr{I}_{2,3} \cup \mathscr{I}_{3,3}}^{\text{Fault tolerance set}} = \{(7,4), (6,5), (5,6), (4,7), (6,4), (5,5), (4,6), (5,4), (4,5), (4,4)\} \tag{19}$$

The resulting set of pairs $\mathscr{V}$ would have 11 elements, listed in Table 1. The second column shows the measured values of $\beta^*_{(\mathbf{s},\mathbf{t})}$ for a simple example solving an advection-diffusion equation (more details in Sect. 6.1). The values corresponding to pairs $\{(4,7), (4,6)\}$ and $\{(5,6), (4,6)\}$ are particularly high, which could indicate that solution $u_{(4,6)}$ has been affected by SDC. A measurement is "high" if it deviates too much from the model. This means we should perform *robust regression* on the values we measure, that is, trying to fit the measured values $\beta^*_{(\mathbf{s},\mathbf{t})}$ to the model $\tilde{\beta}_{(\mathbf{s},\mathbf{t})}$, but with the constraint that some of the values could be outliers. This translates into finding the values of the functions C_1 and C_2 at all mesh sizes h_i that appear

Table 1 Measurements of $\beta^*_{(\mathbf{s},\mathbf{t})}$ with one solution affected by SDC, namely, $u_{(4,6)}$

Pair $(\mathbf{s}, \mathbf{t})$	$\beta^*_{(\mathbf{s},\mathbf{t})}$
(4, 5) (4, 4)	0.0275
(4, 7) (4, 6)	**0.2180**
(4, 7) (5, 6)	−0.0029
(5, 4) (4, 4)	0.0152
(5, 4) (5, 6)	−0.0498
(5, 5) (4, 5)	0.0158
(5, 6) (4, 6)	**0.2210**
(6, 5) (5, 5)	0.0111
(6, 5) (6, 4)	0.0283
(7, 4) (6, 4)	0.0061
(7, 4) (6, 5)	−0.0222

in the combination technique, namely, $h_i = \{h_{\tau+1}, h_{\tau+2}, \ldots, h_n\}$. If we denote by $\mathbf{c} := (C_1(h_{\tau+1}), \ldots, C_1(h_n), C_2(h_{\tau+1}), \ldots, C_2(h_n))$ our vector of $2 \cdot (n - \tau)$ unknowns, the robust least squares problem that results is

$$\mathbf{c}_{\min} \leftarrow \min_{\mathbf{c}} \sum_{(\mathbf{s},\mathbf{t}) \in \mathscr{V}} \rho\left(\tilde{\beta}_{(\mathbf{s},\mathbf{t})}(\mathbf{c}) - \beta^*_{(\mathbf{s},\mathbf{t})}\right), \tag{20}$$

$$\tilde{\beta}_{(\mathbf{s},\mathbf{t})} = \beta^*_{(\mathbf{s},\mathbf{t})} + e_{(\mathbf{s},\mathbf{t})}. \tag{21}$$

$e_{(\mathbf{s},\mathbf{t})}$ is the difference between the measurements and the model. Since the model $\tilde{\beta}_{(\mathbf{s},\mathbf{t})}$ is linear, it can be written as the matrix-vector product $\tilde{\beta}_{(\mathbf{s},\mathbf{t})}(\mathbf{c}) := \mathbf{X} \cdot \mathbf{c}$, where the matrix $\mathbf{X} \in \mathbb{R}^{|\mathscr{V}| \times 2 \cdot (n-\tau)}$ contains the corresponding coefficients h_i^p. The role of the loss function ρ is to attenuate the effect of outliers, and it should have the following properties [14]:

$$\begin{aligned}
&\rho(e) \geq 0 \\
&\rho(0) = 0 \\
&\rho(-e) = \rho(e) \\
&\rho(e_1) \geq \rho(e_2) \quad \text{for} \quad |e_1| > |e_2|
\end{aligned} \tag{22}$$

The choice $\rho(e) = e^2$ leads to the ordinary least squares problem, but the 2-norm is not robust to outliers. Some choices for the loss function that are robust to outliers include

- Huber's function:

$$\rho(e) = \begin{cases} e, & e \leq 1 \\ \sqrt{e} - 1, & e > 1 \end{cases}$$

- Cauchy's function:

$$\rho(e) = \ln(1 + e)$$

- arctan function:

$$\rho(e) = \arctan(e).$$

There are well-established algorithms to solve the minimization problem (20), such as the Trust Region Reflective algorithm [4] or the Iteratively Reweighted Least Squares (IRLS) method [22]. In this paper we opt for the latter, which is implemented in the GSL library [14]. The main idea behind using robust regression is that outlier measurements will have large residuals $\mathbf{r} = \beta^*_{(\mathbf{s},\mathbf{t})} - \mathbf{X} \cdot \mathbf{c}_{\min}$, but in order to have an absolute criterion for outliers, the residuals have to be normalized by a scaling factor to make them dimensionless. One way to do this is described in [28]. The idea is to calculate a preliminary scale estimate σ^0 given by

$$\sigma^0 = 1.4826 \left(1 + \frac{5}{|\mathscr{V}| - 2 \cdot (n - \tau)}\right) \sqrt{\text{med } \mathbf{r}^2}. \tag{23}$$

Here, $|\mathscr{V}|$ is the number of measurements we have (equal to the number of pairs we are comparing) and $2 \cdot (n - \tau)$ corresponds to the number of unknowns in the minimization problem (number of C_i functions). One then calculates a weight w_i for each residual as follows:

$$w_i = \begin{cases} 1, & \text{if } \;|r_i/s^0| \leq 2.5 \\ 0, & \text{otherwise} \end{cases}$$

With these weights we then compute a more robust scale estimate σ^* given by

$$\sigma^* = \sqrt{\frac{\sum_{i=1}^{|\mathscr{V}|} w_i r_i^2}{\sum_{i=1}^{|\mathscr{V}|} w_i - 2 \cdot (n - \tau)}}.$$

One can finally use this scale estimate to compute the *standardized residuals*

$$\hat{\mathbf{r}} = \frac{\mathbf{r}}{\sigma^*}. \tag{24}$$

A common heuristic is to label the i-th measurement as outlier if $|\hat{r}_i| > 2.5$.

5.2 Method 2: Comparing Combination Solutions via their Function Values Directly

Instead of measuring the value of $\beta^*_{(\mathbf{s},\mathbf{t})}$ (the maximum difference between pairs of combination solutions), one could also look at the function values $u_{\mathbf{i}}(x_{\mathbf{l},\mathbf{j}})$ directly. We proceed as in the previous section by first finding the grid point $x^*_{\mathbf{l},\mathbf{j}}$ where the value of $\beta_{(\mathbf{s},\mathbf{t})}$ is largest (Eq. (17)). But now we take a look at the function values $u_{\mathbf{i}}(x^*_{\mathbf{l},\mathbf{j}})$ *for all combination solutions* $u_{\mathbf{i}}$ *containing the grid point* $x^*_{\mathbf{l},\mathbf{j}}$. Since we can expect the different values of $u_{\mathbf{i}}(x^*_{\mathbf{l},\mathbf{j}})$ to be somewhate similar across the combination solutions (despite having different discretizations), we try to fit them to a constant $\tilde{u}$:

$$u_{\min} \leftarrow \min_{\tilde{u}} \sum_{\mathbf{l}' \geq \mathbf{l}} \rho\left(u_{\mathbf{l}'}(x^*_{\mathbf{l},\mathbf{j}}) - \tilde{u}\right). \tag{25}$$

(A grid point $x_{\mathbf{l},\mathbf{j}}$ can be found in all combination solutions $u_{\mathbf{l}'}$ for which $\mathbf{l}' \geq \mathbf{l}$.) The residuals $r_{\mathbf{i}} = u_{\mathbf{i}}(x^*_{\mathbf{l},\mathbf{j}}) - u_{\min}$ can be normalized as in the previous section (Eq. (24)) to determine if any of them corresponds to an outlier measurement, with $|\mathcal{V}|$ substituted by the number of measurements we have (number of combination solutions containing the grid point $x^*_{\mathbf{l},\mathbf{j}}$) and $2 \cdot (n - \tau)$ substituted by 1 (number of unknowns in the minimization problem—in this case, only $\tilde{u}$). One advantage of this method is that it doesn't explicitly rely on the ESA, which could make it more useful in cases where it is not clear whether the ESA holds.

5.3 Cost and Parallelization

The two methods we have described require two main steps: searching for the grid point $x^*_{\mathbf{l},\mathbf{j}}$ where $\beta_{(\mathbf{s},\mathbf{t})}$ is largest for all pairs of solutions and solving the modified least squares problems (20) and (25).

Let us begin with the cost of calculating all values of $\beta_{(\mathbf{s},\mathbf{t})}$, that is, the maximum difference over all pairs of combination solutions from the set $\mathcal{V}$. The number of pairs we have depends on the number of combination solutions $u_{\mathbf{i}}$ resulting from the truncated combination technique (13). For a given level n and truncation parameter

τ, the total number of combination solutions is given by [1]

$$\left|\bigcup_{q=0}^{d+1} \mathscr{I}_{q,\tau}\right| = \sum_{q=0}^{d+1} \frac{(n+(d-2)-q-\tau)!}{(n-q-\tau-1)!(d-1)!}. \tag{26}$$

For large n there are $\mathscr{O}(d \cdot n^{d-1})$ such grids. If we compare each combination solution to its d nearest neighbors, then we have $|\mathscr{V}| = \mathscr{O}(d^2 \cdot n^{d-1})$ pairs to compare. Each comparison requires a simple traversal of the two grids (each of which has $\mathscr{O}(2^n)$ grid points), which is not expensive, especially compared to the actual work of solving N time steps of a PDE on each of the $|\mathscr{I}_{q,\tau}|$ grids.

The cost of solving any of the minimization problems (20) and (25) is also small. In the former case, the regression model matrix $\mathbf{X}$ is very small (of size $|\mathscr{V}| \times 2 \cdot (n-\tau)$), and the minimization problem usually takes $10^1 - 10^2$ iterations to converge (using IRLS). In the latter case, we are only fitting a scalar value, so the cost is negligible.

We can use the parallel framework described in Sect. 4 to accelerate the computation of $\beta_{(\mathbf{s},\mathbf{t})}$ (and $\beta^*_{(\mathbf{s},\mathbf{t})}$). Since each solution $u_{\mathbf{i}}$ is parallelized using domain decomposition, each process within a process group can calculate its local value of $\beta_{(\mathbf{s},\mathbf{t})}$ for all pairs of solutions. The local values can then be reduced within the group using *MPI_MAX*. Both methods are implemented at a group level, so there is no global communication involved, which also means that each process group should contain enough combination solutions $u_{\mathbf{i}}$ for the statistics to be significant. As we will see in our tests, this is not a problem since one only needs approximately 5 or more measurements (so roughly 5 grids per group). These numbers will become more clear in Sect. 6 where we describe typical simulation scenarios in various dimensions.

5.4 Detection Rates

It is difficult to estimate the detection rate of our methods a priori since the rate depends on various different factors, namely:

- The loss function used for the minimization problem
- The specific minimization algorithm used to solve (20) and (25)
- The threshold for the standardized residuals
- The number of combination technique solutions involved

So instead of trying to deduce theoretical detection rates, we will study fault scenarios extensively and report the detection quality of our algorithms.

[1] This can be calculated using, for example, the stars and bars method.

6 Numerical Tests

6.1 Experimental Setup

To test our algorithm we used a d-dimensional advection-diffusion equation

$$\partial_t u - \Delta u + \mathbf{a} \cdot \nabla u = f \quad \text{in } \Omega \times [0, T) \tag{27}$$
$$u(\cdot, t) = 0 \quad \text{in } \partial\Omega$$

as implemented in [20], with $\Omega = [0, 1]^d$, $t = [0, 0.05]$, $\mathbf{a} = \mathbf{1}^T$ and $u(\cdot, 0) = e^{-100 \sum_{i=1}^{d} (x_i - 0.5)^2}$ using the framework `DUNE-pdelab` [3]. This means we simulate a Gaussian function centered in the middle of the domain at $t = 0$, traveling diagonally with a constant velocity in every dimension. The spatial domain is discretized using the finite volume element method on rectangular grids, and we use a simple explicit Euler scheme in time.

We carried out experiments in up to 5 dimensions. In all cases we simulated 50 time steps using $\Delta t = 10^{-3}$ and we combined the solutions every 10 time steps according to Eq. (13). The robust linear regression problems (20) and (25) were solved using the GNU Scientific Library [14] with a Cauchy loss function. The library implements the IRLS algorithm.

6.2 SDC Injection

In [21] we proposed a way to simulate SDC in the combination technique. It is based on the suggestions in [11] and [12], where the authors argue against injecting perturbed values into the floating point data (for example, in the form of random bit flips) to study how an algorithm reacts. Instead, they propose trying to identify the worst-case scenarios that a solver could face, and to come up with numerical bounds or additional algorithmic steps to guarantee stability in these cases. Although it is often difficult to identify the worst-case scenarios (and simulating these may lead to olverly pessimistic conclusions), we think it's the more robust approach. See [21] for a more detailed discussion.

For our simulations, the fault injection looks as follows. First we choose a combination solution $u_{\mathbf{i}}$ to inject SDC into. For our experiments we choose a solution with the highest resolution (any $u_{\mathbf{i}}$ with $|\mathbf{i}|_1 = n + d - 1$), for a reason that will later become evident. To simulate the effect of SDC, we choose a grid point $x_{\mathbf{l},\mathbf{j}}$ in $u_{\mathbf{i}}$ and alter the function value at that point in one of the following ways:

1. $\tilde{u}_{\mathbf{i}}(x_{\mathbf{l},\mathbf{j}}) = u_{\mathbf{i}}(x_{\mathbf{l},\mathbf{j}}) \times 10^{-300}$ (very small)
2. $\tilde{u}_{\mathbf{i}}(x_{\mathbf{l},\mathbf{j}}) = u_{\mathbf{i}}(x_{\mathbf{l},\mathbf{j}}) \times 10^{-0.5}$ (slightly smaller)
3. $\tilde{u}_{\mathbf{i}}(x_{\mathbf{l},\mathbf{j}}) = u_{\mathbf{i}}(x_{\mathbf{l},\mathbf{j}}) \times 10^{+150}$ (very large)

We investigate two injection scenarios: one where SDC is injected in the middle of the domain ($x_i = 0.5,\ i = 1, \ldots, d$ in the unit cube) and another where the injection is done near the middle of the domain, namely at $x_i = 0.5 - h_i,\ i = 1, \ldots, d$. In the former case, the chosen grid point exists in all combination solutions $u_{\mathbf{i}}$, whereas in the latter case, the grid point appears only in the solution affected. This means that if SDC is injected into a combination solution $u_{\mathbf{i}}$ of the finest resolution (which is what we do), it should be easier to detect when injected in the middle of the domain (since the grid point is common to all solutions) than in the latter case (since the grid point appears only in that grid). It is this latter case that proves most challenging since SDC can potentially go undetected. Since the solution of our PDE is a Gaussian that travels away from the center, the function values in the middle of the domain vary from 1 to almost 0, which allows us to investigate all possible magnitudes of SDC.

Finally, we choose a time iteration where the fault is to be injected and we inject SDC only once in the entire simulation. Our goal is to detect the wrong solution before performing the next combination and exclude it from the set of solutions by assigning it a coefficient of zero and computing alternative combination coefficients, as discussed in Sect. 4.

6.3 Results: Detection Rates and Errors

We investigated the quality of our two detection methods in 2, 3 and 5 dimensions. We are primarily interested in the percentage of cases where SDC is detected, as well as the quality of the combination technique after detecting and removing the wrong solution.

Figure 2 shows simulation results with $n = 7$ and a truncation parameter $\tau = 2$ for the 2D truncated combination technique (13), which results in 14 combination solutions. For each of the three magnitudes of SDC from the previous subsection we ran 50 independent simulations, such that at the i-th simulation we injected SDC at the i-th time step. At the end of each simulation we calculated the l_2 relative error of the combination solution compared to a full grid solution of level 7 (i.e., with $(2^7 + 1) \times (2^7 + 1)$ grid points), $e = \frac{\|u_n^{(c)} - u_{\text{ref}}\|_2}{\|u_{\text{ref}}\|_2}$. When SDC was injected in the middle of the domain (Fig. 2a), Method 1 detected the SDC in 88%, 64% and 100% of the times for the three magnitudes of SDC, while for Method 2 the rates were 98%, 84% and 100%. When we injected SDC near the middle of the square domain (Fig. 2b), the detection rates were 22%, 10% and 100% for Method 1 and 0%, 0% and 100% for Method 2. But these low detection rates are not a bad result, since the error introduced by the SDC is so small that it is tolerable to not detect it: it affects the combination solution almost imperceptibly. When SDC was large (10^{+150}), SDC was always detected.

For the simulations in 3D we used once again $n = 7$ and a truncation parameter $\tau = 2$, which results in 10 combination solutions. For each magnitude of SDC we

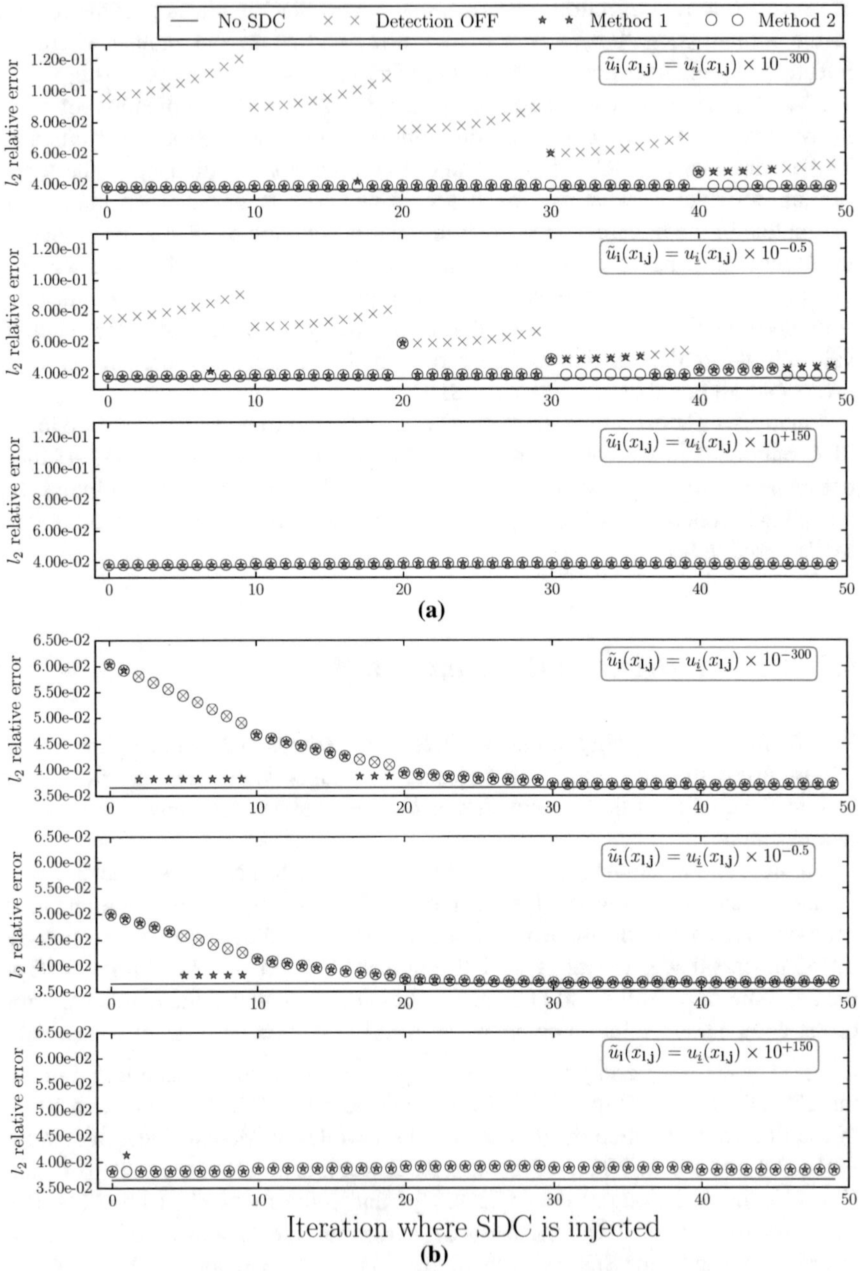

Fig. 2 l_2 relative error of the 2D combination technique with simulated SDC injected in (**a**) the middle of the domain and (**b**) near the middle of the domain

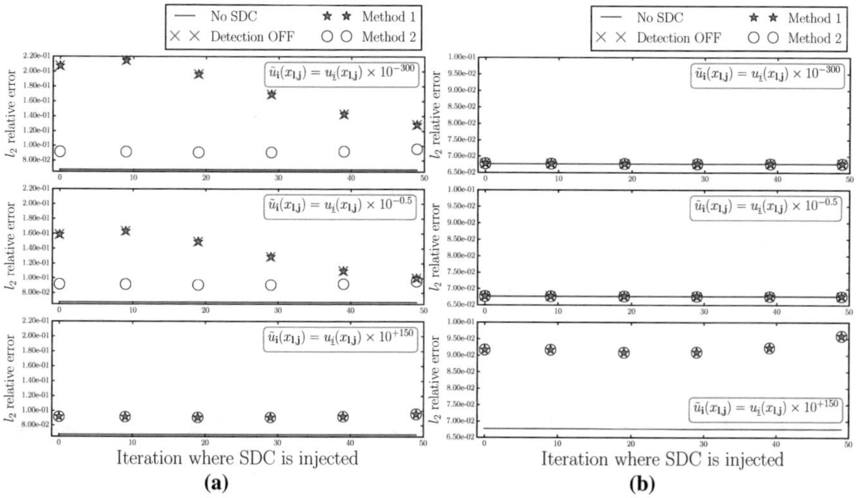

Fig. 3 l_2 relative error of the 3D combination technique with simulated SDC injected in (**a**) the middle of the domain and (**b**) near the middle of the domain

ran 6 independent simulations, injecting SDC at iterations 0, 9, 19, 29, 39 and 49, respectively. We calculated the error of the solution compared to a full grid solution of level $n = 5$ at the end of each simulation. The results are very similar to the 2D case, with the exception that Method 1 performs more poorly when SDC is of moderate magnitude (10^{-300} or $10^{-0.5}$). Method 2 remained robust, detecting all instances of SDC that would have otherwise led to large errors. Figure 3 summarizes our results.

Finally, for our detection tests in 5D we used $n = 11$ and a truncation parameter $\tau = 2$, which resulted in 21 combination solutions. Injection was done as in the 3D case and we compared the combination technique with a full grid solution of level $n = 5$. Due to the very high cost of running these tests, we only simulated one scenario, injecting SDC in the middle of the domain and detecting using Method 2 (Fig. 4a). We used 2 process groups, each with 1024 processes. As in the 2D and 3D cases, SDC is not detected when its effect is too small. In all other cases, it is detected and fixed.

6.4 Results: Scaling

To test the parallel performance of our algorithm, we measured the time needed to compare all pairs of grids and solve the robust regression problem of Method 2. We used a five dimensional scenario with $n = 15$ and $\tau = 2$, which results in 126 combination solutions. For the parallelization we used 8 process groups, doubling the number of processes per group from 256 until 4096. Our time measurements

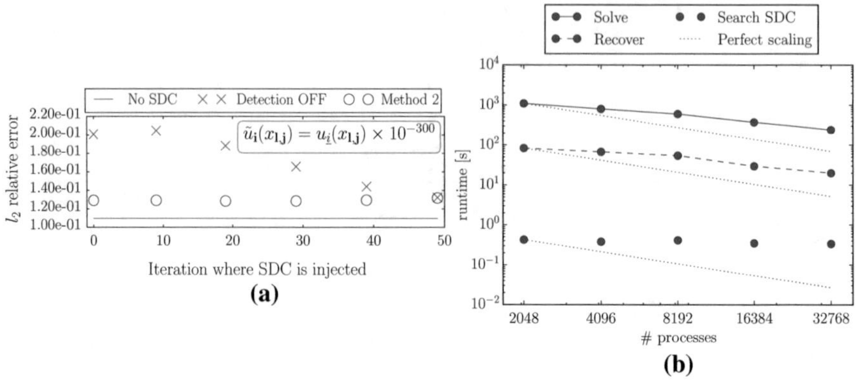

Fig. 4 (**a**) l_2 relative error of the 5D combination technique with simulated SDC injected in the middle of the domain; (**b**) scaling experiments

can be seen in Fig. 4b (*Search SDC*). Once the SDC is detected, the wrong solution is removed and the combination technique is adapted. The time to perform these operations is shown in the plot as *Recover*. As expected, this time is negligible compared to the time required for one time step of the solver (*Solve*, which is 3–4 orders of magnitude larger), but the time remains fairly constant with increasing number of processors. For this simulation scenario, we would still need to increase the number of processors by several orders of magnitude for the *Recover* step to start playing a role. The main cost of the recovery step comes from reinitializing the wrong task, as we reported in [20], and this cost depends on the solver. In the case of DUNE, initializing a grid does not seem to scale.

6.5 Dealing with False Positives

Increasing the dimension of the problem gave rise to an increasing number of false positives: combination solutions marked as outliers when they were not. This happened in scenarios where the function value at a grid point was almost exactly the same across many combination solutions, but only slightly different in others, as illustrated in the measurements in Table 2.

If outliers are penalized strongly, such slight variations are assigned very large standardized residuals and they are marked as outliers.

To solve this problem, we propose the following approach. Consider again the six measurements from Table 2. We first combine the six values according to their classical combination coefficients,

$$u_c = \sum_{\mathbf{i}} c_{\mathbf{i}} u_{\mathbf{i}}(0.5, 0.5).$$

Table 2 Example of values clustering

$\mathbf{i}$	$u_{\mathbf{i}}(0.5, 0.5)$
(3, 3, 3, 3, 3)	0.94203
(3, 3, 4, 3, 3)	0.94203
(4, 3, 3, 3, 3)	0.94203
(3, 3, 3, 3, 4)	**0.94571**
(3, 4, 3, 3, 4)	0.94203
(3, 3, 3, 4, 3)	0.94203

Grid with index (3, 3, 3, 3, 4) can be erroneously marked as wrong if outliers are penalized strongly

We then remove the suspect value as if it were an outlier (we assign it a zero coefficient) and we combine the rest of the values with their alternative combination coefficients (computed according to the Fault Tolerant Combination Technique),

$$u'_c = \sum_{\mathbf{i}} c'_{\mathbf{i}} u_{\mathbf{i}}(0.5, 0.5).$$

If the suspect value is indeed an outlier, the relative difference between u_c and u'_c should be large, since u_c is affected by the outlier but u'_c is not. On the other hand, if the suspect value is only slightly different (and thus not really an outlier), then u_c and u'_c should be very similar. So we can compute

$$e_{\text{rel}} = \frac{|u_c - u'_c|}{|u_{\min}|},$$

and if this error is small (say, $< 5\%$), we conclude that the suspect value is not actually an outlier. $u_{\min}$ is the robust value obtained from the regression problem (25). We used this strategy in all scenarios described earlier.

It is important to keep in mind that the integration scheme we have been using is explicit, which means that the error introduced by SDC at a given time step propagates slowly and locally. Implicit schemes, on the contrary, could spread the error to the whole domain after only one time step. It is not clear how large the error of the combination solution would be if all values were affected by a small factor. If no wrong values are detected, the small errors could accumulate, leading to a large l_2 error. On the other hand, if all values in one combination solution are wrong, there is a chance that the largest such error could be large enough to be detected, especially considering that errors are amplified during hierarchization. Such a scenario should be included in future experiments.

7 Extensions to Quantities of Interest

The two methods described in this paper are based on observing the function values of the different combination solutions. However, we are often not interested in the function values themselves, but rather in a Quantity of Interest (QoI) Q corresponding to the numerical solution $u_{\mathbf{i}}$, or $Q(u_{\mathbf{i}})$. We can also apply the techniques of robust regression presented in this paper if we have a mathematical model of the discrete QoI (analogous to Eq. (18)). That is, if we have a model for

$$Q(u) - Q(u_{\mathbf{i}}),$$

we can perform a regression analysis on the measured QoIs.

As an example, consider the QoI given by the integral of the solution field over the entire domain [31]:

$$Q(u) = \int_{\Omega} u(\mathbf{x}) \mathrm{d}\mathbf{x}. \tag{28}$$

We can calculate this integral numerically for each of the combination solutions $u_{\mathbf{i}}$ and perform robust regression to find outliers. If we use, say, a trapezoidal rule Q_h to calculate (28) numerically, we know that the error will be of order 2 in $h = 1/N$, with $N = \prod_{j=1}^{d}(2^{i_j} + 1)$,

$$Q(u) - Q_h(u_{\mathbf{i}}) = C \cdot h^2 + \mathcal{O}(h^3). \tag{29}$$

We could then try to fit our measurements of $Q_h(u_{\mathbf{i}})$ to, say, a polynomial model of powers of h,

$$Q_h(u_{\mathbf{i}}) \approx \tilde{Q}(\mathbf{c}, h) := \sum_{j=0}^{p} c_j h^j. \tag{30}$$

The robust minimization problem would then be

$$\min_{\mathbf{c}} \sum_{\mathbf{i}} \rho \left(Q_h(u_{\mathbf{i}}) - \tilde{Q}(\mathbf{c}, h) \right). \tag{31}$$

The advantage now is that since we look at an integrated quantity, it matters little where the SDC occurs in the solution field.

Initial tests with a linear advection equation in 2D show promising results. Injecting SDC of magnitude $10^{-0.5}$ into one solution affects its QoI, as seen in Table 3. (Here we used an example for which the exact integral is $Q(u) = \pi$.) Trying to fit the measurements of $Q_h(u_{\mathbf{i}})$ to model (30) with order $p = 2$ reveals the outlier QoI (corresponding to solution $u_{(6,3)}$).

Table 3 Detecting outlier quantities of interest with a simple polynomial model: grid (6,3) is identified

$\mathbf{i}$	$Q_h(u_{\mathbf{i}})$	$\hat{r}_{\mathbf{i}}$
(7, 3)	3.14175	0.01204
(5, 4)	3.14268	1.32745
(4, 6)	3.14215	0.67156
(6, 4)	3.14215	0.67156
(4, 5)	3.14268	1.32745
(4, 3)	3.14245	0.23318
(4, 4)	3.14344	2.19886
(5, 5)	3.14222	0.79327
(6, 3)	**3.13748**	**7.66325**
(3, 6)	3.14189	0.00742
(3, 7)	3.14175	0.01204
(3, 5)	3.14215	0.01368
(3, 4)	3.14245	0.23318
(5, 3)	3.14215	0.01368

8 Conclusion

There has been an increasing interest in studying the effect of SDC in large parallel systems. This has resulted in different proposals to overcome them. Checksums, for example, are quite common but they can be very cumbersome to implement and computationally expensive, sometimes requiring checkpointing, additional checkpoint processes, and a fault-tolerant MPI implementation ([29], Sect. 5.4.2 includes an extensive list of examples.) Data replication can work for some problems (see [30]), but the overhead is not negligible. We believe that a one-size-fits-all solution that is both scalable and robust is not likely to be developed, and that we can do better by focusing on the particular features of our algorithms.

We proposed two methods to detect SDC using the combination technique. Both are based on robust regression, but the minimization problems are different. In the first algorithm we try to fit the difference of the function values for a set of pairs of combination solutions to an error expansion. This method performs very well in 2D but becomes less reliable as the dimension increases. The second method involves fitting the function values directly, and we showed that this method is robust in up to 5D but false positives have to be taken care of. Our scaling experiments with up to 32k cores showed that the cost of searching and recovering from SDC is very small. We believe that focusing instead on the Quantities of Interest could be a more promising approach since for some QoIs it is irrelevant where SDC occurs in the data. This will be the topic of future work.

References

1. L. Bautista-Gomez, F. Cappello, Detecting silent data corruption for extreme-scale MPI applications, in *Proceedings of the 22nd European MPI Users' Group Meeting* (ACM, New York, 2015), p. 12
2. E. Berrocal, L. Bautista-Gomez, S. Di, Z. Lan, F. Cappello, Lightweight silent data corruption detection based on runtime data analysis for HPC applications, in *Proceedings of the 24th International Symposium on High-Performance Parallel and Distributed Computing, HPDC '15* (ACM, New York, 2015), pp. 275–278
3. M. Blatt, A. Burchardt, A. Dedner, C. Engwer, J. Fahlke, B. Flemisch, C. Gersbacher, C. Gräser, F. Gruber, C. Grüninger et al., The distributed and unified numerics environment, version 2.4. Archive Numer. Softw. **4**(100), 13–29 (2016)
4. M.A. Branch, T.F. Coleman, Y. Li, A subspace, interior, and conjugate gradient method for large-scalebound-constrained minimization problems. Tech. Rep., Cornell University, 1995
5. H.J. Bungartz, M. Griebel, Sparse grids. Acta Numer. **13**, 147–269 (2004).
6. H.J. Bungartz, M. Griebel, D. Röschke, C. Zenger, Pointwise convergence of the combination technique for Laplace's equation. Technische Universität München. Institut für Informatik (1993)
7. F. Cappello et al., Toward exascale resilience: 2014 update. Supercomput. Front. Innov. **1**(1), 4–27 (2014)
8. Z. Chen, Online-ABFT: an online algorithm based fault tolerance scheme for soft error detection in iterative methods, in *ACM SIGPLAN Notices*, vol. 48 (ACM, New York, 2013), pp. 167–176
9. C. Constantinescu, I. Parulkar, R. Harper, S. Michalak, Silent data corruption–myth or reality? in *IEEE International Conference on Dependable Systems and Networks With FTCS and DCC, 2008. DSN 2008* (IEEE, New York, 2008), pp. 108–109
10. J. Elliott, M. Hoemmen, F. Mueller, Evaluating the impact of SDC on the GMRES iterative solver, in *2014 IEEE 28th International Parallel and Distributed Processing Symposium* (IEEE, New York, 2014), pp. 1193–1202
11. J. Elliott, M. Hoemmen, F. Mueller, Resilience in numerical methods: a position on fault models and methodologies (2014). arXiv preprint arXiv:1401.3013
12. J. Elliott, M. Hoemmen, F. Mueller, A numerical soft fault model for iterative linear solvers, in *Proceedings of the 24th International Symposium on High-Performance Parallel and Distributed Computing* (ACM, New York, 2015), pp. 271–274
13. D. Fiala, F. Mueller, C. Engelmann, R. Riesen, K. Ferreira, R. Brightwell, Detection and correction of silent data corruption for large-scale High-Performance Computing, in *Proceedings of the International Conference on High Performance Computing, Networking, Storage and Analysis* (IEEE Computer Society Press, Washington, 2012), p. 78
14. M. Galassi, J. Davies, J. Theiler, B. Gough, G. Jungman, P. Alken, M. Booth, F. Rossi, R. Ulerich, GNU scientific library reference manual (2015). Library available online at http://www.gnu.org/software/gsl
15. J. Garcke, Sparse grids in a nutshell, in *Sparse Grids and Applications* (Springer, Berlin, 2013), pp. 57–80
16. M. Griebel, M. Schneider, C. Zenger, A combination technique for the solution of sparse grid problems, in *Iterative Methods in Linear Algebra* (1992), pp. 263–281
17. B. Harding, Fault tolerant computation of hyperbolic partial differential equations with the sparse grid combination technique. Ph.D. thesis, 2016
18. B. Harding et al.: Fault tolerant computation with the sparse grid combination technique. SIAM J. Sci. Comput. **37**(3), C331–C353 (2015)
19. M. Heene, D. Pflüger, Scalable algorithms for the solution of higher-dimensional PDEs, in *Software for Exascale Computing-SPPEXA 2013–2015* (Springer, Berlin, 2016), pp. 165–186
20. M. Heene, A.P. Hinojosa, H.J. Bungartz, D. Pflüger, A massively-parallel, fault-tolerant solver for high-dimensional PDEs, in *Euro-Par 2016: Parallel Processing Workshops* (2016)

21. A.P. Hinojosa et al., Handling silent data corruption with the sparse grid combination technique, in *Proceedings of the SPPEXA Workshop*. Lecture Notes in Computational Science and Engineering (Springer, Berlin, 2016)
22. P.W. Holland, R.E. Welsch, Robust regression using iteratively reweighted least-squares. Commun. Stat. Theory Methods **6**(9), 813–827 (1977)
23. F. Jenko et al., Electron temperature gradient driven turbulence. Phys. Plasmas **7**(5), 1904–1910 (2000). http://www.genecode.org/
24. C. Kowitz, D. Pflüger, F. Jenko, M. Hegland, The combination technique for the initial value problem in linear gyrokinetics, in *Sparse Grids and Applications*. Lecture Notes in Computational Science and Engineering, vol. 88 (Springer, Heidelberg, 2012), pp. 205–222
25. B. Mohr, W. Frings, Jülich Blue Gene/P extreme scaling workshop 2009. Tech. Rep., Technical report FZJ-JSC-IB-2010-02 (2010). http://juser.fz-juelich.de/record/8924/files/ib-2010-02.ps.gz
26. A. Pan, J.W. Tschanz, S. Kundu, A low cost scheme for reducing silent data corruption in large arithmetic circuits, in *IEEE International Symposium on Defect and Fault Tolerance of VLSI Systems, 2008. DFTVS'08* (IEEE, New York, 2008), pp. 343–351
27. C. Reisinger, Analysis of linear difference schemes in the sparse grid combination technique. IMA J. Numer. Anal. **33**(2), 544–581 (2012)
28. P.J. Rousseeuw, A.M. Leroy, *Robust Regression and Outlier Detection*, vol. 589 (Wiley, New York, 2005)
29. M. Snir, R.W. Wisniewski, J.A. Abraham, S.V. Adve, S. Bagchi, P. Balaji, J. Belak, P. Bose, F. Cappello, B. Carlson, et al. Addressing failures in exascale computing. Int. J. High Perform. Comput. Appl. **28**, 129–173 (2014)
30. H.J. van Dam, A. Vishnu, W.A. De Jong, A case for soft error detection and correction in computational chemistry. J. Chem. Theory Comput. **9**(9), 3995–4005 (2013)
31. M. Wakefield, Bounds on quantities of physical interest. Ph.D. thesis, University of Reading, 2003

Sparse Grid Quadrature Rules Based on Conformal Mappings

P. Jantsch and C. G. Webster

Abstract In this work, we demonstrate the extension of quadrature approximations, built from conformal mapping of interpolatory rules, to sparse grid quadrature in the multidimensional setting. In one dimension, computation of an integral involving an analytic function using these transformed quadrature rules can improve the convergence rate by a factor approaching $\pi/2$ versus classical interpolatory quadrature (Hale and Trefethen, SIAM J Numer Anal 46:930–948, 2008). For the computation of high-dimensional integrals with analytic integrands, we implement the transformed quadrature rules in the sparse grid setting, and we show that in certain settings, the convergence improvement can be exponential with growing dimension. Numerical examples demonstrate the benefits and drawbacks of the approach, as predicted by the theory.

1 Introduction and Background

Standard interpolatory quadrature methods, such as Gauss–Legendre and Clenshaw–Curtis, tend to have points which cluster near the endpoints of the domain. As seen in the well-known interpolation example of Runge, this can mitigate the spurious effects of the growth of the polynomial basis functions at the boundary. However, this clustering can be problematic and inefficient in some situations. Gauss–Legendre and Clenshaw–Curtis grids, with n quadrature points on $[-1, 1]$, distribute asymptotically as $\frac{1}{\pi\sqrt{1-x^2}}$ [33]. Hence these clustered grids

P. Jantsch
University of Tennessee, Knoxville, TN, USA
e-mail: jantsch@math.utk.edu

C. G. Webster (✉)
University of Tennessee, Knoxville, TN, USA

Oak Ridge National Laboratory, Oak Ridge, TN, USA
e-mail: webstercg@ornl.gov

J. Garcke et al. (eds.), *Sparse Grids and Applications – Miami 2016*,
Lecture Notes in Computational Science and Engineering 123,
https://doi.org/10.1007/978-3-319-75426-0_6

may have a factor of $\pi/2$ fewer points near the middle of the domain, compared with a uniform grid. This may have unintended negative effects, and the issue is compounded when considering integrals over high-dimensional domains.

For numerical integration of an analytic function in one dimension, the convergence of quadrature approximations based on orthogonal polynomial interpolants depends crucially on the size of the region of analyticity, which we denote by Σ. More specifically, they depend on $\rho \geq 1$, the parameter yielding the largest Bernstein ellipse E_ρ contained within the region of analyticity Σ. The Bernstein ellipse is defined as the open region in the complex plane bounded by the curve

$$\left\{z \in \mathbb{C} : z = (u + u^{-1})/2,\ u = \rho e^{\mathrm{i}\theta},\ 0 \leq \theta \leq 2\pi\right\}. \tag{1}$$

This gives some intuition as to why the most stable quadrature rules place more nodes toward the boundary of the domain $[-1, 1]$; since the boundary of E_ρ is close to $\{\pm 1\}$, the analyticity requirement is weaker near the endpoints of the domain. More specifically, to be analytic in E_ρ, the radius of the Taylor series of f at $\{\pm 1\}$ is only required to be $\rho - 1/\rho$, while the radius of the Taylor series centered at 0 is required to be at least $\rho + 1/\rho$.

On the other hand, the appearance of the Bernstein ellipse in the analysis is not tied fundamentally to the integrand, but only to the choice of polynomials as basis functions. Thus, we may consider other types of quadrature rules which still take advantage of the analyticity of the integrand. Using non-polynomial functions as a basis for the rule may improve the convergence rate of the approximation. Much research has gone into investigating ways to find the optimal quadrature rule for a function analytic in Σ, and to overcome the aforementioned "$\pi/2$-effect", including end-point correction methods [2, 15, 18], non-polynomial based approximation [4–6, 25, 34], and the transformation methods [7, 10, 13, 16, 17, 19, 24, 26, 30, 30] which map a given set of quadrature points to a less clustered set. In this paper, we consider the transformation approach, based on the concept of conformal mappings in the complex plane. Many such transformations have been considered in the literature, especially for improper integrals where the integrand has endpoint singularities, e.g., [11, 30]. Our interest here is in analytic functions which have singularities in the complex plane away from the endpoints of the interval. We consider the transformations from [13], which offer the following benefits: (1) practical and implementable maps; and (2) simple concepts leading to theorems which may precisely quantify their benefits in mitigating the effect of the endpoint clustering.

Our contribution to this line of research is to implement and analyze the application of the transformed rules to sparse grid quadratures in the high-dimensional setting. For high-dimensional integration over the cube $[-1, 1]^d$, the endpoint clustering means that a simple tensor product quadrature rule may use $(\pi/2)^d$ too many points. On the other hand, we show that for sparse Smolyak quadrature rules [27] based on tensorization of transformed one-dimensional quadrature, this effect may be mitigated to some degree. Even in the sparse grid setting, the use of mapped quadratures is not new. The paper [11] uses a similar method to generate new quadrature rules. In their setting, the goal is to compute integrals where the integrand has boundary singularities. In contrast, our objective is to analyze the rules for integrands which are analytic but may have singularities in the complex plane away from $[-1, 1]^d$.

The remainder of the paper is outlined next. First, we introduce the one-dimensional transformed quadrature rules in Sect. 2, and in Sect. 2.2 describe how to use them in the construction of sparse grid quadrature rules for integration of multidimensional functions. In Sect. 3, we provide a brief analysis of the corresponding mapped method to show that the improvement in the convergence rate to a d-dimensional integral is $(\pi/2)^{1/\xi(d)}$, where $\xi(d)^{-1} \geq d$, and provide numerical tests for the sparse grid transformed quadrature rules in Sect. 4. We conclude this effort with some remarks on the benefits and limitations of the method in Sect. 5.

2 Transformed Quadrature Rules

In this section, we introduce one-dimensional transformed quadrature rules, based on the conformal mappings described in [13], applied to classical polynomial interpolation based rules. These rules will be used as a foundation for sparse tensor product quadrature rules for computing high-dimensional integrals, introduced in later sections.

To begin, suppose we want to integrate a given function f over the domain $[-1, 1]$, and assume this function admits an analytic extension in a region $[-1, 1] \subset \Sigma \subset \mathbb{C}$. Given a set of points $\{x_j\}_{j=1}^n$, an interpolatory quadrature rule is defined from the Lagrange interpolant of f, which is the unique degree $n-1$ polynomial matching f at each of the abcissas x_j, i.e.,

$$L_n[f](x) = \sum_{j=1}^{n} f(x_j) l_j^n(x), \quad \text{where} \quad l_j^n(x) = \prod_{\substack{i=1 \\ i \neq j}}^{n} \frac{x - x_i}{x_j - x_i}.$$

The quadrature approximation of the integral of f, denoted $Q_n[f]$, is then defined by

$$\int_{-1}^{1} f(x)\, dx \approx \int_{-1}^{1} L_n[f](x)\, dx = \sum_{j=1}^{n} c_j f(x_j) =: Q_n[f], \tag{2}$$

with weights given explicitly as

$$c_j = \int_{-1}^{1} l_j^n(x)\, dx. \tag{3}$$

Now, according to the Cauchy integral theorem, since f has an analytic extension, we can evaluate the integral along any (complex) path contained in Σ with endpoints $\{\pm 1\}$. Next, let g be a conformal mapping satisfying the conditions:

$$g(\pm 1) = \pm 1, \text{ and } g\left([-1, 1]\right) \subset \Sigma. \tag{4}$$

According to the argument above, the integral can be rewritten as the path integral from -1 to 1, with the path parameterized by the map g, i.e.,

$$\int_{-1}^{1} f(x)\,dx = \int_{-1}^{1} f(g(s))g'(s)\,ds.$$

Applying our original quadrature rule to the latter integral,

$$\int_{-1}^{1} f(g(s))g'(s)\,ds \approx \sum_{j=1}^{n} \underbrace{c_j g'(x_j)}_{:=\tilde{c}_j} f(\underbrace{g(x_j)}_{:=\tilde{x}_j}) =: \widetilde{Q}_n[f], \tag{5}$$

we obtain a new quadrature rule with transformed weights $\{\tilde{c}_j\}_{j=1}^n$ and points $\{\tilde{x}_j\}_{j=1}^n$.

Equation (5) provides the motivation for the choice of the conformal mapping g. Specifically, the Taylor series for f, centered at points $x \in [-1, 1]$ which are close to the boundary, may have a radius which extends beyond the largest Bernstein ellipse in which f is analytic. We may then hope to find a g such that a Bernstein ellipse is conformally mapped onto the whole region where f is analytic, where classical convergence theory yields the convergence rate for $(f \circ g) \cdot g'$. In addition to (4), it is especially advantageous to have g map $[-1, 1]$ onto itself, i.e.,

$$g([-1, 1]) = [-1, 1]. \tag{6}$$

In this case, the transformed weights and points remain real-valued, and we avoid evaluations of f with complex inputs.

We now turn our attention to several specific conformal mappings which satisfy the conditions (4), along with the extra condition (6). For more details on the derivation and numerical implementation of the maps, see [13]. The first mapping we consider applies to functions which admit an analytic extension at every point on real line; in other words, functions which have only complex singularities. In this case, the natural transformations to consider are ones that conformally map the interior of a Bernstein ellipse (1) to a strip about the real line. Specifically, we define a map which takes the Bernstein ellipse with shape parameter ρ to the complex

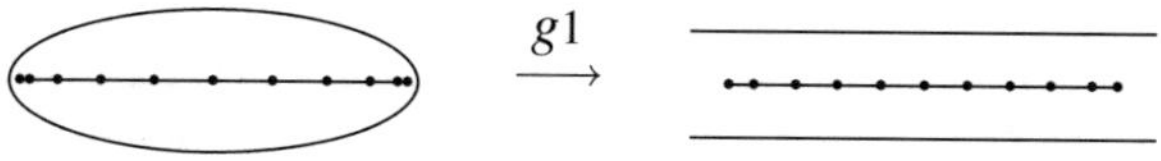

Fig. 1 The mapping (7) takes the Bernstein ellipse $E_{1.4}$ (left) to a strip of half-width $2(1.4 - 1)/\pi \approx 0.255$

strip with half-width $\frac{2}{\pi}(\rho - 1)$, as shown in Fig. 1. We can do this through the use of Jacobi elliptic functions [1, 8]. First, fixing a value $\rho > 1$, we define the parameter $0 < m < 1$ through

$$m^{1/4} = 2\sum_{j=1}^{\infty} \rho^{-4(j-\frac{1}{2})^2} \Bigg/ \left(1 + 2\sum_{j=1}^{\infty} \rho^{-4j^2}\right),$$

and the associated parameter $K = K(m)$,

$$K = \int_0^{\pi/2} \frac{d\theta}{\sqrt{1 - m\sin\theta}},$$

which is an incomplete Jacobi elliptic integral of the first kind [1]. Finally, we define the mapping in terms of the elliptic sine function $sn(\cdot; m)$:

$$g_1(z) = \tanh^{-1}\left(m^{1/4}\text{sn}\left(\frac{2K}{\pi \sin^{-1}(z)}; m\right)\right) / \tanh\left(m^{1/4}\right). \tag{7}$$

We'll refer to this map as the "*strip map*" in the following.

According to (5), we also need to know the derivative of g_1, given by

$$g_1'(z) = \frac{2Km^{1/4}}{\pi\sqrt{1 - z^2}} \frac{\text{cn}(\omega(z); m)\text{dn}(\omega(z); m)}{(1 - m^{1/2}\text{sn}(\omega(z); m))} / \tanh\left(m^{1/4}\right), \tag{8}$$

with $\omega(z) = 2K\sin^{-1}(z)/\pi$. Here we have also made use of the elliptic cosine function, cn, and elliptic amplitude function, dn [1]. For our applications, we also require the values of g_1' at the endpoints of the interval, which are given by

$$g_1'(\pm 1) = 4K^2m^{1/4}\left(1 + m^{1/2}\right)/\pi^2 \tanh\left(m^{1/4}\right).$$

Again we refer the reader to [13] for additional details.

Another way to change the endpoint clustering, and transform the quadrature rule under a conformal map, is to use an appropriately normalized truncation of the power series for $\sin^{-1}(z)$. The map $\frac{2}{\pi}\sin^{-1}(z)$ perfectly eliminates the clustering of the Gauss–Legendre and Clenshaw–Curtis points, but since it has singularities at ± 1, it is useless for our purposes. On the other hand, by considering a *truncation* of

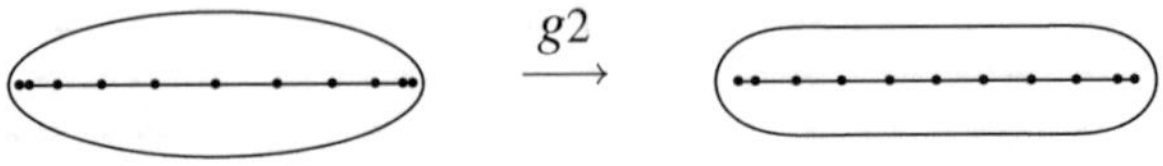

Fig. 2 The mapping (9), with $M = 4$, takes the Bernstein ellipse $E_{1.4}$ (left) to a pill-shaped region with sides of length ≈ 0.255

the power series

$$\sin^{-1}(z) = \sum_{k=1}^{\infty} \frac{\Gamma(k+1/2)}{\Gamma(1/2)} \frac{z^{2k+1}}{(2k+1)k!},$$

we define a more desirable mapping. To this end, for $M \geq 1$, we define

$$g_2(z) = c(M) \sum_{k=1}^{M} \frac{\Gamma(k+1/2)}{\Gamma(1/2)} \frac{z^{2k+1}}{(2k+1)k!}, \tag{9}$$

with a constant $c(M) \in (0, 1)$ appropriately chosen so that $g_2(\pm 1) = \pm 1$. This mapping is much simpler to implement than the previous mapping. We will call this map the "*pill map*", since it maps the Bernstein ellipse to a pill-shaped region about $[-1, 1]$ with flatter sides. In Fig. 2, we plot the image of the ellipse E_ρ with $\rho = 1.4$, under the mapping (9) with $M = 4$. The region on the right has almost flat sides, with width a little bigger than $\frac{2}{\pi}(1.4 - 1) \approx 0.255$.

2.1 *Standard One-Dimensional Quadrature Rules*

Here we give a brief summary of some standard interpolatory-type quadrature rules, to which we will apply the mappings of the previous section. Only the nodes are discussed here, as the weights for each method will be defined according to (3). For an overview of the theory of interpolatory quadrature, see [33, Ch. 19].

The first quadrature rule is based on the extrema of the Gauss–Chebychev polynomials. For a given number of points n, these are given by:

$$x_{n,j} = \cos\left(\frac{(j-1)\pi}{n}\right), \quad 1 \leq j \leq n. \tag{10}$$

If we choose the number of nodes $n = n(l)$ to grow according to $n(1) = 1, n(l) = 2^{l-1} + 1, l > 1$, this generates a nested sequence known as the Clenshaw–Curtis nodes.

Another set of points of interest are the well-known Gaussian abscissa, which are the roots of orthogonal polynomials with respect to a given measure. Here we consider the sequence of Gauss–Legendre nodes, which consists of the roots of the sequence of polynomials orthogonal to the uniform measure on $[-1, 1]$, i.e., the n roots of the polynomials

$$P_n(x) = \frac{d^n}{dx^n}\left[(x^2-1)^n\right], \quad n \geq 0. \tag{11}$$

With the introduction of a weight into the integral from (2), other families of orthogonal polynomials can be used. The main advantage of Gauss points is their high degree of accuracy, i.e., the one-dimensional quadrature rules built from n Gauss points integrate exactly polynomials of degree $2n - 1$.

Remark 1 Gauss–Legendre points do not form a nested sequence, which may lead to inefficiency in the high-dimensional quadrature setting. In fact, without nestedness of the one-dimensional sequence, the sparse grid rule described in the following section may not even be interpolatory. Even so, we only require the *one-dimensional rule* to be interpolatory to apply the conformal mapping theory in the multidimensional setting. We also remark that nested quadrature sequences based on the roots of orthogonal polynomials, the so-called Gauss–Patterson points, are also available, but we do not consider these types of rules herein.

The final set of nodes we consider are known as the Leja points. Leja points satisfy a recursive definition, that is, given a point $x_1 \in [-1, 1]$, for $n \geq 2$ define

$$x_n = \underset{x\in[-1,1]}{\arg\max} \prod_{j=1}^{n-1} |x - x_j|, \tag{12}$$

where we typically take $x_1 = 0$. Of course, there may be several minimizers to (12), so for computational purposes, we simply choose the minimizer closest to the left endpoint. In the interpolation setting, Leja sequences are known to have good properties for approximation in high-dimensions [20], and there has been much research related to the stability properties of such nodes when used for Lagrange interpolation [14, 31]. The lack of symmetry of the sequence may not be ideal for all applications, and certain symmetric "odd Leja" constructions may be used in these cases [29]. On the other hand, the points here have the added benefit of being a nested sequence and grow one point at a time, and furthermore have asymptotic distribution which is that same as that of Gauss and Clenshaw–Curtis nodes [20].

2.2 Sparse Quadrature for High Dimensional Integrals

For the numerical approximation of high-dimensional integrals over product domains, it is natural to consider simple tensor products of one-dimensional quadrature rules. Unfortunately, these rules suffer from the curse of dimensionality, as the number of points required to accurately compute the integral grows exponentially with the underlying dimension of the integral; i.e., a rule using n points in each dimension requires n^d points. For certain smooth integrands, we can mitigate this effect by considering sparse combinations of tensor products of these one-dimensional rules, i.e., sparse grid quadrature. It is known that sparse grid rules can asymptotically achieve approximately the same order of accuracy as full tensor product quadrature, but use only a fraction of the number of quadrature nodes [9, 22, 23, 27].

Rather than the one-dimensional integral from before, we let $d > 1$ be the dimension and define $\Gamma := [-1, 1]^d$. In addition, by letting $\boldsymbol{x} = (x_1, \dots, x_d)$ be an arbitrary element of Γ, we consider the problem of approximating the integral

$$I^d[f] = \int_\Gamma f(\boldsymbol{x})\, d\boldsymbol{x}, \tag{13}$$

using transformed quadrature rules. To define the sparse grid rules, we first denote by $\{I_{p(l)}\}_{l\geq 1}$ a sequence of given one-dimensional quadrature operators using $p(l)$ points. Here $I_{p(l)}$ may be a standard interpolatory quadrature $Q_{p(l)}$ from (2) or its conformally transformed version $\widetilde{Q}_{p(l)}$ from (5). With $I_0 := 0$, define the difference operator

$$\Delta_l := I_{p(l)} - I_{p(l-1)}.$$

Then given a set of multiindices $\Lambda_w \subset \mathbb{N}_0^d$, we define the sparse grid quadrature operator to be

$$I_{N_w}[f] = \sum_{\boldsymbol{l}\in\Lambda_w} \bigotimes_{i=1}^{d} \Delta_{p(l_i)}[f] = \sum_{\boldsymbol{l}\in\Lambda_w} \bigotimes_{i=1}^{d} \left(I_{p(l_i)} - I_{p(l_i-1)}\right)[f], \tag{14}$$

where we refer to the natural number w as the *level* of the sparse grid rule, and N_w is the total number of points in Γ used by the sparse grid. The choice of multiindex set Λ_w may vary based on the problem at hand. We only require that it be *downward closed*, i.e., if $\boldsymbol{l} \in \Lambda_w$, then $\nu_i \leq l_i$ for all $i = 1, \dots, d$ implies $\boldsymbol{\nu} \in \Lambda_w$. The index set may be anisotropic, i.e., dimension dependent, or if appropriate error indicators are defined, it may even be chosen adaptively. Some typical choices are given in Table 1, but for simplicity, we consider only standard isotropic Sparse Smolyak grids. For more information on anisotropic rules, see [21].

Table 1 The functions $p : \mathbb{N}_+ \to \mathbb{N}_+$ and index sets Λ_w, with the corresponding polynomial subspaces

Polynomial space	$p(l)$	Λ_w
Tensor product	$p(l) = l$	$\max\limits_{1 \le i \le d} (l_i - 1) \le w$
Total degree	$p(l) = l$	$\sum_{i=1}^{d} (l_i - 1) \le w$
Hyperbolic cross	$p(l) = l$	$\prod_{i=1}^{N} (l_i - 1) \le w$
Sparse Smolyak	$p(l) = 2^{l-1} + 1,\ l > 1$	$\sum_{i=1}^{d} (l_i - 1) \le w$

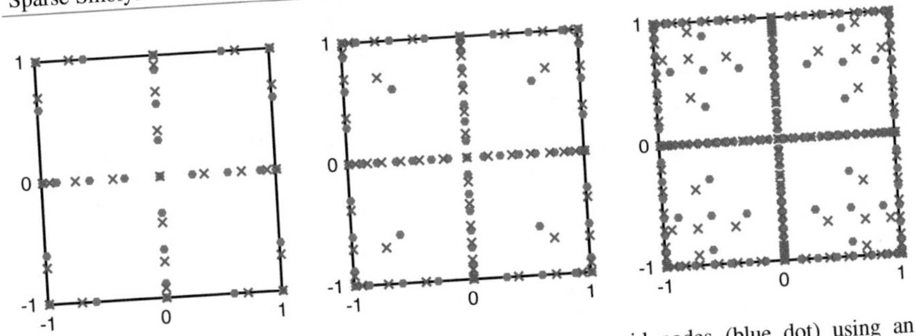

Fig. 3 Location of the two-dimensional transformed sparse grid nodes (blue dot) using an underlying Clenshaw–Curtis rule, compared to standard Clenshaw–Curtis sparse grids (red x)

The effect of the conformal mapping on the placement of the nodes used by the sparse quadrature rule (14) is similar to the one-dimensional case. In Fig. 3, we have plotted the nodes of a two-dimensional Clenshaw–Curtis sparse grid with the transformation map (9), using $\rho = 1.4$, versus a traditional Clenshaw–Curtis sparse grid. Note how the clustering of the nodes toward the outer boundary of the cube is diminished.

3 Comparison of the Transformed Sparse Grid Quadrature Method

In this section we investigate the potential improvement in convergence for computation of high-dimensional integrals using the sparse grid quadrature method based on the transformed rules. The different mappings (7) and (9), since they have different properties, will be considered separately. Furthermore, the focus of this section will be on the transformation of Gauss–Legendre rules, though we remark that starting from a one-dimensional convergence result such as the following theorem, the rest of the analysis is similar for the Clenshaw–Curtis case. We begin by quoting the following one-dimensional result stated from [13], establishing the convergence of the transformed Gauss–Legendre rule for an analytic integrand.

Theorem 1 *For some $\rho > 1$, let f be analytic and uniformly bounded by $\gamma > 0$ in the ellipse E_ρ. Then for $n \geq 1$, the Gauss–Legendre quadrature rule has the error bound*

$$|I[f] - I_n[f]| \leq \frac{64\gamma}{15(1-\rho^{-2})}\rho^{-2n}. \tag{15}$$

Now taking a specific region of analyticity and a given conformal map, we apply Theorem 1 to quantify the benefit of the transformation method. We start by considering functions analytic in the strip S_ε of half width ε about the real line, and the Gauss–Legendre rule transformed under the map (7).

Theorem 2 *([13, Theorem 3.1]) Let f be analytic in a strip S_ε of half width ε about the real line, and g_1 the conformal map (7) mapping $E_{1+\frac{\pi}{2}\varepsilon} \to S_\varepsilon$. Then for $n \geq 1$, and any $\tilde{\varepsilon} < \varepsilon$, the transformed Gauss–Legendre quadrature rule has the error bound*

$$\left|I[f] - \widetilde{Q}_n[f]\right| \leq \frac{64\gamma_1}{15(1-(1+\frac{\pi}{2}\tilde{\varepsilon})^{-2})}\left(1+\frac{\pi}{2}\tilde{\varepsilon}\right)^{-2n}, \tag{16}$$

where $\gamma_1 = \sup_{s \in E_{1+\frac{\pi}{2}\tilde{\varepsilon}}} |f(g_1(s))g_1'(s)|$.

This theorem follows by an application of Theorem 1 to the integral of $(f \circ g)g'$ using (5). We must take $\tilde{\varepsilon} < \varepsilon$, since otherwise the value of γ_1 will be infinite due to the behavior of g'. Furthermore, since the closure of $g(E_{1+\frac{\pi}{2}\tilde{\varepsilon}})$ is contained in the open set S_ε, we have $\gamma_1 < \infty$ without assuming the boundedness of f. However, we do not lose much from this assumption, and this theorem shows that we can achieve savings of almost a factor of $\pi/2$ for functions analytic in a strip S_ε.

For the mapping (9), the results are somewhat more complicated, due to the fact that the properties of the map depend crucially on the chosen degree M of the truncation, and for a given M we may not be able to realize the full factor of $\pi/2$. From a practical standpoint, this is not much worse than the case of the strip mapping (7), since full information about the analyticity of the integrand may not be available, and hence it may be difficult to tune the parameter of the mapping to the integral at hand. Thus, what we have in the case of the map (9) is a more precise result with all the parameters specified. The following result from [13] will apply to functions which are analytic in the ε-neighborhood of $[-1, 1]$, denoted U_ε. Then we have the following theorem.

Theorem 3 *([13, Theorem 6.1]) Let $\varepsilon \leq 0.8$, and let f be analytic in a ε-neighborhood U_ε of $[-1, 1]$. Let g_2 be the conformal map (9), truncated at degree $M = 4$. Then for $n \geq 1$, the transformed Gauss–Legendre quadrature rule has the error bound*

$$\left|I[f] - \widetilde{Q}_n[f]\right| \leq \frac{64\gamma_2}{15(1-(1+1.3\varepsilon)^{-2})}(1+1.3\varepsilon)^{-2n}, \tag{17}$$

where $\gamma_2 = \sup_{s \in E_{1+1.3\varepsilon}} |f(g_2(s))g_2'(s)|$.

From the one dimensional results of Theorems 2 and 3, for the maps (7) and (9), resp., we are able to fully quantify the benefits of the TQ rules applied to sparse grid quadrature in high dimensions. The following theorems give the expected sub-exponential convergence rate for a sparse grid quadrature approximation of an analytic integrand based on the Gauss–Legendre points. These results are in accord with other subexponential rates for sparse multidimensional polynomial approximation obtained in [3, 12, 22, 32]. Recall that we are considering only isotropic sparse Smolyak constructions, according to the last row in Table 1.

Theorem 4 *Let f be analytic in $\prod_{i=1}^{d} S_\varepsilon$ for some $\varepsilon > 0$, and let g_1 be the conformal mapping* (7)*. Then for any $\tilde{\varepsilon} < \varepsilon$, the sparse quadrature* (14) *built from transformed Gauss–Legendre quadrature rules satisfies the following error bound in terms of the number of quadrature nodes:*

$$|I^d[f] - I_{N_w}[f]| \leq C(\tilde{\varepsilon}, f, \gamma_1, d) \exp\left(-\log\left(1 + \frac{\pi}{2}\tilde{\varepsilon}\right) \frac{2d}{2^{1/d}} N_w^{\xi(d)}\right). \tag{18}$$

where γ_1 is defined as in Theorem 2, and

$$\xi(d) = \frac{\log(2)}{d(\zeta + \log(d))}, \tag{19}$$

with the constant $\zeta = 1 + \log(2)(1 + \log_2(1.5)) \approx 2.1$.

Theorem 5 *For some $0 < \varepsilon \leq 0.8$, let f be analytic in $\prod_{i=1}^{d} U_\varepsilon$, and let g_2 be the conformal mapping* (9) *truncated at degree $M = 4$. Then the sparse quadrature* (14) *built from transformed Gauss–Legendre quadrature rules satisfies the following error bound in terms of the number of quadrature nodes:*

$$|I^d[f] - I_{N_w}[f]| \leq C(\varepsilon, f, \gamma_2, d) \exp\left(-\log(1 + 1.3\varepsilon) \frac{2d}{2^{1/d}} N_w^{\xi(d)}\right), \tag{20}$$

with γ_2 as in Theorem 3, and $\xi(d)$ as in (19)*.*

Sketch of Proof From the one dimensional results of Theorems 2 and 3, resp., the proof of the results above follows from well-known sparse grid analysis techniques and estimates on the number of quadrature nodes [22]. Specifically, we may follow along the lines of the proof of [22, Theorem 3.19], with the one-dimensional convergence estimates [22, p. 2230] replaced by (16) and (17), resp, and noting that, e.g.,

$$\left(1 + \frac{\pi}{2}\tilde{\varepsilon}\right)^{-2\cdot 2^i} = e^{-\sigma 2^i}, \qquad \sigma = 2\log\left(1 + \frac{\pi}{2}\tilde{\varepsilon}\right).$$

The rest of the proof is an application of Lemmas 3.4 and 3.5 from the aforementioned [22], along with the estimate on the number of quadrature points given by Lemma 3.17. □

We remark again that it is not necessary to use the same ε in each dimension, but we make that choice for clarity of presentation. As mentioned in Sect. 2.2, in the case that the integrand f has dimension-dependent smoothness, *anisotropic* sparse grid methods are available.

We now make a few remarks on the improvements of Corollaries 4 and 5 over sparse grids based on traditional interpolatory quadrature methods. First, note that for functions $f \in C(\Gamma)$ which admit an analytic extension in either $\prod_{i=1}^{d} S_\varepsilon$ or $\prod_{i=1}^{d} U_\varepsilon$, the largest (isotropic) polyellipse in which f is analytic has the shape parameter $\rho = 1 + \varepsilon$. Hence, the convergence rate of typical sparse grid Gauss–Legendre quadrature, using N abscissa, is

$$|I[f] - \mathscr{I}_N[f]| = \mathscr{O}\left(\exp\left(-\log(1+\varepsilon)\frac{2d}{2^{1/d}} N_w^{\xi(d)}\right)\right). \tag{21}$$

Thus, the improvement in convergence rate is multiplied exponentially in the sparse grid case, i.e., in the case of Corollary 4, the number of points required to reach a certain tolerance is reduced by a factor approaching $(\pi/2)^{\xi(d)^{-1}}$, with ξ as in (19). To see this, let N_{SGTQ} and N_{SG} be the necessary number of points for the right-hand sides of (18) and (21), respectively, to be less than a given tolerance. Then, we may calculate that

$$\frac{N_{SG}}{N_{SGTQ}} = \left(\frac{\log(1+\frac{\pi}{2}\varepsilon)}{\log(1+\varepsilon)}\right)^{\xi(d)^{-1}} \xrightarrow{\varepsilon\to 0} \left(\frac{\pi}{2}\right)^{\xi(d)^{-1}}. \tag{22}$$

The constants are ignored in the calculation, though the transformed quadrature may have slightly improved constant versus the standard case. We also note that $\xi(d)^{-1} \geq d$, so the improvement is exponential in the dimension, and that as $\varepsilon \to \infty$, i.e., for functions which are analytic in a large region containing $[-1, 1]$, the improvement factor degrades to 1. In the case of the sparse grid quadrature approximation transformed by (9), we use (20), so the improvement is $1.3^{\xi(d)^{-1}}$, as long as $\varepsilon \leq 0.8$. As mentioned in the work [13], the factor of 1.3 is still less than $\pi/2 \approx 1.57$, but for smaller ε and large truncation parameter M this can improved to $3/2$; see [13, Theorem 6.2].

4 Numerical Tests of the Sparse Grid Transformed Quadrature Rules

In this section we test the sparse grid transformed quadrature rules on a number of multidimensional integrals, and compare the performance versus standard rules. The transformed rules we consider are based on the conformal mapping of Gauss–Legendre, Clenshaw–Curtis, and Leja quadrature nodes, which are describe in Sect. 2.1. We transform these rules using both of the conformal mappings (7) and (9), using the Matlab code provided in [13] to generate the one-dimensional quadrature sequences. The Tasmanian sparse grid toolkit [28, 29] is used for the implementation of the full sparse grid quadrature rule. For the Clenshaw–Curtis rule, we use a standard Smolyak sparse grid with doubling rule; see Table 1. For the Gauss–Legendre and Leja sequences, we use isotropic total degree index sets with linear point growth $p(l) = l, l \geq 1$. For each of the rules, we will consider error versus the total number of sparse grid points, N_w.

4.1 Comparison of Maps

For the first test, we compare the sparse grid methods with the transformed quadratures to traditional quadrature approximations for computing the integral of three test functions over the cube $[-1, 1]^3$ in three dimensions. In each case, we compare the different maps (7) and (9) for the generation of the transformed one-dimensional quadrature from the Clenshaw–Curtis, Gauss–Legendre, and Leja rules. The chosen mapping parameters are $\rho = 1.4$ with (7) and truncation parameter $M = 4$ for (9).

In Fig. 4, we plot the results for approximating the integral over $[-1, 1]^3$ of the function

$$f(x, y, z) = \frac{1}{(1 + 5x^2)(1 + 5y^2)(1 + 5z^2)}. \tag{23}$$

This function has complex singularities at points $z \in \mathbb{C}^3$ where at least one coordinate $z_j = \frac{1}{\sqrt{5}}\mathrm{i}$, and is hence analytic in the complex hyper-strip $\prod_{i=1}^{3} S_{1/\sqrt{5}}$. As expected, the quadrature generated according to the mapping (7) performs the best here, though the chosen parameter $\rho = 1.4$ is somewhat less than the optimal,

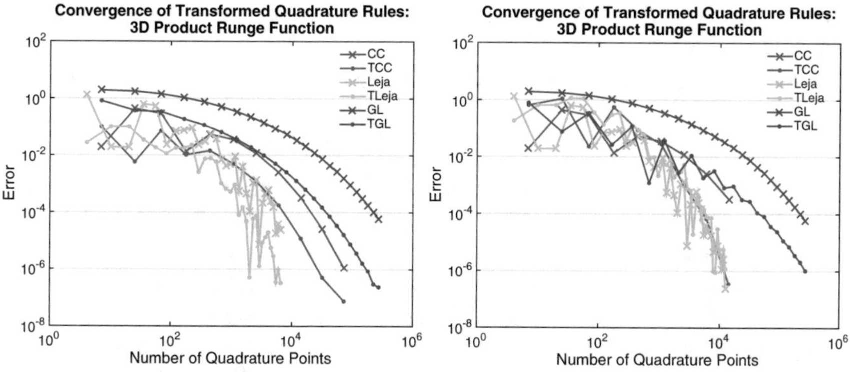

Fig. 4 Comparison of sparse grid quadrature rules for computing the integral of (23) over the cube $[-1, 1]^3$, using the conformal maps (7) (left), and (9) (right)

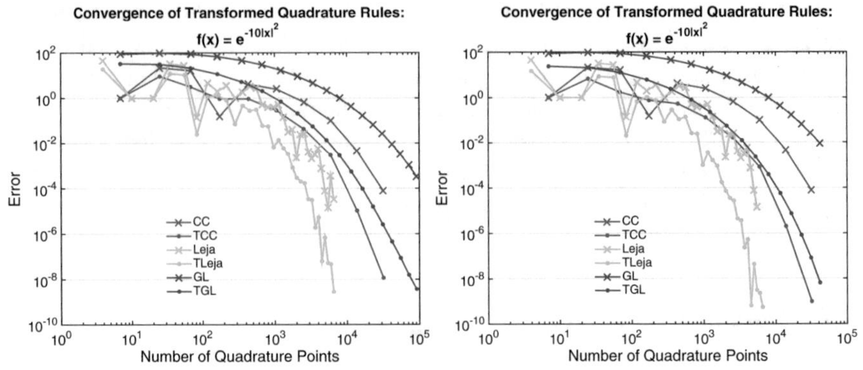

Fig. 5 Comparison of sparse grid quadrature rules for computing the integral of (24) over the cube $[-1, 1]^3$, using the conformal maps (7) (left), and (9) (right)

since the value $\frac{2}{\pi}(1.4 - 1) \approx 0.255 < 1/\sqrt{5}$. Regardless, the transformed sparse grid approximations again perform better than their classical counterparts, gaining up to two orders of magnitude in the error for Clenshaw–Curtis and Gauss rules. Note that on the right-hand plot, the transformation (9) does not work well with the Leja rule. The results for the standard quadrature are repeated in each plot for ease of comparison.

Figure 5 again shows the results for approximating the integral of the function

$$f(x, y, z) = \exp^{-10(x^2+y^2+z^2)}, \tag{24}$$

over the cube $[-1, 1]^3$. This function is entire, but grows rapidly in the complex hyperplane away from $[-1, 1]^3$. The left-hand plot shows the performance of the sparse grid transformed quadratures using the transformation (7), while the right-hand plot uses (9). In each case, the sparse quadrature approximations using mapped rules outperform traditional sparse grid quadrature, and there is only a slight difference in the performance of the transformed rules corresponding to the different mappings.

Finally, in Fig. 6, we plot results for approximating the integral of the function

$$f(x, y, z) = \cos(1 + x^2 + y^2 + z^2), \tag{25}$$

over the cube, $[-1, 1]^3$. This function is entire and does not grow too quickly away from the unit cube in the complex hyperplane $\mathbb{C}^3$. On the other hand, by fixing the parameters in the conformal mapping, the convergence rate of the transformed sparse grid rules is restricted by the analyticity of the composition $(f \circ g)g'$. In other words, the conformal mapping technique cannot take full advantage of the analyticity of the function f. Thus, we see that the rules based on holomorphic mappings are inferior for computing the integral of this function, though the transformed Leja rule using (7) is somewhat competitive, at least up to the computed level.

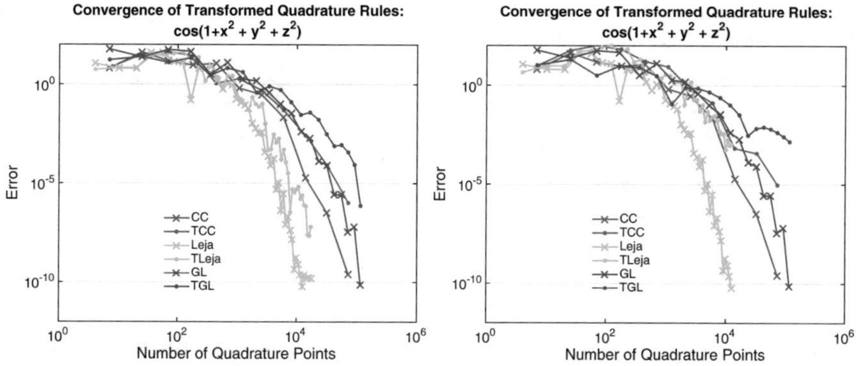

Fig. 6 Comparison of sparse grid quadrature rules for computing the integral of (25) over the cube $[-1, 1]^3$, using the conformal maps (7) (left), and (9) (right)

4.2 *Effect of Dimension*

Next we investigate the effect of increasing the dimension d of the integral problem, and see whether the holomorphic transformation idea indeed decreases the computational cost with growing dimension. The test integral for this experiment is

$$\int_{[-1,1]^d} \prod_{i=1}^{d} \left(\frac{1}{1 + 5x_i^2} \right) dx. \tag{26}$$

In Table 2 we compare the number of points used to estimate the integral (26) in $d = 2, 4, 6$ dimensions, up to the given error tolerance. We use both the Clenshaw–Curtis and the Leja rules, with their corresponding transformed versions. We do not include the results for the Gauss–Legendre method, since as in Fig. 4, the GL method performs much worse than the others for this test function. Here we implement only the map (7) with $\rho = 1.7$, which maps the interior of the ellipse (1) to a strip of half-width $\frac{1}{\pi}(1.7 - 1) \approx 1/\sqrt{5}$. This integral has simple product structure, so we compare the computed sparse grid approximation to the "true" integral value computed to high precision. As expected, the sparse grid rules using transformed quadrature need far fewer points to compute the value of the integral up to a given tolerance, as compared with standard sparse grid rules. As the dimension increases, because of the doubling rule p from Table 1, the number of points grows rapidly from one level to the next. Thus, a certain grid may vastly undershoot or overshoot the optimal number of points needed to achieve a certain error. Furthermore, it may be the case that the convergence has not yet reached the asymptotic regime for such a large tolerance 10^{-2}, and so we claim from Table 2 that the transformed sparse grid rules may work well even before the convergence is governed by the asymptotic theory.

Table 2 Comparison of the number of points used by a given sparse grid quadrature rule to approximate the integral (26) to the given tolerance

Dimension	Tol	CC	TCC	Ratio	Leja	TLeja	Ratio
2	10^{-7}	1537	705	2.18	666	435	1.53
4	10^{-5}	1,507,329	271,617	5.55	73,815	20,475	3.61
6	10^{-2}	6,436,865	127,105	50.64	593,775	12,376	47.98

5 Conclusions

In this work, we have demonstrated the application of the transformed quadrature rules of [13] to isotropic sparse grid quadrature in high dimensions, and showed that in certain situations we are able to speed up convergence of a transformed sparse approximation by a factor approaching $(\pi/2)^{\xi(d)^{-1}}$, where $\xi(d)^{-1} \approx d \log d$. We applied the rules to several test integrals, and experimented with different conformal mappings g, and found that the sparse grid quadratures with conformally mapped rules outperformed the standard sparse grid rules based on one-dimensional interpolatory quadrature by a significant amount for several example integrands. For entire functions, or functions which are analytic and grow slowly in a large region around $[-1, 1]$, the transformation method fails to beat standard quadrature rules, since the convergence of the transformed quadrature is dictated by the chosen mapping parameter. However, the transformed rules perform especially well for functions which are analytic only in a small neighborhood around $[-1, 1]$, even if the mapping parameter is not tuned exactly to the region of analyticity.

References

1. M. Abramowitz, I.A. Stegun, *Handbook of Mathematical Functions: with Formulas, Graphs, and Mathematical Tables*, vol. 55 (Courier Corporation, North Chelmsford, 1964)
2. B.K. Alpert, Hybrid Gauss-trapezoidal quadrature rules. SIAM J. Sci. Comput. **20**(5), 1551–1584 (1999)
3. J. Beck, F. Nobile, L. Tamellini, R. Tempone, Convergence of quasi-optimal stochastic galerkin methods for a class of pdes with random coefficients. Comput. Math. Appl. **67**(4), 732–751 (2014)
4. G. Beylkin, K. Sandberg, Wave propagation using bases for bandlimited functions. Wave Motion **41**(3), 263–291 (2005)
5. J.P. Boyd, Prolate spheroidal wavefunctions as an alternative to Chebyshev and Legendre polynomials for spectral element and pseudospectral algorithms. J. Comput. Phys. **199**(2), 688–716 (2004)
6. Q.-Y. Chen, D. Gottlieb, J.S. Hesthaven, Spectral methods based on prolate spheroidal wave functions for hyperbolic PDEs. SIAM J. Numer. Anal. 43(5), 1912–1933 (2005)
7. P. Favati, G. Lotti, F. Romani, Bounds on the error of Fejér and Clenshaw–Curtis type quadrature for analytic functions. Appl. Math. Lett. **6**(6), 3–8 (1993)

8. H.E. Fettis, Note on the computation of Jacobi's nome and its inverse. Computing **4**(3), 202–206 (1969)
9. T. Gerstner, M. Griebel, Numerical integration using sparse grids. Numer. Algoritm. **18**(3–4), 209–232 (1998)
10. M. Götz, Optimal quadrature for analytic functions. J. Comput. Appl. Math. **137**(1), 123–133 (2001)
11. M. Griebel, J. Oettershagen, Dimension-adaptive sparse grid quadrature for integrals with boundary singularities, in *Sparse Grids and Applications-Munich 2012* (Springer, Cham, 2014), pp. 109–136
12. M. Griebel, J. Oettershagen, On tensor product approximation of analytic functions. J. Approx. Theory **207**, 348–379 (2016)
13. N. Hale, L.N. Trefethen, New quadrature formulas from conformal maps. SIAM J. Numer. Anal. **46**(2), 930–948 (2008)
14. P. Jantsch, C.G. Webster, G. Zhang, On the Lebesgue constant of weighted Leja points for Lagrange interpolation on unbounded domains. IMA J. Numer. Anal. (2018). https://doi.org/10.1093/imanum/dry002
15. S. Kapur, V. Rokhlin, High-order corrected trapezoidal quadrature rules for singular functions. SIAM J. Numer. Anal. **34**(4), 1331–1356 (1997)
16. D. Kosloff, H. Tal-Ezer, Modified Chebyshev pseudospectral method with $O(N^{-1})$ time step restriction. J. Comput. Phys. **104**, 457–469 (1993)
17. M. Kowalski, A.G. Werschulz, H. Woźniakowski, Is Gauss quadrature optimal for analytic functions? Numer. Math. **47**(1), 89–98 (1985)
18. J. Ma, V. Rokhlin, S. Wandzura, Generalized Gaussian quadrature rules for systems of arbitrary functions. SIAM J. Numer. Anal. **33**(3), 971–996 (1996)
19. M. Mori, An IMT-type double exponential formula for numerical integration. Publ. Res. Inst. Math. Sci. **14**(3), 713–729 (1978)
20. A. Narayan, J.D. Jakeman, Adaptive Leja sparse grid constructions for stochastic collocation and high-dimensional approximation. SIAM J. Sci. Comput. **36**(6), A2952–A2983 (2014)
21. F. Nobile, R. Tempone, C.G. Webster, An anisotropic sparse grid stochastic collocation method for partial differential equations with random input data. SIAM J. Numer. Anal. **46**(5), 2411–2442 (2008)
22. F. Nobile, R. Tempone, C.G. Webster, A sparse grid stochastic collocation method for partial differential equations with random input data. SIAM J. Numer. Anal. **46**(5), 2309–2345 (2008)
23. E. Novak, K. Ritter, High dimensional integration of smooth functions over cubes. Numer. Math. **75**(1), 79–97 (1996)
24. K. Petras, Gaussian versus optimal integration of analytic functions. Constr. Approx. **14**(2), 231–245 (1998)
25. D. Slepian, H.O. Pollak, Prolate spheroidal wave functions, Fourier analysis and uncertainty–I. Bell Labs Tech. J. **40**(1), 43–63 (1961)
26. R.M. Slevinsky, S. Olver, On the use of conformal maps for the acceleration of convergence of the trapezoidal rule and sinc numerical methods. SIAM J. Sci. Comput. **37**(2), A676–A700 (2015)
27. S. Smolyak, Quadrature and interpolation formulas for tensor products of certain classes of functions. Soviet Math. Dokl. **4**, 240–243 (1963)
28. M. Stoyanov, User manual: Tasmanian sparse grids v4.0. Technical Report ORNL/TM-2015/596. Oak Ridge National Laboratory (2017). https://tasmanian.ornl.gov/manuals.html
29. M.K. Stoyanov, C.G. Webster, A dynamically adaptive sparse grids method for quasi-optimal interpolation of multidimensional functions. Comput. Math. Appl. **71**(11), 2449–2465 (2016)
30. H. Takahasi, M. Mori, Quadrature formulas obtained by variable transformation. Numer. Math. **21**(3), 206–219 (1973)
31. R. Taylor, V. Totik, Lebesgue constants for Leja points. IMA J. Numer. Anal. **30**(2), 462–486 (2010)

32. H. Tran, C.G. Webster, G. Zhang, Analysis of quasi-optimal polynomial approximations for parameterized PDEs with deterministic and stochastic coefficients. Numer. Math. **137**, 451–493 (2017)
33. L.N. Trefethen, *Approximation Theory and Approximation Practice* (SIAM, Philadelphia, 2013)
34. H. Xiao, V. Rokhlin, N. Yarvin, Prolate spheroidal wavefunctions, quadrature and interpolation. Inverse Prob. **17**(4), 805–838 (2001)

Solving Dynamic Portfolio Choice Models in Discrete Time Using Spatially Adaptive Sparse Grids

Peter Schober

Abstract In this paper, I propose a dynamic programming approach with value function iteration to solve Bellman equations in discrete time using spatially adaptive sparse grids. In doing so, I focus on Bellman equations used in finance, specifically to model dynamic portfolio choice over the life cycle. Since the complexity of the dynamic programming approach—and other approaches—grows exponentially in the dimension of the (continuous) state space, it suffers from the so called *curse of dimensionality*. Approximation on a spatially adaptive sparse grid can break this curse to some extent. Extending recent approaches proposed in the economics and computer science literature, I employ local linear basis functions to a spatially adaptive sparse grid approximation scheme on the value function. As economists are interested in the optimal choices rather than the value function itself, I discuss how to obtain these optimal choices given a solution to the optimization problem on a sparse grid. I study the numerical properties of the proposed scheme by computing Euler equation errors to an exemplary dynamic portfolio choice model with varying state space dimensionality.

1 Introduction

The individual's lifetime planning of consumption and investment decision can be formulated as an expected utility maximization problem. Economists solve these dynamic portfolio choice problems according to the Bellman principle of optimality [1]. It allows to characterize the solution as a value function and reduces the multi-period optimal choice problem to a sequence of one-period maximization problems. Since in the last period the consumption decision is known, and hence the value function, the problem is solved backwards in time using the previous value

P. Schober (✉)
Goethe University Frankfurt, Chair of Investment, Portfolio Management and Pension Finance, Theodor-W.-Adorno-Platz 3, 60323, Frankfurt am Main, Germany
e-mail: schober@finance.uni-frankfurt.de

J. Garcke et al. (eds.), *Sparse Grids and Applications – Miami 2016*,
Lecture Notes in Computational Science and Engineering 123,
https://doi.org/10.1007/978-3-319-75426-0_7

function evaluated at the current choices. The optimal consumption and investment choices can then be inferred from the value function. These optimal choices are subject to studies by economists, e.g., by simulating asset price scenarios, evaluating the optimal choices for each scenario, and computing moments of the resulting wealth distribution.

The first generation of dynamic portfolio choice models was solved analytically in two companion papers by Robert Merton [35] for the continuous time case and Paul Samuelson [40] for the discrete time case. In current research, dynamic portfolio choice models are increasingly calibrated realistically and therefore require numerical solution methods. Recently, new risk sources have been introduced, e.g., labor income risk [12], health risk [15], and uncertain family transitions [26, 33]. Other studies focus on the addition of endogenous decisions, e.g., decisions on housing [11], labor supply or retirement age [10, 30], as well as individual (variable) life annuities [22–24].

The numerical solution by value function iteration using discrete time dynamic programming and likewise approaches (such as iterations on the optimal choices rather than the value function) suffer from the so-called *curse of dimensionality*. That is, the compute time of an ε-approximation of the true solution grows exponentially as it is linear in the number of grid points of the discretized state space—and the number of grid points grows exponentially, $O(N^d)$, where d is the number of dimensions and N the maximal number of grid points in one coordinate direction. In addition, stochastic risk factors imply the necessity to compute expected utility within the maximization problem. Here, the computation of the expectation can also suffer from the curse of dimensionality when numerical quadrature rules are applied. This makes models with a high-dimensional state spaces and/or many stochastic risk factors practically *intractable* in the sense that the *time-to-result* is unacceptably high.

Besides exploiting the exponential growth of computing power to solve ever larger dynamic programming problems [9, 25], *sparse grids* can break the curse of dimensionality to some extent [19, 41, 45]. Compared to basis constructions using tensor product approaches on a nodal grid, a classical sparse grid uses a hierarchical formulation of piecewise linear basis functions in one dimension, which are then extended to a d-dimensional basis also via tensor products. The use of more sophisticated basis functions, like higher-order polynomials with local support [5], wavelets [6], or B-splines [37, 43]—which in addition to their local support are globally smooth and continuously differentiable—is also possible on a sparse grid. In a sparse grid the number of grid points grows like $O(N(\log N)^{d-1})$, while the accuracy is only slightly deteriorated. The accuracy obtained with piecewise linear basis functions, for example, is $O(N^{-2}(\log N)^{d-1})$ with respect to the L_2- and L_∞-norm, if the solution has bounded second mixed derivatives. This way, the curse of dimensionality is overcome to some extent. The sparse grid approach can be extended to non-smooth solutions by adaptive refinement methods [20, 38], some of which also exist for global polynomial basis function defined on certain grid sequences [42].

Spatially adaptive sparse grids with local linear basis functions have been employed to solve the Hamilton-Jacobi-Bellman equation in discrete time using a Semi-Lagrangian scheme [2, 17]. More recently, they have been combined with a hybrid CPU/GPU parallelization to solve economic problems, namely (1) an international real business cycle model with smooth and non-smooth choice functions and (2) a menu-cost problem with value function iteration approximating the value function on an adaptive sparse grid [4]. Earlier, sparse grids have been used for high dimensional approximation with global polynomials to compute unknown equilibrium asset demand functions [32]. Global polynomial-based approximation on a sparse grid has also been applied to time and fixed point iterations in a standard representative agent stochastic growth model as well as a multi-country model [29, 34] and to high-dimensional quadrature [21, 44].[1] Economists compare the various sparse grid approaches to numerical methods based on full Cartesian grids in terms of run time, convergence (using Euler equation errors) and the Degrees of Freedom used in the approximation of the choice functions.

In this paper, I focus on the solution of Bellman equations used to model dynamic portfolio choice over the life cycle in discrete time. Following Bokanowski et al. [2] and Brumm et al. [4], I propose a spatially adaptive sparse grid dynamic programming scheme using piecewise linear basis functions and a hierarchical surplus-based refinement criterion. Whereas the sparse grid serves to break the curse of dimensionality, spatially adaptive refinement allows to adapt the approximation of the value function to its functional form in the course of the iterative solution procedure. Also, manually choosing a discretization of the state space that yields an appropriate approximation error without spending too many grid points can be time consuming and spatially adaptive refinement relieves the researcher from this task. Besides combining and adopting these approaches to dynamic portfolio choice models, I also discuss how approximations for the optimal choices can be constructed from the solution to the optimization problem, when the optimization problem is solved on a sparse grid. I complement this approach by the use of global polynomial-based numerical quadrature rules on a sparse grid to compute expectations over the stochastic risk factors.

I analyze the numerical properties of the proposed scheme using a transaction costs problem [7, 8] whose dimensionality depends on the number of modeled assets for which transaction costs accrue. To measure the numerical error of the solution to the dynamic portfolio choice problem, I compute the deviation from the Euler equations when plugging in the approximations of the optimal choices. I then compute the pointwise absolute error of the value function approximated on a sparse grid compared to a reference solution on a full grid to separate out the error in the value function. I find that the spatially adaptive sparse grid dynamic programming scheme provides similar Euler errors compared to the full grid reference solution and the pointwise errors of the value function converge with decreasing refinement tolerances.

The overall Euler errors remain rather high, which is an indicator that the applicability of the presented approach is limited by the choice of only piecewise

[1]Multi-country real business cycle models are popular example problems among economists as they allow to vary the dimensionality of the problem simply by varying the number of countries in the model.

continuously differentiable basis functions, which pose a problem for gradient-based optimization methods. Non-convergence or failure of the optimizer can result in outlier grid points and hence locally non-smooth value functions or optimal choices, misdirecting the scheme when using absolute values of hierarchical surpluses as the refinement criterion. The use of globally smooth and continuously differentiable local basis functions, such as B-splines [43], promises to overcome this problem while maintaining the advantages of local spatial adaptivity compared to globally defined, smooth basis functions (such as polynomials).

The paper is structured as followed: In the next section, I introduce the class of discrete time life cycle models I investigate and describe a current, non-sparse grid solution method using dynamic programming with value function iteration. Section 3 introduces sparse grids and explains spatial adaptivity. I present my spatially adaptive sparse grid dynamic programming scheme in Sect. 4. The details of the numerical examples and the results are depicted in Sect. 5. I conclude in Sect. 6.

2 Dynamic Portfolio Choice Models

In discrete time dynamic portfolio choice models the investor seeks to maximize expected life time utility u from consumption c_t

$$\mathscr{E}_0\left[\sum_{t=0}^{T}\rho^t u\left(c_t\left(\boldsymbol{p}_t,\boldsymbol{s}_t,\theta_t\right)\right)\right], \tag{1}$$

with consumption depending on the investor's choice $\boldsymbol{p}_t \in \mathbb{R}^k$, her continuous state $\boldsymbol{s}_t \in \mathbb{R}^d$, and discrete state $\theta_t \in \Theta$, where Θ is the finite set of all possible states (such as "alive", "dead", "healthy", "sick", etc.). Here, $\rho < 1$ denotes the time discount factor. I assume the utility function to be of *Constant Relative Risk Aversion* type[2]:

$$u\left(c_t\right) = \frac{1}{1-\gamma}c_t^{1-\gamma}. \tag{2}$$

This problem can be reformulated as an optimization problem in terms of the value function $j_t, t \in \{0,\ldots,T\}$, with known terminal utility v

$$j_t(\boldsymbol{s}_t,\theta_t) = \max_{\boldsymbol{p}_t}\left\{u(c_t) + \rho\mathscr{E}_t\left[j_{t+1}\left(\boldsymbol{f}_{t+1}\left(\boldsymbol{p}_t,\boldsymbol{s}_t,\theta_t,\boldsymbol{\omega}_{t+1}\right),\theta_{t+1}\right)\right]\right\}, \tag{3}$$

$$j_T(\boldsymbol{s}_T,\theta_T) = v(\boldsymbol{s}_T,\theta_T), \tag{4}$$

[2]With this choice of utility function the approach does not lose generality, as it can be applied to other established utility functions, such as Epstein-Zin utility, in the same way.

subject to the constraints

$$g_j\left(\boldsymbol{p}_t, \boldsymbol{s}_t, \theta_t\right) = 0\,, \quad j = 1, \ldots, m\,, \tag{5}$$

$$h_j\left(\boldsymbol{p}_t, \boldsymbol{s}_t, \theta_t\right) \leq 0\,, \quad j = 1, \ldots, l\,. \tag{6}$$

I assume the state transition between the discrete states to be *Markovian* and given by the transition probabilities $\pi_t(\theta_{t+1}|\theta_t)$. The randomness in the continuous state transition between t and $t+1$ is independent from the discrete state transition and captured by $\boldsymbol{\omega}_{t+1} \in \Omega$, where $\Omega \subset \mathbb{R}^q$ denotes the sample space. The random variable $\boldsymbol{f}_{t+1} : \mathbb{R}^k \times \mathbb{R}^d \times \Theta \times \Omega \mapsto \mathbb{R}^d$ then describes the continuous state dynamics between t and $t+1$. The corresponding expected value of $j_{t+1}(\boldsymbol{f}_{t+1}, \theta_{t+1})$ is:

$$\mathscr{E}_t\left[j_{t+1}(\boldsymbol{f}_{t+1}, \theta_{t+1})\right] = \sum_{\theta_{t+1}\in\Theta} \pi_t(\theta_{t+1}|\theta_t) \int_{\Omega} j_{t+1}\left(\boldsymbol{f}_{t+1}, \theta_{t+1}\right) d\Phi_t\left(\boldsymbol{\omega}_{t+1}|\boldsymbol{s}_t, \theta_t\right)\,.$$

Here, $\Phi_t(\cdot|\boldsymbol{s}_t, \theta_t)$ denotes the conditional distribution of $\boldsymbol{\omega}_{t+1}$. The equality constraints (5) and the inequality constraints (6) can be non-linear.

A general way to solve problem (3)–(6) numerically is to use a discrete time dynamic programming approach stepping backwards in time iterating over the value function.[3] Therefore, the three ingredients are approximation, optimization and integration.

2.1 Approximation

Amongst others, one common method to approximate the value function defined on the continuous state space using finitely many points is *collocation* (see [27] for this and alternative approaches). For a given discrete state $\theta_t \in \Theta$ one discretizes the associated continuous state space $G(\theta_t) \subset \mathbb{R}^d$ by a rectangular grid of mesh size $\boldsymbol{h} = ((b_1 - a_1)/n_1, \ldots, (b_d - a_d)/n_d)^\top \in \mathbb{R}^d$, where $\boldsymbol{n} \in \mathbb{N}^d$ is the number of grid points, $\boldsymbol{a} \in \mathbb{R}^d$ is a chosen lower, and $\boldsymbol{b} \in \mathbb{R}^d$ the respective upper boundary for every dimension of the state space.[4] Given a multi-index $\boldsymbol{i} \in \mathbb{N}^d$ from the index set $I_{\boldsymbol{n}}(\theta_t) = \{\boldsymbol{i} \in \mathbb{N}^d \mid \boldsymbol{1} \leq \boldsymbol{i} \leq \boldsymbol{n}\}$, the points of the grid $G_{\boldsymbol{n},\boldsymbol{h},\boldsymbol{a}}(\theta_t) = \{\boldsymbol{s}^{\boldsymbol{i}} \in \mathbb{R}^d \mid \boldsymbol{i} \in I_{\boldsymbol{n}}(\theta_t)\}$ are given by $\boldsymbol{s}^{\boldsymbol{i}} = \boldsymbol{a} + \boldsymbol{i} \cdot \boldsymbol{h}$, where "$\leq$", "+", and "$\cdot$" and have to be

[3] For an extensive discussion of dynamic programming in economics, see [28, 39].

[4] In many models the continuous state space is the same for every discrete state. However, this doesn't have to be the case. Especially the discretization, e.g., the choice of upper and lower boundaries, may vary depending on the discrete state.

read element-wise. The approximating (or, equivalently, interpolating) function on this grid can be defined as

$$j_t(\boldsymbol{s}_t, \theta_t) \approx \mathscr{A}_{\boldsymbol{n}}[j_t](\boldsymbol{s}_t, \theta_t) = \sum_{\boldsymbol{i} \in I_{\boldsymbol{n}}(\theta_t)} c_{\boldsymbol{i},\theta_t} \phi_{\boldsymbol{i}}(\boldsymbol{s}_t),$$

where the basis functions $\phi_{\boldsymbol{i}}$ can be global polynomials or Ansatz functions with local support and the coefficients $c_{\boldsymbol{i},\theta_t}$ are chosen in such a way that the approximation fits the known function values at all grid points, $\mathscr{A}_{\boldsymbol{n}}[j_t](\boldsymbol{s}_t^{\boldsymbol{i}}, \theta_t) = j_t(\boldsymbol{s}_t^{\boldsymbol{i}}, \theta_t), \forall \boldsymbol{s}_t^{\boldsymbol{i}} \in G_{\boldsymbol{n},\boldsymbol{h},\boldsymbol{a}}(\theta_t), \forall \theta_t \in \Theta$.

2.2 *Optimization*

The terminal period's value function (4) is given by $j_T(\boldsymbol{s}_T^{\boldsymbol{i}}, \theta_T) = v(\boldsymbol{s}_T^{\boldsymbol{i}}, \theta_T)$, where v is a known function of the continuous and discrete state space. To determine the optimal solution of the penultimate period $j_{T-1}(\boldsymbol{s}_{T-1}^{\boldsymbol{i}}, \theta_{T-1})$ and all earlier periods $t \in \{0, \ldots, T-2\}$ at all grid points $\boldsymbol{s}_t^{\boldsymbol{i}}$ and in all discrete states θ_t, numerical optimization routines, which solve (3) over the real-valued vector $\boldsymbol{p}_t$ subject to the constraints (5)–(6) can be used, e.g., *sequential quadratic programming (SQP)*. Let me define the objective function as

$$\tilde{j}_t\left(\boldsymbol{p}_t, \boldsymbol{s}_t, \theta_t\right) := u(\boldsymbol{p}_t, \boldsymbol{s}_t, \theta_t) + \rho \mathscr{E}_t\left[j_{t+1}\left(\boldsymbol{f}_{t+1}\left(\boldsymbol{p}_t, \boldsymbol{s}_t, \theta_t, \boldsymbol{\omega}_{t+1}\right), \theta_{t+1}\right)\right]. \quad (7)$$

Then, SQP routines use the linearization of the Lagrangian of (3) with Kuhn-Tucker multipliers for the equality ($\boldsymbol{\lambda}_t \in \mathbb{R}_+^m$) and inequality ($\boldsymbol{\mu}_t \in \mathbb{R}_+^l$) constraints

$$\begin{aligned} \mathscr{L}_t\left(\boldsymbol{p}_t, \boldsymbol{\lambda}_t, \boldsymbol{\mu}_t\right) = \tilde{j}_t\left(\boldsymbol{p}_t, \boldsymbol{s}_t, \theta_t\right) &- \sum_{j=1}^{m} \lambda_t^j g_j\left(\boldsymbol{p}_t, \boldsymbol{s}_t, \theta_t\right) \\ &- \sum_{j=1}^{l} \mu_t^j h_j\left(\boldsymbol{p}_t, \boldsymbol{s}_t, \theta_t\right) \end{aligned} \quad (8)$$

to find the optimum $(\boldsymbol{p}_t^*, \boldsymbol{\lambda}_t^*, \boldsymbol{\mu}_t^*)$.

SQP optimization routines require information on the gradient and the Hessian of the objective function and thus any approximation of the objective function to be twice continuously differentiable. Since this is true for the utility function (2), this is the case iff $\mathscr{A}_{\boldsymbol{n}}[j_t](\cdot, \theta_t) \in C_2(G(\theta_t))$ for each $\theta_t \in \Theta$ and each $t \in \{1, \ldots, T\}$ (assuming that I can interchange differentiation and integration in the computation of the expected value). If this is not the case, *derivative free optimizers* like simulated annealing, pattern search, and genetic algorithms might present alternative solution methods.

2.3 Integration

Within the optimization routine, the state dynamics mapping $\boldsymbol{f}_{t+1}$ and the Markovian transition probabilities are applied to each grid point $\boldsymbol{s}_t^i$ and discrete state θ_t. Then, the expectation operator $\mathscr{E}_t[j_{t+1}(\boldsymbol{f}_{t+1}(\boldsymbol{p}_t, \boldsymbol{s}_t^i, \theta_t, \boldsymbol{\omega}_{t+1}), \theta_{t+1})]$ can be approximated by

$$\begin{aligned} &\sum_{\theta_{t+1}\in\Theta} \pi_t(\theta_{t+1}|\theta_t) \int_{\Omega} j_{t+1}\left(\boldsymbol{f}_{t+1}\left(\boldsymbol{p}_t, \boldsymbol{s}_t^i, \theta_t, \boldsymbol{\omega}_{t+1}\right), \theta_{t+1}\right) d\Phi_t(\boldsymbol{\omega}_{t+1}|\boldsymbol{s}_t^i, \theta_t) \\ &\approx \sum_{\theta_{t+1}\in\Theta} \pi_t(\theta_{t+1}|\theta_t) \sum_{q} \mathscr{A}_{\boldsymbol{n}}[j_{t+1}]\left(\boldsymbol{f}_{t+1}\left(\boldsymbol{p}_t, \boldsymbol{s}_t^i, \theta_t, \boldsymbol{\omega}_{t+1}^q\right), \theta_{t+1}\right) w_{t+1}^q . \end{aligned} \tag{9}$$

The set of tuples $\{(\boldsymbol{\omega}_{t+1}^q, w_{t+1}^q) \mid q \in Q_t(\boldsymbol{s}_t^i, \theta_t)\}$ are the evaluation points and weights of a quadrature rule. Note that $j_{t+1}(\boldsymbol{f}_{t+1}(\boldsymbol{p}_t, \boldsymbol{s}_t^i, \theta_t, \boldsymbol{\omega}_{t+1}), \theta_{t+1})$ is only known at the grid points $\boldsymbol{s}_{t+1}^i$, and since in general $\boldsymbol{f}_{t+1}(\boldsymbol{p}_t, \boldsymbol{s}_t^i, \theta_t, \boldsymbol{\omega}_{t+1}^q)$ does not correspond with a grid point, one has to evaluate the approximation $\mathscr{A}_{\boldsymbol{n}}[j_{t+1}](\boldsymbol{f}_{t+1}(\boldsymbol{p}_t, \boldsymbol{s}_t^i, \theta_t, \boldsymbol{\omega}_{t+1}^q), \theta_{t+1})$ within the numerical integral (9).

2.4 Dynamic Programming

Algorithm 1 formalizes the discrete time dynamic programming approach outlined in Sects. 2.1–2.3. A solution to problem (3)–(6) is characterized by the value function values $j_t(\boldsymbol{s}_t^i, \theta_t)$ and the optimal choices $\boldsymbol{p}_t(\boldsymbol{s}_t^i, \theta_t)$ at all grid points $\boldsymbol{s}_t^i \in G_{\boldsymbol{n},\boldsymbol{h},\boldsymbol{a}}(\theta_t)$ of the grid $G_{\boldsymbol{n},\boldsymbol{h},\boldsymbol{a}}(\theta_t)$. In principle, the number of grid points $\boldsymbol{n}$, the mesh width $\boldsymbol{h}$, and the lower boundary $\boldsymbol{a}$ can be chosen differently for each time step t and in each discrete state $\theta_t \in \Theta$. Let me denote the time t solution obtained by means of the dynamic programming approach by the set of tuples $S_t = \{(\boldsymbol{s}_t^i,\ j_t(\boldsymbol{s}_t^i, \theta_t),\ \boldsymbol{p}_t(\boldsymbol{s}_t^i, \theta_t)) \mid \boldsymbol{s}_t^i \in G_{\boldsymbol{n},\boldsymbol{h},\boldsymbol{a}}(\theta_t),\ \theta_t \in \Theta\}$.[5]

The innermost loop iterates over all grid points in $G_{\boldsymbol{n},\boldsymbol{h},\boldsymbol{a}}(\theta_t)$. Here, for each grid point one optimization problem has to be solved to determine the value function and the optimal choices at this grid point. The computational effort to solve a single optimization problem within the optimization routine involves computing the expectation according to Eq. (9). On the one hand, the number of quadrature nodes used to compute the expectation grows exponentially with the dimensionality of the sample space Ω. On the other hand, the computational effort of the evaluation of the approximation $\mathscr{A}_{\boldsymbol{n}}[j_{t+1}]$ at the states $\boldsymbol{f}_{t+1}$ implied by the quadrature nodes

[5] Using the SQP method described in Sect. 2.2, the optimal solution is additionally characterized by the $m+l$ Kuhn-Tucker multipliers $(\lambda_t^1, \ldots, \lambda_t^m, \mu_t^1, \ldots, \mu_t^l)^\top$ at all grid points and all discrete states.

Algorithm 1: Discrete time dynamic programming

Data:
Final solution $S_T = \{(\boldsymbol{s}_T^i,\ j_T(\boldsymbol{s}_T^i, \theta_T),\ \boldsymbol{p}_T(\boldsymbol{s}_T^i, \theta_T)) \mid \boldsymbol{s}_T^i \in G_{\boldsymbol{n},\boldsymbol{h},\boldsymbol{a}}(\theta_T),\ \theta_T \in \Theta\}$
Grids $G_{\boldsymbol{n},\boldsymbol{h},\boldsymbol{a}}$
Result:
Solutions $S_t = \{(\boldsymbol{s}_t^i,\ j_t(\boldsymbol{s}_t^i, \theta_t),\ \boldsymbol{p}_t(\boldsymbol{s}_t^i, \theta_t)) \mid \boldsymbol{s}_t^i \in G_{\boldsymbol{n},\boldsymbol{h},\boldsymbol{a}}(\theta_t),\ \theta_t \in \Theta\}, t \in \{0, \dots, T-1\}$
for $t := T-1$ **to** 0 **do**
    $S_t := \emptyset$;
    for $\theta_t \in \Theta$ **do**
        for $\boldsymbol{s}_t^i \in G_{\boldsymbol{n},\boldsymbol{h},\boldsymbol{a}}(\theta_t)$ **do**
            $\boldsymbol{p}_t(\boldsymbol{s}_t^i, \theta_t) := \arg\max_{\boldsymbol{p}_t} \left\{\tilde{j}_t\left(\boldsymbol{p}_t, \boldsymbol{s}_t^i, \theta_t\right)\right\}$ subject to (5)–(6);
            $j_t(\boldsymbol{s}_t^i, \theta_t) := \tilde{j}_t\left(\boldsymbol{p}_t(\boldsymbol{s}_t^i, \theta_t), \boldsymbol{s}_t^i, \theta_t\right)$;
            $S_t := S_t \bigcup (\boldsymbol{s}_t^i,\ j_t(\boldsymbol{s}_t^i, \theta_t),\ \boldsymbol{p}_t(\boldsymbol{s}_t^i, \theta_t))$;

is conditional on the approximation method used. For example, using Lagrange polynomials as (global) basis functions ϕ_i, the evaluation of the approximation off the grid is costly in terms of arithmetic operations since the Lagrange polynomial for every $\boldsymbol{i} \in \boldsymbol{I_n}(\theta_t)$ has to be evaluated in order to compute the approximated value off the grid (whereas the coefficients c_i are easily obtained as the value function values at the grid points $\boldsymbol{s}_t^i \in G_{\boldsymbol{n},\boldsymbol{h},\boldsymbol{a}}(\theta_t)$).

The number of optimization problems that have to be solved grows exponentially with the dimensionality d of the state space G as the number of grid points needed by conventional methods to approximate the value function on the discretized state space $G_{\boldsymbol{n},\boldsymbol{h},\boldsymbol{a}}(\theta_t)$ grows exponentially in d. Additionally, the computation of the approximation coefficients c_i also depend on the state space dimension for certain approximation methods. For example, using cubic splines as the basis functions ϕ_i, a linear system of equations, whose dimensionality depends on the number of grid points, must be solved to determine the coefficients of the approximation in each time step of Algorithm 1.

In consequence, this approach suffers from the curse of dimensionality. It is eminent to break the exponential growth in the number of optimization problems that have to be solved, i.e., by using approximation on a sparse grid. Additionally, quadrature on a sparse grid can break the curse of dimensionality inherent in the computation of the expected value necessary to solve a single optimization problem [21, 25].

Lastly, choosing the right discretization of the state space $G(\theta_t)$, i.e., the number of grid points $\boldsymbol{n}$, the mesh width $\boldsymbol{h}$, and the lower boundaries $\boldsymbol{a}$, that yields an appropriate approximation error without unnecessary many grid points, can be a time consuming task in practice. This is especially true when the functional form of the value function is different across times and discrete states.

2.5 Optimal Choices

The solution of the discrete time dynamic programming algorithm 1 contains the optimal choices $\boldsymbol{p}_t(\boldsymbol{s}_t^i, \theta_t) \in \mathbb{R}^k$ at each grid point $\boldsymbol{s}_t^i$ and for each discrete state θ_t. Ultimately, these optimal choices are subject to analysis by economists. For example, by setting up a Monte-Carlo simulation to simulate n agents in the portfolio choice model and computing the resulting consumption distribution's moments to analyze the expected utility (1) or directly by evaluating the optimal choices at points of interest.

In any case, the optimal choices are needed for an arbitrary state $\boldsymbol{s}_t$. It is straightforward to again construct an approximating function for any $p_t^u(\boldsymbol{s}_t, \theta_t)$, $u = 1, \ldots, k$, and evaluate the approximation $\mathscr{A}_{\boldsymbol{n}}[p_t^u](\boldsymbol{s}_t, \theta_t)$. Since in many models the constraints (5)–(6) are linear, a linear approximation of the optimal choices ensures that the approximated choices also fulfill the constraints—at least for full grid approximations.

3 Sparse Grid Approximation

The goal is to approximate a function $f \in C_2(G)$ in one and later multiple dimensions by linear splines on a grid. For simplification, for an $\boldsymbol{x} \in G = [0, 1]^d$ let $f(\boldsymbol{x}) \in \mathbb{R}$ and f be zero on the boundary. An approximation on an arbitrary subset $[a_1, b_1] \times \cdots \times [a_d, b_d] \subset \mathbb{R}^d$ can be found via the affine transformation $\boldsymbol{x}(\boldsymbol{y}) = (\boldsymbol{y} - \boldsymbol{a})/(\boldsymbol{b} - \boldsymbol{a})$, where $\boldsymbol{a} = (a_1, \ldots, a_d)^\top, \boldsymbol{b} = (b_1, \ldots, b_d)^\top$ and "/" is to be read element-wise. A non-zero boundary requires additional basis functions on the boundary but does not change the general ideas on hierarchical bases and sparse subspace selection as well as spatially adaptive refinement presented in Sects. 3.1 and 3.2, respectively (see [16]).

3.1 Hierarchical Bases and Sparse Subspace Selection

For a given level l, let $h_l = 2^{-l}$ be the mesh width of a grid with grid points $x_{l,i} = i \cdot h_l$, where i is from the index set $I_l^{\text{nod}} = \{i \in \mathbb{N} \mid 1 \leq i \leq 2^l - 1\}$. A *nodal basis* on this grid is given by the one-dimensional hat functions

$$\phi_{l,i}(x) = \begin{cases} 1 - \left|\frac{x - x_{l,i}}{h_l}\right| & \text{if } x \in [x_{l,i} - h_l, x_{l,i} + h_l] \\ 0 & \text{else} \end{cases} \tag{10}$$

that span the space

$$V_l = \operatorname{span}\left\{\phi_{l,i} | i \in I_l^{\text{nod}}\right\}.$$

A continuous, piecewise approximation $\mathscr{A}_l^{\text{nod}}[f](x) \in C_0(G)$ of f can be obtained by the weighted sum of these basis functions

$$f(x) \approx \mathscr{A}_l^{\text{nod}}[f](x) = \sum_{i \in I_l^{\text{nod}}} c_{l,i}\phi_{l,i}(x)$$

with the weights (from now on called coefficients)

$$c_{l,i} = f\left(x_{l,i}\right) .$$

Equivalently, the space V_l can be spanned by a *hierarchical basis*. Therefore, define the index set $I_l^{\text{hier}} = \{i \in \mathbb{N} \mid 1 \le i \le 2^l - 1, i \text{ odd}\}$ for which the hat functions (10) span the hierarchical subspaces

$$W_k = \text{span}\left\{\phi_{k,i} | i \in I_k^{\text{hier}}\right\} .$$

These hierarchical subspaces can be combined to span the nodal space by

$$V_l = \bigoplus_{k=1}^{l} W_k .$$

Figure 1 compares the nodal and the hierarchical basis for V_3.

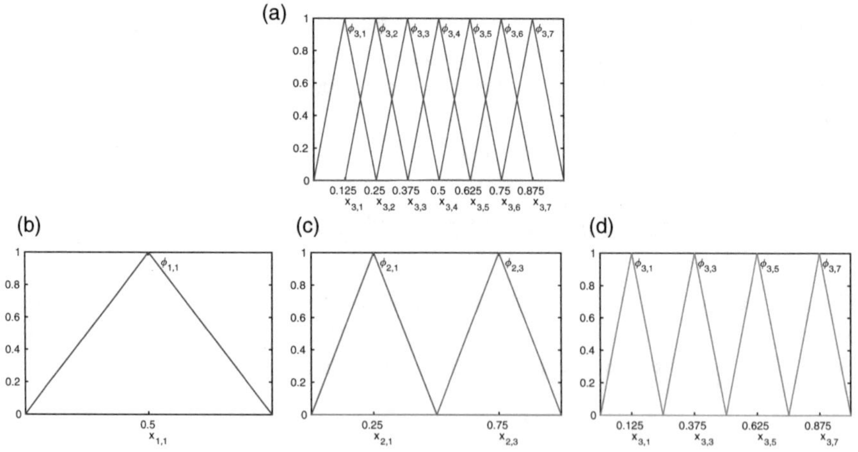

Fig. 1 The (**a**) nodal basis V_3 and the (**b**)–(**d**) basis functions of the subspaces W_k, $k = 1, 2, 3$, (**b**)–(**d**) that span the same space V_3

The approximation $\mathscr{A}_l^{\text{hier}}[f](x) \in C_0(G)$ of f is then obtained via

$$f(x) \approx \mathscr{A}_l^{\text{hier}}[f](x) = \sum_{k=1}^{l} \underbrace{\sum_{i \in I_k^{\text{hier}}} c_{k,i}\phi_{k,i}(x)}_{:=\Delta_k[f]},$$

where $\Delta_k[f]$ is the hierarchical "increment" of $\mathscr{A}_{I_l^{\text{hier}}}[f]$ on level k and the coefficients are computed such that the approximation matches the function values at the grid points:

$$c_{l,i} = \frac{-f\left(x_{l,i} - h_l\right) + 2f\left(x_{l,i}\right) - f\left(x_{l,i} + h_l\right)}{2} = \underbrace{\begin{bmatrix} -\frac{1}{2} & 1 & -\frac{1}{2} \end{bmatrix}_{l,i} f}_{:=\mathscr{D}_{l,i}[f]}. \tag{11}$$

These coefficients happen to be the second (numerical) derivative of f multiplied by $-h_l^2/2$ and can be expressed by the operator form of the finite difference stencil $\mathscr{D}_{l,i}$ [6, 16, 45]. The coefficients (11) are called *hierarchical surpluses*, as Fig. 2 depicts.

In d dimensions the hierarchical basis on the dimension-wise equidistant grid with mesh widths $\boldsymbol{h_l} = 2^{-\boldsymbol{l}}$ and grid points $\boldsymbol{x_{l,i}} = \boldsymbol{i} \cdot \boldsymbol{h_l}$ with $\boldsymbol{i}$ from the multi-index set $I_{\boldsymbol{l}}^{\text{hier}} = \{\boldsymbol{i} \in \mathbb{N}^d \mid \mathbf{1} \leq \boldsymbol{i} \leq 2^{\boldsymbol{l}} - \mathbf{1}, i_j \text{ odd for } j = 1, \ldots, d\}$ is constructed via a tensor product approach, where the piecewise d-dimensional basis is defined by

$$\phi_{\boldsymbol{l},\boldsymbol{i}}(\boldsymbol{x}) := \prod_{j=1}^{d} \phi_{l_j,i_j}(x_j). \tag{12}$$

Accordingly, the hierarchical subspaces given by

$$W_{\boldsymbol{k}} = \text{span}\left\{\phi_{\boldsymbol{k},\boldsymbol{i}} | \boldsymbol{i} \in I_{\boldsymbol{k}}^{\text{hier}}\right\}$$

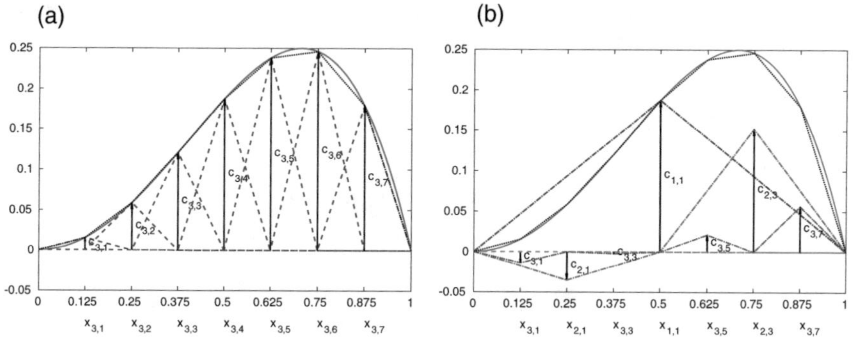

Fig. 2 The dotted lines represent the (**a**) nodal basis approximation $\mathscr{A}_3^{\text{nod}}[f]$ and the (**b**) hierarchical basis approximation $\mathscr{A}_3^{\text{hier}}[f]$, respectively, of $f(x) = -x^4 + x^2$ on [0, 1]

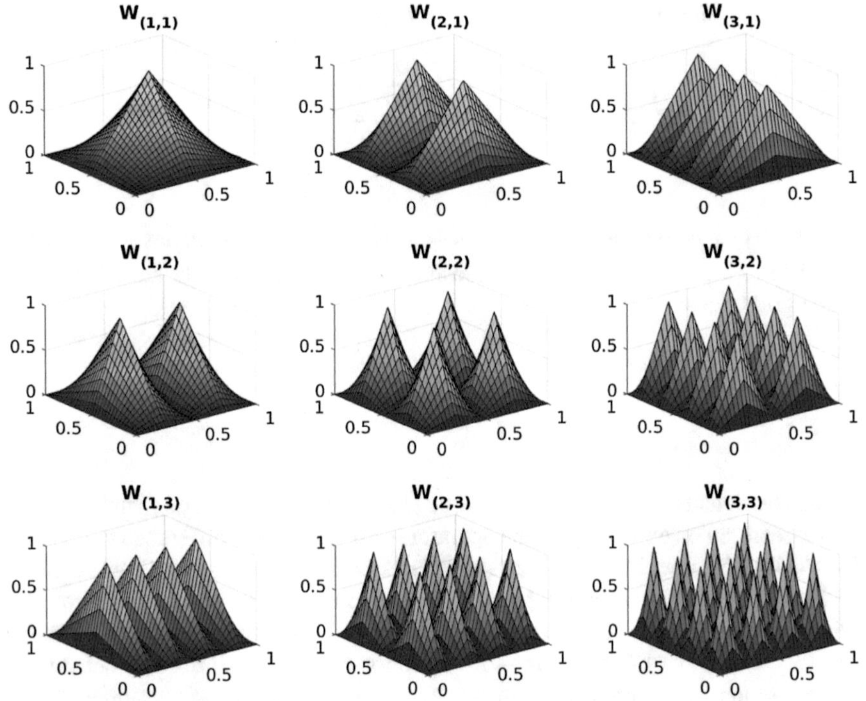

Fig. 3 The hierarchical basis in two dimension that spans $V_{(3,3)}$

span the full grid space

$$V_{\boldsymbol{l}} = \left(\bigoplus_{k_1=1}^{l_1} \cdots \bigoplus_{k_d=1}^{l_d} \right) W_{\boldsymbol{k}} = \bigoplus_{\boldsymbol{k} \leq \boldsymbol{l}} W_{\boldsymbol{k}} .$$

Figure 3 shows the two-dimensional hierarchical basis (12). Finally, $f : [0, 1]^d \to \mathbb{R}$ can be approximated by

$$\begin{aligned} f(\boldsymbol{x}) \approx \mathscr{A}_{\boldsymbol{l}}^{\text{hier}}[f](\boldsymbol{x}) &= \left(\mathscr{A}_{l_1}^{\text{hier}} \otimes \cdots \otimes \mathscr{A}_{l_d}^{\text{hier}} \right) [f](\boldsymbol{x}) \\ &= \left(\sum_{k_1=1}^{l_1} \Delta_{k_1}[f] \otimes \cdots \otimes \sum_{k_d=1}^{l_d} \Delta_{k_d}[f] \right) (\boldsymbol{x}) \\ &:= \sum_{k_1=1}^{l_1} \cdots \sum_{k_d=1}^{l_d} \left(\Delta_{k_1} \otimes \cdots \otimes \Delta_{k_d} \right) [f](\boldsymbol{x}) \\ &= \sum_{\boldsymbol{k} \leq \boldsymbol{l}} \underbrace{\left(\Delta_{k_1} \otimes \cdots \otimes \Delta_{k_d} \right) [f](\boldsymbol{x})}_{:= \Delta_{\boldsymbol{k}}[f]} \end{aligned}$$

with the definition

$$\Delta_{\boldsymbol{k}}[f](\boldsymbol{x}) := \sum_{\boldsymbol{i} \in I_{\boldsymbol{k}}^{\text{hier}}} c_{\boldsymbol{k},\boldsymbol{i}} \phi_{\boldsymbol{k},\boldsymbol{i}}(\boldsymbol{x}) \,. \tag{13}$$

As in the one-dimensional case, the coefficients can be computed by the d-dimensional finite difference stencil

$$c_{\boldsymbol{l},\boldsymbol{i}} := \underbrace{\left(\prod_{j=1}^{d} \begin{bmatrix} -\frac{1}{2} & 1 & -\frac{1}{2} \end{bmatrix}_{l,i} \right) f}_{:=\mathscr{D}_{\boldsymbol{l},\boldsymbol{i}}[f]} \,, \tag{14}$$

which includes second mixed derivatives.

The idea of sparse grids is to leave out those subspaces that contribute little to the overall approximation error. The selection of these subspaces can be done a priori, that means, without knowledge of the approximated function f, with respect to a given norm $\|\cdot\|$ (e.g., L_2 or L_∞) and as long as f has bounded second mixed derivatives, $\|\partial^{2d}/(\partial x_1^2 \cdots \partial x_d^2) f\| < \infty$ (see [6] for details). As a result, the sparse grid space for a given level l is constructed by

$$V_l^{\mathrm{S}} := \bigoplus_{|\boldsymbol{k}|_1 \le l+d-1} W_{\boldsymbol{k}} \tag{15}$$

with $|\boldsymbol{x}|_1 := \sum_{i=1}^{d} x_i$, $\boldsymbol{x} \in \mathbb{N}^d$. Figure 4 depicts the full grid space and the sparse grid space for level $l = 3$. The corresponding sparse grid approximation is given by

$$f(\boldsymbol{x}) \approx \mathscr{A}_l^{\mathrm{S}}[f](\boldsymbol{x}) = \sum_{|\boldsymbol{k}|_1 \le l+d-1} \Delta_{\boldsymbol{k}}[f](\boldsymbol{x}) \,.$$

Whereas in the isotropic full grid space $V_l = V_{(l,\ldots,l)}$ the number of grid points grows exponentially as $O(2^{ld})$, the number of grid points in the sparse grid space V_l^{S} only grows like $O(2^l l^{d-1})$. The L_2 and L_∞ error of the approximation is only slightly deteriorated by a logarithmic factor from $O(2^{-2l})$ to $O(2^{-2l} l^{d-1})$ [6].

3.2 *Spatially Adaptive Refinement*

The selection of subspaces as chosen in (15) is only a priori optimal, that is, if the approximated function f fulfills the smoothness condition of bounded second mixed derivatives and no further information on f is available [20, 38]. However, in the hierarchical basis formulation (13), the absolute value $|c_{\boldsymbol{l},\boldsymbol{i}}|$ of the hierarchical surpluses contains information on the smoothness of f as obvious by

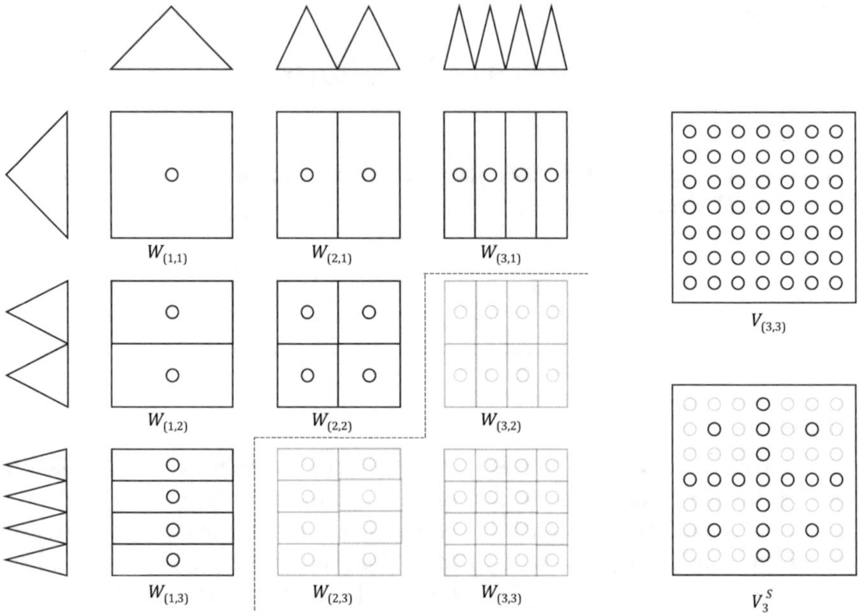

Fig. 4 Grids $G_{\boldsymbol{k}}$ of the subspaces $W_{\boldsymbol{k}}$, $\boldsymbol{k} \leq (3, 3)$, associated in the construction of the full grid space $V_{(3,3)}$ (all grids) and the sparse grid space V_3^S (only black grids)

their definition (14). It is fairly straightforward to enhance the approximation by adding extra grid points where the absolute value of the coefficients is larger than a chosen tolerance, $|c_{\boldsymbol{l},\boldsymbol{i}}| > \epsilon$—so called *surplus-based refinement* [37]. If one relates the absolute value of the coefficient to the volume of its associated basis function, one ends up at the *surplus volume-based* refinement criterion:

$$|c_{\boldsymbol{l},\boldsymbol{i}}| > \frac{\epsilon}{\|\phi_{\boldsymbol{l},\boldsymbol{i}}\|_2}. \tag{16}$$

Given this criterion, one could add all $2d$ children $\{x_{\tilde{\boldsymbol{k}},\tilde{\boldsymbol{i}}} \mid \tilde{\boldsymbol{k}} = \boldsymbol{k} + \mathbf{1}_j,\ \tilde{\boldsymbol{i}} = \boldsymbol{i} + \boldsymbol{i} \cdot \mathbf{1}_j \pm \mathbf{1},\ j = 1, \ldots, d\}$ of the grid point $x_{\boldsymbol{k},\boldsymbol{i}}$ (Fig. 5). Here, $\mathbf{1}_j$ denotes the unit vector in dimension j. A grid point is called *refinable* when at least one child does not exist. Given the tree-like structure of hierarchical grids it is sensible to include all up to d parents of the children added, if they are not already included in the grid, all up to d parents of the parents, if they are not already included, and so on [37].

It should be noted that the absolute values of the hierarchical surpluses merely provide an indicator on where the approximation error could be large and surplus (volume)-based refinement does not constitute a reliable error estimator.

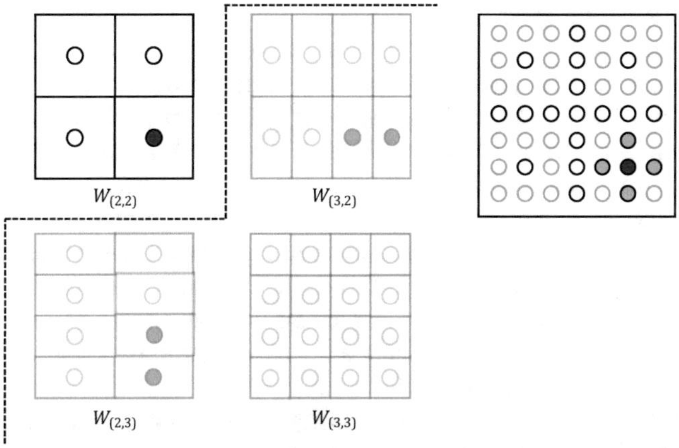

Fig. 5 Spatially adaptive refinement by adding all $2d$ children (green) of the refinable grid point (blue)

4 Spatially Adaptive Sparse Grid Dynamic Programming Scheme

To break the exponential growth in the number of optimization problems that have to be solved in the dynamic programming algorithm 1, my idea is to use a sparse grid for the approximation of the value function. This should reduce the computational effort needed to obtain a solution to the dynamic portfolio choice problem considerably, especially in higher dimensions of the continuous state space. Additionally, my approach employs spatial adaptivity to construct value function approximations for every time step and every discrete state. This allows to start at a coarse grid level with few grid points and to only add grid points that presumably reduce the approximation error considerably. Also, spatial adaptivity can save a substantial amount of time spent by researchers to find appropriate grid choices that allow for accurate approximations at every time step and discrete state. Although my approach differs in certain aspects and focuses on dynamic portfolio choice models, spatially adaptive sparse grid approximations for value function iterations have already been picked up in similar manner in the computational science [2] and economics literature [4]. I present my spatially adaptive sparse grid dynamic programming algorithm for dynamic portfolio choice problems in Sect. 4.1.

In dynamic portfolio choice models the optimal choices often have kinks due to binding constraints or must fulfill non-negativity constraints (such as short selling and/or budget constraints). One way to resolve these kinks and increase the approximation quality is adaptive Delaunay triangulation [3]. Unfortunately, non-differentiability poses a problem for a sparse grid, which requires bounded second mixed derivatives and sparse grid approximations may become negative even though the approximated function is strictly positive. Hence the question comes up how to treat the optimal choices when using a sparse grid approximation of the value function. Again, Delaunay triangulation can be used, if one treats the optimal choices at the sparse grid points as scattered data and the dimension of the problem is comparably low. In my as of yet unpublished work together with Yannick Dillschneider and Raimond Maurer we show that this approach delivers accurate approximations of the optimal choices and a significant reduction of effort to compute the solution for dynamic portfolio choice models of $d \leq 3$ with annuities when the interest rate is stochastic. However, Delaunay triangulation also suffers from the curse of dimensionality. Alternatively, spatially adaptive refinement on the optimal choice approximations on a sparse grid can again be used to construct approximations that fulfill constraints such as non-negativity and resolve kinks appropriately, which is the topic of Sect. 4.2.

To approximate the value function and the optimal choices, I use linear basis functions. In consequence, the hierarchical surplus provides an indicator for refinement as described in Sect. 3. Section 4.3 discusses implications of the choice of basis functions on the optimization routine and how to treat extrapolation.

4.1 Spatially Adaptive Sparse Grid Dynamic Programming

For a given level l and the associated sparse grid space V_l^{S} let me denote the grid by $G_l = \{s^{\boldsymbol{k},\boldsymbol{i}} \mid \boldsymbol{i} \in I_{\boldsymbol{k}}^{\mathrm{hier}},\ \boldsymbol{k} \in \mathbb{N}^d,\ |\boldsymbol{k}|_1 \leq l + d - 1\}$. Furthermore, as spatially adaptively refined grids are created in every time step t and for every discrete state $\theta_t \in \Theta$, let me denote the adaptively refined grid that was constructed starting from level l for a given discrete state θ_t at a given time t by $G_l^t(\theta_t)$.

The spatially adaptive sparse grid dynamic programming algorithm 2 is initialized with the refined grid $G_l^T(\theta_T)$ for all discrete states $\theta_T \in \Theta$ at terminal time T and the corresponding values of the terminal value function and the optimal choices at these grid points. Altogether, they form the final solution S_T. Starting from T, the algorithm iterates backwards in time to determine all solutions S_t at the times $t \in \{T - 1, \ldots, 0\}$. For a given time step t, the grids associated with the previously computed solution S_{t+1} are refined for each discrete state θ_{t+1} to construct the approximation of the value function. The grids of the solution S_t are then initialized as a copy of the refined grids from the solution in $t + 1$. For all of these grid points the time t optimization problems are solved at all discrete states. Therefore, the refined approximations of the value functions for each discrete state are used within the objective function $\tilde{j}_t$ that is maximized to determine the value function values and optimal choices at the grid points of time step t, see Eqs. (7) and (9). Note that I assume that the set of discrete states Θ remains the same at all times and only the transition probabilities $\pi_t(\theta_{t+1}|\theta_t)$ may vary. Also, the solution in t could be initialized with a copy of the base level grids instead of taking the grids from the previous solution. Thus, I assume that the grids of the solution in t and $t + 1$ are likely to be similar.

Algorithm 2: Spatially adaptive sparse grid dynamic programming

Data:
Base level l
Final solution $S_T = \{(s_T^{k,i},\ j_T(s_T^{k,i}, \theta_T),\ p_T(s_T^{k,i}, \theta_T)) \mid s_T^{k,i} \in G_l^T(\theta_T),\ \theta_T \in \Theta\}$
Refinement tolerance ϵ
Result:
Solutions $S_t = \{(s_t^{k,i},\ j_t(s_t^{k,i}, \theta_t),\ p_t(s_t^{k,i}, \theta_t)) \mid s_t^{k,i} \in G_l^t(\theta_t),\ \theta_t \in \Theta\}, t \in \{0, \ldots, T-1\}$

```
for t := T − 1 to 0 do
    if t + 1 < T then
        S_{t+1} := RefineGridsInSolution(S_{t+1}, S_{t+2}, ε);
    S_t := ∅;
    for θ_t ∈ Θ do
        G_l^t(θ_t) := G_l^{t+1}(θ_t);
        // Note that Θ is constant over time
        for s_t^{k,i} ∈ G_l^t(θ_t) do
            p_t(s_t^{k,i}, θ_t) := arg max_{p_t} { j̃_t(p_t, s_t^{k,i}, θ_t) } subject to (5)–(6);
            // Here, the approximation on the previously refined
               grid is evaluated
            j_t(s_t^{k,i}, θ_t) := j̃_t(p_t(s_t^{k,i}, θ_t), s_t^{k,i}, θ_t);
            S_t := S_t ∪ (s_t^{k,i}, j_t(s_t^{k,i}, θ_t), p_t(s_t^{k,i}, θ_t));
```

In fact, to refine the grids of S_{t+1} for all discrete states, Algorithm 3 starts at the base grid G_l. Note that, by definition of the grids G_l^{t+1} in S_{t+1} as the copy of the grids G_l^{t+2} in S_{t+2} and subsequent computation of the value function values and optimal choices for the grid points of S_{t+1}, the grid points of the base grid G_l as well as the value function values and optimal choices at these grid points are known. That is, $\{(s_{t+1}^{k,i},\ j(s_{t+1}^{k,i}, \theta_{t+1}),\ p(s_{t+1}^{k,i}, \theta_{t+1})) \mid s_{t+1}^{k,i} \in G_l(\theta_{t+1}),\ \theta_{t+1} \in \Theta\} \subset S_{t+1}$. The base grid is then refined by adding all $2d$ children $s_{t+1}^{\tilde{k},\tilde{i}}$ as defined in Sect. 3. Starting from the base grid ensures that a minimum coverage of the domain by grid points is maintained. If the value function value and the optimal choices for an added grid point are already contained in the solution S_{t+1}, they are kept in the solution. This implies that the added grid point was already part of the solution S_{t+1}. If not, i.e., if the added grid point does not yet exist in the grid of S_{t+1} for a given $\theta_{t+1} \in \Theta$, the value function and the optimal choices for these grid points have to be added to the solution S_{t+1} by solving the optimization problem at these grid points. This, of course, requires the solution S_{t+2}, which has been refined already. At the end, the approximation of the value function is fitted to all grid points and value function values. This process is repeated until no more grid points are refined given the refinement tolerance ϵ. All grid points that were part of the initial solution to

Algorithm 3: Refine grids in solution

Data: Solutions S_{t+1}, S_{t+2}, refinement tolerance ϵ
Result: Solution $\widehat{S}_{t+1}$ with refined grids
for $\theta_{t+1} \in \Theta$ **do**
    $\widehat{G}_l^{t+1}(\theta_{t+1}) := G_l$;
    $\widehat{S}_{t+1} := \{(\boldsymbol{s}_{t+1}^{\boldsymbol{k},\boldsymbol{i}},\ j(\boldsymbol{s}_{t+1}^{\boldsymbol{k},\boldsymbol{i}}, \theta_{t+1}),\ \boldsymbol{p}(\boldsymbol{s}_{t+1}^{\boldsymbol{k},\boldsymbol{i}}, \theta_{t+1})) \mid \boldsymbol{s}_{t+1}^{\boldsymbol{k},\boldsymbol{i}} \in \widehat{G}_l^{t+1}(\theta_{t+1}),\ \theta_{t+1} \in \Theta\} \subset S_{t+1}$;
    repeat
        $G^{\text{tmp}} := \widehat{G}_l^{t+1}(\theta_{t+1})$;
        for $\boldsymbol{s}_{t+1}^{\boldsymbol{k},\boldsymbol{i}} \in$ `Leafs`$(\widehat{G}_l^{t+1}(\theta_{t+1}))$ **do**
            if $|c_{\boldsymbol{k},\boldsymbol{i}}| \cdot \|\phi_{\boldsymbol{k},\boldsymbol{i}}\| > \epsilon$ **then**
                $\widehat{G}_l^{t+1}(\theta_{t+1}) := \widehat{G}_l^{t+1}(\theta_{t+1}) \bigcup$ `RefinePoint`$(\boldsymbol{s}_{t+1}^{\boldsymbol{k},\boldsymbol{i}})$;
                `// Add all children and all missing parents`
        for $\boldsymbol{s}_{t+1}^{\tilde{\boldsymbol{k}},\tilde{\boldsymbol{i}}} \in \left\{\widehat{G}_l^{t+1}(\theta_{t+1}) \backslash G^{\text{tmp}}\right\}$ **do**
            if $\boldsymbol{s}_{t+1}^{\tilde{\boldsymbol{k}},\tilde{\boldsymbol{i}}} \in G_l^{t+1}$ **then**
                $X := (\boldsymbol{s}_{t+1}^{\tilde{\boldsymbol{k}},\tilde{\boldsymbol{i}}},\ j_{t+1}(\boldsymbol{s}_{t+1}^{\tilde{\boldsymbol{k}},\tilde{\boldsymbol{i}}}, \theta_{t+1}),\ \boldsymbol{p}_{t+1}(\boldsymbol{s}_{t+1}^{\tilde{\boldsymbol{k}},\tilde{\boldsymbol{i}}}, \theta_{t+1})) \in S_{t+1}$;
            else
                $\boldsymbol{p}_{t+1}(\boldsymbol{s}_{t+1}^{\tilde{\boldsymbol{k}},\tilde{\boldsymbol{i}}}, \theta_{t+1}) := \arg\max_{\boldsymbol{p}_{t+1}} \left\{\tilde{j}_{t+1}(\boldsymbol{p}_{t+1}, \boldsymbol{s}_{t+1}^{\tilde{\boldsymbol{k}},\tilde{\boldsymbol{i}}}, \theta_{t+1})\right\}$ subject to (5)–(6);
                $j_{t+1}(\boldsymbol{s}_{t+1}^{\tilde{\boldsymbol{k}},\tilde{\boldsymbol{i}}}, \theta_{t+1}) := \tilde{j}_{t+1}(\boldsymbol{p}_{t+1}(\boldsymbol{s}_{t+1}^{\tilde{\boldsymbol{k}},\tilde{\boldsymbol{i}}}, \theta_{t+1}), \boldsymbol{s}_{t+1}^{\tilde{\boldsymbol{k}},\tilde{\boldsymbol{i}}}, \theta_{t+1})$;
                $X := (\boldsymbol{s}_{t+1}^{\tilde{\boldsymbol{k}},\tilde{\boldsymbol{i}}},\ j_{t+1}(\boldsymbol{s}_{t+1}^{\tilde{\boldsymbol{k}},\tilde{\boldsymbol{i}}}, \theta_{t+1}),\ \boldsymbol{p}_{t+1}(\boldsymbol{s}_{t+1}^{\tilde{\boldsymbol{k}},\tilde{\boldsymbol{i}}}, \theta_{t+1}))$;
                `// If the added grid point is not already in the grids of solution` S_{t+1}`, compute the value function and the optimal choices at this grid point using` S_{t+2}
            $\widehat{S}_{t+1} := \widehat{S}_{t+1} \bigcup X$;
        `// Fit approximation to the updated grid points and value function values`
    until $\widehat{G}_l^{t+1}(\theta_{t+1}) \backslash G^{\text{tmp}} = \emptyset$;
    `// ... there are no grid points to be refined given` ϵ

be refined, but have not been added in the refinement process, are discarded.[6] This ensures that the number of grid points in the solutions is not monotonously growing as the algorithm is stepping backwards in time.

Overall, Algorithms 2 and 3 guarantee that the value function approximation for each time step and every discrete state is spatially adaptively refined and the value function values and optimal choices at all of these grid points are part of the respective solution. Discarding points that are not needed for a good approximation

[6]Aside from additionally added parents, this is similar to removing all grid points that are leafs below level l in the hierarchical grid structure and fulfill the criterion $|c_{l,i}| \leq \epsilon / \|\phi_{l,i}\|_2$, so called *coarsening*.

of the value function within the refinement procedure minimizes the number of optimization problems that have to be solved.

4.2 Optimal Choices

Given the solution of the spatially adaptive sparse grid dynamic programming algorithm 2 optimal choice approximations on a sparse grid can be constructed and individually adapted, again by spatially adaptive refinement. This is sensible when the optimal choices have high curvature in parts of the domain where the value function is comparably flat and, hence, the solution to the optimization problem contains only few grid points in these parts.

Since $\boldsymbol{p}_t(s_t, \theta_t) \in \mathbb{R}^k$, there are k optimal choices to approximate for a given time t and discrete state $\theta_t \in \Theta$. Consistent with the notion of a solution to the optimization problem S_t, I define the uth optimal choice at time t by a set of tuples $P_t^u = \{(\boldsymbol{s}_t^{k,i}, p_t^u(\boldsymbol{s}_t^{k,i}, \theta_t)) \mid \boldsymbol{s}_t^{k,i} \in G_l^{t,u}(\theta_t), \theta_t \in \Theta\}$, $t \in \{0, \ldots, T\}$. For each $t \in \{0, \ldots, T\}$, Algorithm 4 takes the optimal choices $p_t^u(\boldsymbol{s}_t^{k,i}, \theta_t)$ at all grid points from the solution S_t to construct an initial P_t^u, $u = 1, \ldots, k$. Since these grids are of hierarchical structure, it is possible to construct a sparse grid approximation of the optimal choices. These initial optimal choices are subsequently refined using surplus-based refinement for a given choice dependent refinement constant ϵ_u. Since the intersection of the grid points added in all refined optimal choices grids is frequently large, it is beneficial to compute the optimal choices at all newly added grid points for the union set of these points.

Note that in the initialization of the optimal choices from the solution and during the refinement procedure no grid points are discarded. They contain information useful for the approximation and, contrary to the refinement of the value function in Algorithm 3, no costly operation (like optimization) has to be performed on the set of grid points for any optimal choice.[7]

4.3 Choice of Basis Functions, Extrapolation, and Refinement

Let $G \subset \mathbb{R}^d$ be a bounded domain within the continuous state space. If the state space is unbounded, which is often the case, a boundary has to be chosen a priori, since the sparse grid approximation using local basis functions requires a bounded domain. Let the boundary of this domain be δG.

[7]This is true for constructing approximations of the optimal choices using linear basis functions. Using different approximation methods on a sparse grid can involve significant computational effort for fitting the approximation to the function values, e.g., using conventional B-spline basis functions.

Algorithm 4: Generate optimal choices

Data:
Solutions $S_t = \{(\boldsymbol{s}_t^{k,i},\ j_t(\boldsymbol{s}_t^{k,i}, \theta_t),\ \boldsymbol{p}_t(\boldsymbol{s}_t^{k,i}, \theta_t)) \mid \boldsymbol{s}_t^{k,i} \in G_l^t(\theta_t),\ \theta_t \in \Theta\}, t \in \{0, \ldots, T\}$
Refinement tolerances ϵ_u, $u = 1, \ldots, k$
Result:
Optimal choices $P_t^u = \{(\boldsymbol{s}_t^{k,i},\ p_t^u(\boldsymbol{s}_t^{k,i}, \theta_t)) \mid \boldsymbol{s}_t^{k,i} \in G_l^{t,u}(\theta_t),\ \theta_t \in \Theta\}, t \in \{0, \ldots, T\}$

for $t := 0$ **to** T **do**
  for $\theta_t \in \Theta$ **do**
    for $u := 1$ **to** k **do**
      $G_l^{t,u}(\theta_t) := G_l^t(\theta_t)$;
      $P_t^u := \emptyset$;
      for $\boldsymbol{s}_t^{k,i} \in G_l^{t,u}(\theta_t)$ **do**
        $P_t^u := P_t^u \bigcup (\boldsymbol{s}_t^{k,i},\ \cdot,\ p_t^u(\boldsymbol{s}_t^{k,i}, \theta_t),\ \cdot) \in S_t$
    `// Initialize the optimal choice as obtained from the solution and fit the approximation to the values at the grid points`
    repeat
      $G^{\text{tmp}} := \bigcup_{u=1}^k G_l^{t,u}(\theta_t)$;
      for $u := 1$ **to** k **do**
        for $\boldsymbol{s}_t^{k,i} \in$ `Leafs`$(G_l^{t,u}(\theta_t))$ **do**
          if $|c_{\boldsymbol{k},\boldsymbol{i}}| \cdot \|\phi_{\boldsymbol{k},\boldsymbol{i}}\| > \epsilon_u$ **then**
            $G_l^{t,u}(\theta_t) := G_l^{t,u}(\theta_t) \bigcup$ `RefinePoint`$(\boldsymbol{s}_t^{k,i})$;
            `// Add all children and all missing parents`
      $G^{\text{union}} := \bigcup_{u=1}^k G_l^{t,u}(\theta_t)$;
      for $\boldsymbol{s}_t^{\tilde{k},\tilde{i}} \in \left\{G^{\text{union}} \backslash G^{\text{tmp}}\right\}$ **do**
        `// Compute optimal choices for the union of all added grid points`
        $\boldsymbol{p}_t(\boldsymbol{s}_t^{\tilde{k},\tilde{i}}, \theta_t) := \arg\max_{\boldsymbol{p}_t} \left\{\tilde{j}_t(\boldsymbol{p}_t, \boldsymbol{s}_t^{\tilde{k},\tilde{i}}, \theta_t)\right\}$ subject to (5)–(6);
        for $u := 1$ **to** k **do**
          if $\boldsymbol{s}_t^{\tilde{k},\tilde{i}} \in G_l^{t,u}(\theta_t)$ **then**
            $P_t^u := P_t^u \bigcup \{(\boldsymbol{s}_t^{\tilde{k},\tilde{i}},\ p_t^u(\boldsymbol{s}_t^{\tilde{k},\tilde{i}}, \theta_t))\}$;
            `// Here, take only the grid points that have been added to the grid for optimal choice` u
      `// Fit all` k `approximations against the updated grid points and optimal choices`
    until $G^{\text{union}} \backslash G^{\text{tmp}} = \emptyset$;

Firstly, the choice of the domain may introduce an approximation error. If the boundary is chosen too tight, a relevant part of the state space may lay outside the domain. If the boundary is chosen too loose, a high discretization level of the sparse grid approximation is necessary to obtain a reasonable approximation error.

As already mentioned in Sect. 2, the state space mapping $\boldsymbol{f}_{t+1}$ does not correspond with a grid point and may map to points outside the domain G. If the value function is defined outside the bounded domain, the value function has to be extrapolated, e.g., by nearest neighbor extrapolation or other means.[8] To approximate a function value $f(\boldsymbol{x}_\mathrm{o})$ outside the domain $G \subset \mathbb{R}^d$ at a point $\boldsymbol{x}_\mathrm{o} \notin G$, I extrapolate linearly. Let $\boldsymbol{x}_\mathrm{b} \in \delta G$ be the closest point on the boundary of the domain, i.e., the point that solves $\min_{\boldsymbol{x} \in \delta G} \|\boldsymbol{x}_\mathrm{o} - \boldsymbol{x}\|_2$ with the standard euclidean norm $\|\cdot\|_2$. Furthermore, let $\boldsymbol{h} \in \mathbb{R}^d$ be an extrapolation accuracy. Then, the linearly extrapolated approximation is computed as

$$\begin{aligned}\mathscr{A}[f](\boldsymbol{x}_\mathrm{o}) =& \mathscr{A}[f](\boldsymbol{x}_\mathrm{b}) \\ &+ \sum_{j=1}^{d} \left(\frac{\mathscr{A}[f](\boldsymbol{x}_\mathrm{b} - \boldsymbol{h} \cdot \mathbf{1}_j) - \mathscr{A}[f](\boldsymbol{x}_\mathrm{b})}{h_j} (x_{\mathrm{o}_j} - x_{\mathrm{b}_j}) \mathbb{1}_{\{x_{\mathrm{o}_j} > x_{\mathrm{b}_j}\}} \right) \\ &+ \sum_{j=1}^{d} \left(\frac{\mathscr{A}[f](\boldsymbol{x}_\mathrm{b} + \boldsymbol{h} \cdot \mathbf{1}_j) - \mathscr{A}[f](\boldsymbol{x}_\mathrm{b})}{h_j} (x_{\mathrm{b}_j} - x_{\mathrm{o}_j}) \mathbb{1}_{\{x_{\mathrm{o}_j} < x_{\mathrm{b}_j}\}} \right),\end{aligned}$$

with $\mathbb{1}$ denoting the indicator function.

As the value function and the optimal choices are non-zero on the boundary, I employ linear basis functions with grid points on the boundary, see [16]. For the adaptive refinement, I use hierarchical surplus volume-based refinement with the linear boundary basis functions, as the theory behind hierarchical surplus volume-base refinement is well studied [20, 37, 45].[9]

[8]The extrapolated value can be crucial. Consider the following simple model: In period $t \ll T$ of investment horizon T an investor can choose how much to invest from her wealth w_t in a stock giving a risky return r_{t+1} from t to $t+1$ and how much to consume c_t. The investor tries to maximize CRRA utility from consumption over the time horizon. Her wealth in $t+1$ is thus $w_{t+1} = (w_t - c_t) \cdot r_{t+1}$. The wealth depends on the distribution of returns and is unbounded, such that an upper boundary w_u of the bounded state space $[0, w_u] \subset \mathbb{R}$ has to be chosen in order to compute a numerical solution to the problem. Assume that her decision in t at the boundary point $w_t = w_u$ shall be optimized. Also, it should be clear that higher values of wealth w_t correspond with higher value function values. For any return realization $r_{t+1} > w_t/(w_t - c_t)$ the next period's wealth w_{t+1} lies outside the bounded domain. With nearest neighbor extrapolation, the value function value for w_{t+1} is the same as for w_u. Investing in the risky stock is not rewarded beyond w_u. With linear extrapolation, investing into the risky stock is rewarded (assuming a positive slope of the value function) and her consumption c_t would be considerably lower compared to the nearest neighbor extrapolation.

[9]I also tested the modified linear basis functions from [37], which extrapolate linearly towards the boundary by folding up the outermost basis functions instead of placing grid points on the boundary. Extrapolation outside the boundary then boils down to extending the outermost basis functions to the outside of the domain [2]. The multi-dimensional modified basis is again the product of the one-dimensional modified basis functions and thus extrapolation in the corners of the d-dimensional cube is d-linear. Conveniently, the modified basis functions make up for the fact that in higher dimensions the majority of the grid points lies on the boundaries. However, the extrapolation towards the boundary—especially in the corners—led to the value function not

Since the optimization is performed on an approximation of the value function using linear hat functions, the basis suffers from the global non-differentiability of piecewise continuously differentiable basis functions. Any gradient-based numerical optimization routine is likely to converge poorly if the optimized function has discontinuous derivatives. However, it might still find a global optimum more effectively than a derivative-free optimizer as long as the value function is convex and monotone. For the numerical examples in this paper, I prefer to use a gradient-based optimizer and terminate the optimization after a certain number of steps over an alternative, derivative-free optimizer (such as pattern search or simulated annealing). To circumnavigate problems with non-differentiability on a sparse grid with linear hat functions, various approaches exist [14, 31]. Local basis functions that are globally smooth and continuously differentiable, such as B-splines [43], pose a promising alternative that still allows for spatially adaptive refinement of the value function.

5 Numerical Example

As numerical example I use a variant of the transaction costs problem analyzed in detail in the dissertation of Yongyang Cai [7] with the parametrization described in the related paper [8]. In the analyzed model, the investor takes transaction costs for multiple stock investment opportunities into account when optimizing her portfolio and consumption choice. These stock investments differ in risk and return characteristics. For each stock in the model, one continuous state variable, two choice variables ("buy" and "sell" amounts), and one stochastic risk factor (the normally distributed logarithmic return shock) is required. The result is a high-dimensional model in terms of the state space, stochastic sample space, and choice variables. It is easy to vary its dimensionality by increasing the number of stock investment options available, see Sect. 5.1.

To separate the numerical error resulting from the optimization and the approximation error in the value function approximation, I employ two error measures. The first measures how much the numerically obtained choices violate the optimality conditions. It thus allows to make a general statement on the quality of the numerical solution to the optimization problem. The second error measure computes the pointwise error of the value function approximation obtained from Algorithm 2 with a spatially adaptively refined sparse grid and a spatially adaptively refined full grid reference solution to the same problem. Since in the Lagrangian optimization setup only the gradient of the value function is relevant, small errors on the policies do not necessarily imply small pointwise errors of the value function. Details are given in Sect. 5.2.

I present and discuss my results in Sect. 5.3.

---

being monotone for certain cases and subsequently to failure of the optimization routine. Hence, the modified basis does not constitute a generally reliable choice of basis functions.

5.1 Transaction Costs Model

In my version of the transaction costs problem I consider a retired investor who maximizes expected utility from consumption (1). Therefore, she tracks the continuous states wealth $w_t \in \mathbb{R}_+$ and fractions of wealth $\boldsymbol{x}_t = (x_t^1, \ldots, x_t^d)^\top \in [0, 1]^d$ invested in stocks $S_1, \ldots, S_d$. The investor can reside in the two discrete states $\theta_t \in \{\text{alive, dead}\}$ and stays alive in the subsequent period with probability π_t. Her choices are how much to buy of stock i for USD amount $\delta_t^{i+} \in \mathbb{R}_+$ with transaction costs $\tau\delta_t^{i+}$ or sell of stock i for USD amount $\delta_t^{i-} \in \mathbb{R}_+$ and transaction costs $\tau\delta_t^{i-}$, where $\tau > 0$ is a percentage cost factor. Additionally, she can invest in a transaction cost free money market account b_t yielding a risk-free return r_f. With $\boldsymbol{\delta}_t^{\pm} \in \mathbb{R}_+^d$ her choices are given by $(\boldsymbol{\delta}_t^+, \boldsymbol{\delta}_t^-, b_t)^\top$. The total transaction costs associated with the implementation of the investor's choices are thus given by:

$$\tau \sum_{i=1}^{d} (\delta_t^{i+} + \delta_t^{i-}) .$$

In line with [7, 8], I assume the logarithms of the returns $\boldsymbol{r}_{t+1}$ of the d stocks are iid normally distributed with mean $\boldsymbol{\mu}$ and covariance matrix Σ, $\log(\boldsymbol{r}_{t+1}) \sim N(\boldsymbol{\mu}, \Sigma)$. Finally, the investor's consumption in period t is the residual of her wealth that is not invested in stocks or bonds, reduced by the transaction costs for rearranging her portfolio in this period:

$$c_t = (1 - \sum_{i=1}^{d} x_t^i) w_t - b_t - (1+\tau) \sum_{i=1}^{d} \delta_t^{i+} - (\tau - 1) \sum_{i=1}^{d} \delta_t^{i-} . \tag{17}$$

Additionally, the investor receives a fixed retirement income α_R. The state dynamics from t to $t+1$ are thus given by:

$$w_{t+1} = \sum_{i=1}^{d} \left(x_t^i w_t + \delta_t^{i+} - \delta_t^{i-} \right) r_{t+1}^i + b_t r_f + \alpha_R ,$$

$$x_{t+1}^i = \frac{\left(x_t^i w_t + \delta_t^{i+} - \delta_t^{i-} \right) r_{t+1}^i}{w_{t+1}} , \quad i = 1, \ldots, d .$$

With u being the CRRA utility (2) the investor faces the optimization problem

$$j_t(w_t, \boldsymbol{x}_t) = \max_{b_t, \boldsymbol{\delta}_t^+, \boldsymbol{\delta}_t^-} u(c_t) + \rho \pi_t \mathscr{E}_t \left[j_{t+1}(w_{t+1}, \boldsymbol{x}_{t+1}) \right] , \tag{18}$$

$$j_T(w_T, \boldsymbol{x}_T) = u \left(\left(1 - \tau \sum_{i=1}^{d} x_T^i \right) w_T \right) , \tag{19}$$

subject to the constraints for each $t \in \{0, \ldots, T\}$

$$b_t + (1+\tau)\sum_{i=1}^{d} \delta_t^{i+} + (\tau - 1)\sum_{i=1}^{d} \delta_t^{i-} \le w_t \left(1 - \sum_{i=1}^{d} x_t^i\right) - c_{\min}\,, \tag{20}$$

$$\delta_t^{i+} \ge 0\,, \qquad i = 1, \ldots, d\,, \tag{21}$$

$$\delta_t^{i-} \in \left[0, x_t^i w_t\right]\,, \qquad i = 1, \ldots, d\,, \tag{22}$$

$$b_t \ge 0\,, \tag{23}$$

$$\sum_{i=1}^{d} x_t^i \le 1\,. \tag{24}$$

Here, a minimum consumption level $c_{\min}$ must be attained and I assume that the final stock holdings $x_T^i w_T, i = 1, \ldots, d$, must be sold before they can be consumed. Also, the investor cannot sell more of stock i than her current holding $x_t^i w_t$.[10] Note that the discrete state is implicitly included in the problem definition (18)–(24) since $\mathscr{E}_t[j_{t+1}(\cdot, \text{dead})] = 0$ and hence $\mathscr{E}_t[j_{t+1}(\cdot, \text{alive})] = \pi_t \mathscr{E}_t[j_{t+1}(\cdot)]$.

The constraint (24) is a constraint on the state space and the resulting eligible subspace in $[0, 1]^d$ is a d-dimensional simplex, not a rectangular domain as needed for the sparse grid approximation. I solve this problem by assuming that any state attained that is not eligible is cropped to an eligible state by selling all stock holdings pro rata until all constraints are satisfied. That is, money is transferred from stocks to wealth, for which the proportionate transaction costs are deducted. The approximation of the value function is then evaluated at this eligible state.

The optimization starts at age $T = 71$, ends at age $t_0 = 65$, and has a fixed period length of $\Delta t = 1$ year. Thus, I assume a 6 years investment horizon to be comparable with the run times reported in [8]. The retirement income is $\alpha_R = 18{,}664$ USD and taken from tables one and two of [12]. For the 1-year conditional mortality π_t I use the 2009 US female mortality table. The risk aversion γ, the log-return distribution parameters $\boldsymbol{\mu}$ and Σ, as well as the transaction costs factor τ are taken from page seven of [8]. I set the time discount factor to $\rho = 0.97$ and $c_{\min} = 0.001$ to ensure minimal consumption is taking place.

I track wealth w_t in 10,000 USD, truncate the infinite domain to $[0.05, 5]$, and transform via $m(w_t) = w_t^{1/3}$ to increase the density of grid points in regions were the value function has a high curvature, that means for lower wealth levels.

[10] To save choice variables one could track the net change $\Delta_t \in \mathbb{R}^d$ in the investor's stock holding. However, the transaction costs function $\tau \sum_{i=1}^{d} |\Delta_t^i|$ would be non-linear and the constraints become non-linear, too; see [7].

5.2 Error Measurement

The optimal choices $\boldsymbol{p}_t(\boldsymbol{s}_t, \theta_t) \in \mathbb{R}^k$ must satisfy the first order conditions of the Lagrangian (8)

$$\nabla_{\boldsymbol{p}_t} u + \rho \mathscr{E}_t \left[\nabla_{\boldsymbol{p}_t} \boldsymbol{f}_{t+1} \cdot \nabla_{\boldsymbol{f}_{t+1}} j_{t+1}\right] - \nabla_{\boldsymbol{p}_t} \boldsymbol{g} \cdot \boldsymbol{\lambda}_t - \nabla_{\boldsymbol{p}_t} \boldsymbol{h} \cdot \boldsymbol{\mu}_t = \boldsymbol{0}$$

and any of these k equations must hold for the optimal choices.[11] In order to eliminate the derivative of the value function $\nabla_{\boldsymbol{s}_{t+1}} j_{t+1}$ with regard to the state $\boldsymbol{s}_{t+1} \in \mathbb{R}^d$, which is not known and would have to be computed from the value function approximation, it is possible to construct a constant matrix $C \in \mathbb{R}^{d \times k}$, which relates $\nabla_{\boldsymbol{s}_t} j_t$ to $\nabla_{\boldsymbol{p}_t} u$ via

$$\nabla_{\boldsymbol{s}_t} j_t = C \cdot \nabla_{\boldsymbol{p}_t} u \,.$$

Thus, I can replace $\nabla_{\boldsymbol{f}_{t+1}} j_{t+1}$ by $C \cdot \nabla_{\boldsymbol{p}_{t+1}} u$. By reformulation I arrive at the $d + 1$ error measures with respect to the d stock returns $\boldsymbol{r}_{t+1}$ and one risk-free return r_f with c_t from Eq. (17):

$$e_{r^i_{t+1}}(w_t, \boldsymbol{x}_t) = \rho \pi_t \mathscr{E}_t \left[\frac{c_{t+1}^{-\gamma}}{c_t^{-\gamma}} r^i_{t+1} \right] - 1 \,, \quad i = 1, \ldots, d \,, \tag{25}$$

$$e_{r_f}(w_t, \boldsymbol{x}_t) = \rho \pi_t \mathscr{E}_t \left[\frac{c_{t+1}^{-\gamma}}{c_t^{-\gamma}} r_f \right] - 1 \,. \tag{26}$$

These errors are frequently called *Euler errors*. Note that these errors are the deviations from the basic pricing equation of asset pricing [13]. For details see the appendix.

To measure the numerical error $e_{X_t}(w_{t+1}, \boldsymbol{x}_{t+1})$ for $X_t \in \{r^1_{t+1}, \ldots, r^d_{t+1}, r_f\}$ at time t, the Euler error equations (25) and (26) are evaluated on an equidistant grid with 23 points in each dimension covering the bounded domain $[0.05, 5] \times \{\boldsymbol{x} \in [0, 1]^d \mid \sum_{i=1}^d x^i_t \leq 1\}$.[12] In case any of the constraints (20)–(23) is binding at a

[11] I define $\nabla_{\boldsymbol{x}} \boldsymbol{f} := (\partial f_j / \partial x_i)_{ij}$.

[12] Using more grid points to compute the error would allow for a better estimation of the error. However, for more than two stocks (three dimensions), the error computation takes unacceptably long. To be consistent with the error estimation when varying the problem dimensionality, I use this rather low number of grid points.

grid point, the Euler error is set to "NaN".[13] Denoting the index set $I_{X_t}^{\text{Err}}$ of the grid points $(w_t^i, \boldsymbol{x}_t^i)$, $\boldsymbol{i} \in I_{X_t}^{\text{Err}}$, with non-NaN errors for X_t and the cardinality of a given finite set by $\#I$, the aggregated error measures $L_1^{X_t}$, $L_2^{X_t}$, and $L_\infty^{X_t}$ can be computed in the usual sense:

$$L_1^{X_t} = \frac{1}{\#I_{X_t}^{\text{Err}}} \sum_{i \in I_{X_t}^{\text{Err}}} \left| e_{X_t}\left(w_t^i, \boldsymbol{x}_t^i\right)\right| ,$$

$$L_2^{X_t} = \left(\frac{1}{\#I_{X_t}^{\text{Err}}} \sum_{i \in I_{X_t}^{\text{Err}}} e_{X_t}\left(w_t^i, \boldsymbol{x}_t^i\right)^2 \right)^{\frac{1}{2}} ,$$

$$L_\infty^{X_t} = \max_{i \in I_{X_t}^{\text{Err}}} \left\{ \left| e_{X_t}\left(w_t^i, \boldsymbol{x}_t^i\right)\right| \right\} .$$

To compute the overall aggregated Euler errors for the optimal choices, I average the respective error measure over all times $t \in \{0, \ldots, T-1\}$ (since for the Euler error in t the period $t+1$ is required). The final Euler errors L_1^{Euler}, L_2^{Euler}, and L_∞^{Euler} are then computed as the average of the time-average errors for all X_t. I consider L_∞^{Euler} errors around 20% and L_1^{Euler} about 5% as economically acceptable. They indicate an average maximum mispricing of the assets of roughly 20% and an average absolute mispricing of around 5%.

The pointwise error for a given state is the absolute difference between the approximation $\mathscr{A}[j_t]$ of the value function of the full grid reference solution and the sparse grid approximation $\mathscr{A}^S[j_t]$:

$$e_{j_t}(w_t, \boldsymbol{x}_t) = \left| \mathscr{A}^S[j_t] - \mathscr{A}[j_t] \right| . \tag{27}$$

I proceed in a similar manner as with the Euler errors and compute the pointwise value function errors (27) at the same grid points as the Euler errors for each point in time $t \in \{0, \ldots, T\}$ and average over all times to determine L_1^{Value}, L_2^{Value}, and L_∞^{Value}. For the pointwise error of the value function there are no binding constraints.

[13] Additionally, I set the Euler error to "NaN", if $w_t x_t^i + \delta_t^{i+} - \delta_t^{i-} = 0$, $i = 1, \ldots, d$. This constraint results from the derivation of the Euler error equation described in the appendix. I consider a constraint to be binding when its absolute value is within the constraint tolerance of the numerical SQP optimizer.

5.3 Results

All solutions have been computed using the certainty equivalent transformation of the value function $\hat{j}_t = ((1-\gamma)j_t)^{1/(1-\gamma)}$, which reduces the curvature of the value function when the utility is of CRRA type.[14] Since this transform is strictly monotone, any maximizer of $\hat{j}_t$ also maximizes j_t. The code is written in MATLAB, where the approximation on sparse and full grids is implemented by a MEX file interface to the sparse grids C++-toolbox SG++ that was originally developed in the course of the dissertation of Dirk Pflüger [37]. Note that the surplus volume in dimensionality d is computed as $\|\phi_{l,i}\|_2 = (2/3)^{d/2} 2^{-|l|_1/2}$, but SG++ uses as $2^{-|l|_1}$ as the denominator in the refinement criterion (16). I do not correct the refinement constant ϵ for this factor. Since for the transaction costs problem the distribution Φ_t is lognormal and state independent, I compute the expectation (9) using Gauß-Hermite quadrature on a sparse grid, thus breaking the curse of dimensionality when including stochastic risk factors. Sparse grid quadrature is used, but not subject to analysis within this paper. For details on sparse grid quadrature in dynamic portfolio choice models, see the appendix of Horneff et al. [25]. The quadrature routine is implemented by a MEX file interface to the TASMANIAN sparse grids C++-toolbox [42]. The optimization is done using the SQP solver SNOPT [18] in the implementation of the Numerical Algorithms Group (www.nag.co.uk). Often convergence of the optimizer cannot be observed due to the discontinuous gradients of the linear approximation. In these cases I stop the optimization after 100 iterations.

Firstly, I solve the transaction costs problem (18)–(24) with one stock (two-dimensional problem) up to three stocks (four-dimensional problem) using Algorithm 2 on a full grid of base level $l = 3$ with spatially adaptive refinement of the value function as described in Algorithm 3 with refinement constant $\epsilon = 5\mathrm{e}{-04}$. Given the solutions to the optimization problem, I then compute approximations for the optimal choices according to Algorithm 4 without spatially adaptive refinement of the optimal choices, which is discussed later in this section. Finally, I compute the Euler errors (25) and (26) using the approximations of the optimal choices as described in Sect. 5.2. The full grid solutions and their Euler errors serve as my reference solutions to compare against the Euler errors of the sparse grid solution and to compute the pointwise error of the value function.

I then solve the two-, three-, and four-dimensional transaction costs problem with Algorithm 2 on a sparse grid. Again, I use the optimal solution obtained this way to compute approximations for the optimal choices according to Algorithm 4 without spatially adaptive refinement. Given these optimal choices, I compute the Euler errors in the same manner as with the full grid solution. Finally, I compute the pointwise error of the value function on the sparse grid compared to the full grid reference solution's value function as given by Eq. (27).

[14]Note that the final condition becomes $\hat{j}_T = ((1-\gamma)/(1-\gamma)c_T^{1-\gamma})^{1/(1-\gamma)} = c_T$.

I determine the Degrees of Freedom (DoF) as the average of the grid points of all grids G_l^t, $t \in \{0, \ldots, T\}$. So if grid G_l^t has $\#G_l^t$ points, I compute the DoF as $\lceil 1/T \sum_{t=0}^{T} \#G_l^t \rceil$. In my implementation, I parallelize over the number of grid points and use a fixed number of cores per problem dimensionality, but more cores for higher dimensionality. For parallelization I use MATLAB's `parfor`. Since the parallel efficiency of `parfor` decreases drastically when too few DoFs are allocated per core [25], I have to adapt the number of cores to the problem dimension to avoid undesired scaling effects.

I repeat these computations for each problem dimensionality with varying base levels $l \in \{2, 3, 4\}$ and refinements constants $\epsilon \in \{5\text{e}{-}01, \ldots, 5\text{e}{-}05\}$ to asses the impact of base level and refinement constant on the solution of the optimization problem. The following Tables 1, 2, and 3 depict the outcome of these experiments. I only report the absolute pointwise errors of the value function, since the relative errors are of the same magnitude as the absolute errors.

For the results of the two-dimensional transaction costs problem in Table 1, the Euler errors are overall low and the average absolute Euler errors L_1^{Euler} are acceptable around the 2.5% level for the full grid and 5% level for the different sparse grid settings with base level $l \geq 3$. For all parametrizations of the sparse grid the L_2^{Euler} and L_∞^{Euler} errors are higher than with the full grid. Increasing the base level of the sparse grid approximation of the value function leads to lower average absolute Euler errors, but does not improve much on the L_2^{Euler} and L_∞^{Euler} errors. Decreasing the refinement tolerance has seemingly little effect. For the pointwise error of the sparse grid value function compared to the reference solution L^{Value}, convergence is visible for both, increasing base level as well as decreasing refinement tolerances. However, a better approximation of the value function on a sparse grid does not necessarily correspond to better Euler errors. With respect to the magnitude of the pointwise errors, note that I do not compare the sparse grid value function approximation to an analytically known or specifically chosen benchmark, but to a possibly error prone numerical reference solution. For refinement tolerance $\epsilon = 5\text{e}{-}05$, all error measures diverge.

The results for the three-dimensional transaction costs problem in Table 2 are of similar nature as for the two-dimensional transaction costs problem. The average absolute Euler errors, however, are nearly twice as high at the 4.8% level for the full grid and 9% level for the sparse grid settings with base level $l = 3$. Again, they are not very sensitive to base level and refinement tolerance variations. The pointwise approximation error stays about the same as in the two-dimensional problem and converges until $\epsilon = 5\text{e}{-}05$, for which Euler errors and pointwise errors start diverging. The DoF are clearly lower in the sparse grid cases and so are the run times compared to the full grid case.

This trend continues for the four-dimensional transaction costs model, see Table 3. The average absolute Euler errors are now at the 7.6% level for the full grid and at the 18% level for the sparse grid with base level $l = 3$. The sparse grid value function converges to its full grid counterpart with increasing base level and decreasing refinement tolerance, but the pointwise errors are slightly higher as

Table 1 Two-dimensional transaction costs problem

l	ϵ	DoF	Run time	L_1^{Euler}	L_2^{Euler}	L_∞^{Euler}	L_1^{Value}	L_2^{Value}	L_∞^{Value}
3	5e−04	109	7.95	2.57e−02	6.19e−02	2.46e−01	–	–	–
2	5e−01	21	3.16	8.95e−02	1.49e−01	4.26e−01	1.78e−02	2.01e−02	3.10e−02
2	5e−02	23	2.45	8.63e−02	1.47e−01	4.25e−01	1.28e−02	1.47e−02	2.49e−02
2	5e−03	34	2.87	7.06e−02	1.47e−01	4.96e−01	2.82e−03	3.46e−03	7.94e−03
2	5e−04	63	4.36	8.74e−02	2.00e−01	6.55e−01	1.65e−04	2.33e−04	6.65e−04
2	5e−05	125	9.32	1.03e−01	2.54e−01	8.89e−01	7.73e−04	8.74e−04	1.79e−03
3	5e−01	49	3.56	5.59e−02	1.24e−01	5.09e−01	3.99e−03	4.57e−03	7.59e−03
3	5e−02	49	3.97	5.59e−02	1.24e−01	5.09e−01	3.99e−03	4.57e−03	7.59e−03
3	5e−03	53	3.75	5.55e−02	1.24e−01	5.06e−01	2.60e−03	3.04e−03	6.23e−03
3	5e−04	77	5.18	6.18e−02	1.41e−01	4.97e−01	5.45e−05	9.32e−05	4.26e−04
3	5e−05	135	10.32	8.75e−02	2.12e−01	7.34e−01	7.47e−04	8.44e−04	1.69e−03
4	5e−01	113	7.96	4.85e−02	1.21e−01	4.99e−01	4.43e−04	5.45e−04	1.36e−03
4	5e−02	113	8.34	4.85e−02	1.21e−01	4.99e−01	4.43e−04	5.45e−04	1.36e−03
4	5e−03	113	8.29	4.85e−02	1.21e−01	4.99e−01	4.43e−04	5.45e−04	1.36e−03
4	5e−04	123	8.94	4.87e−02	1.22e−01	4.98e−01	6.19e−05	1.43e−04	6.93e−04
4	5e−05	172	13.42	5.37e−02	1.31e−01	5.01e−01	7.07e−04	8.00e−04	1.68e−03

Convergence for the spatially adaptive refinement scheme with varying base levels l and refinement tolerances ϵ for the value function. The reference solution on the full grid was refined with refinement tolerance $\epsilon = 5\text{e}{-04}$ and base level $l = 3$ and is given in the first line. Reported are the Euler errors L^{Euler} as well as the pointwise errors of the value function L^{Value} with respect to the reference solution. Run times are reported in seconds with respect to 4 cores

Table 2 Three-dimensional transaction costs problem

l	ϵ	DoF	Run time	L_1^{Euler}	L_2^{Euler}	L_∞^{Euler}	L_1^{Value}	L_2^{Value}	L_∞^{Value}
3	5e−04	789	172.76	4.74e−02	7.36e−02	2.89e−01	–	–	–
2	5e−01	81	12.03	1.37e−01	2.01e−01	5.61e−01	1.76e−02	1.99e−02	3.08e−02
2	5e−02	84	11.53	1.42e−01	2.10e−01	5.92e−01	1.28e−02	1.47e−02	2.51e−02
2	5e−03	108	13.69	1.71e−01	2.66e−01	7.65e−01	2.69e−03	3.34e−03	7.89e−03
2	5e−04	177	21.35	1.97e−01	3.01e−01	8.18e−01	3.84e−04	5.13e−04	1.49e−03
2	5e−05	349	41.23	1.94e−01	3.30e−01	9.28e−01	8.65e−04	1.02e−03	2.56e−03
3	5e−01	225	21.34	8.52e−02	1.48e−01	5.52e−01	4.08e−03	4.67e−03	7.70e−03
3	5e−02	225	27.86	8.52e−02	1.48e−01	5.52e−01	4.08e−03	4.67e−03	7.70e−03
3	5e−03	232	21.60	8.76e−02	1.55e−01	5.63e−01	2.58e−03	3.01e−03	6.15e−03
3	5e−04	285	27.36	9.65e−02	1.62e−01	5.67e−01	1.33e−04	2.00e−04	7.91e−04
3	5e−05	435	61.32	1.61e−01	3.20e−01	9.71e−01	6.58e−04	8.10e−04	2.12e−03
4	5e−01	593	101.23	1.01e−01	1.87e−01	6.81e−01	4.62e−04	5.91e−04	1.86e−03
4	5e−02	593	105.46	1.01e−01	1.87e−01	6.81e−01	4.62e−04	5.91e−04	1.86e−03
4	5e−03	593	106.54	1.01e−01	1.87e−01	6.81e−01	4.62e−04	5.91e−04	1.86e−03
4	5e−04	612	105.71	9.37e−02	1.75e−01	6.53e−01	1.03e−04	1.85e−04	9.58e−04
4	5e−05	720	151.01	8.67e−02	1.71e−01	6.02e−01	7.71e−04	9.09e−04	2.27e−03

Convergence for the spatially adaptive refinement scheme with varying base levels l and refinement tolerances for the value function. The reference solution on the full grid was refined with refinement tolerance $\epsilon = 5\text{e}{-04}$ and base level $l = 3$ and is given in the first line. Reported are the Euler errors L^{Euler} as well as the pointwise errors of the value function L^{Value} with respect to the reference solution. Run times are reported in seconds with respect to 16 cores

Table 3 Four-dimensional transaction costs problem

l	ϵ	DoF	Run time	L_1^{Euler}	L_2^{Euler}	L_∞^{Euler}	L_1^{Value}	L_2^{Value}	L_∞^{Value}
3	5e−04	6699	4247.78	7.56e−02	1.18e−01	5.02e−01	–	–	–
2	5e−01	297	25.51	2.84e−01	3.50e−01	7.70e−01	1.78e−02	2.01e−02	3.25e−02
2	5e−02	302	26.85	2.65e−01	3.35e−01	7.46e−01	1.25e−02	1.44e−02	2.55e−02
2	5e−03	353	39.01	2.95e−01	3.67e−01	8.02e−01	2.81e−03	3.52e−03	9.47e−03
2	5e−04	620	93.21	3.51e−01	4.46e−01	9.80e−01	5.58e−04	8.43e−04	5.33e−03
3	5e−01	945	142.20	1.89e−01	2.82e−01	1.38e+00	3.88e−03	4.56e−03	1.01e−02
3	5e−02	945	149.51	1.79e−01	2.67e−01	1.32e+00	4.00e−03	4.72e−03	1.04e−02
3	5e−03	959	148.69	1.79e−01	2.61e−01	1.26e+00	2.84e−03	3.36e−03	8.48e−03
3	5e−04	1083	176.83	1.81e−01	2.72e−01	1.42e+00	4.44e−04	7.12e−04	5.52e−03
4	5e−01	2769	911.17	1.75e−01	2.74e−01	8.50e−01	9.71e−04	1.42e−03	7.64e−03
4	5e−02	2769	975.12	1.84e−01	2.78e−01	8.25e−01	1.01e−03	1.49e−03	7.80e−03
4	5e−03	2769	985.25	1.84e−01	2.74e−01	8.07e−01	9.18e−04	1.36e−03	7.67e−03
4	5e−04	2806	960.72	1.79e−01	2.72e−01	8.40e−01	7.60e−04	1.15e−03	6.85e−03

Convergence for the spatially adaptive refinement scheme with varying base levels l and refinement tolerances for the value function. The reference solution on the full grid was refined with refinement tolerance $\epsilon = 5\text{e}{-}04$ and base level $l = 3$ and is given in the first line. Reported are the Euler errors L^{Euler} as well as the pointwise errors of the value function L^{Value} with respect to the reference solution. Run times are reported in seconds with respect to 112 cores for the two- and three-dimensional problem, especially for base level $l = 4$. For $\epsilon = 5\text{e}{-}05$ Algorithm 2 does not terminate as excessively many grid points are added, indicating a general failure of the optimization that creates an erratic shape of the value function that is subsequently refined by Algorithm 3. The run time advantage is significant by a factor of nearly 30 for base level $l = 3$ compared to the full grid case and the DoF are reduced by a factor of roughly 7.

Tables 1, 2, and 3 show that on the sparse grid and on the full grid comparable numerical solutions to the transaction costs problem are obtained. Their quality, however, decreases with the dimensionality of the problem as the Euler errors—as a measure of the overall quality of the numerical solution—increase. Since the number of choice variables increases with the dimensionality of the problem (two choices per stock), and hence the dimensionality of the numerically approximated, discontinuous gradient increases, failure of the gradient-based optimization routine is likely. While the Euler errors are mostly insensitive to the chosen base grid level and refinement tolerances for every dimensionality, the sparse grid value function converges to its full grid counterpart with increasing base level and decreasing refinement tolerance. One possible conclusion is that the approximation quality of the value function in the underlying linear basis is not the right choice for the optimization setup. Additionally, divergence of Euler and pointwise errors for refinement tolerances below a certain threshold indicate that the "cost-benefit ratio" of added grid points is poor. That is, the gains in approximation accuracy of the value function by the added basis functions compared to the costs of solving additional optimization problems that might fail and produce outliers within the refinement step, are low. This phenomenon resembles *overfitting*, where, when a function is fitted too closely to in-sample data, its approximation error for out-of-sample data starts to increase again. The error thus does not monotonically decrease, but there exists an optimal size of the fitting data set.

With a sparse grid, significant run time gains can be realized as the DoF involved are lower in higher dimensions, even when grid points are spent on the boundary of the domain.

Given the above observations that reliable optimization results cannot always be obtained in this setup, it is unlikely that additional refinement of the optimal choices improves on the Euler errors. To support my conjecture, I fix the refinement constant of the value function to $\epsilon = 5\text{e}{-}04$ and solve the two-, three-, and four-dimensional transaction costs model for base levels $l \in \{2, 3, 4\}$ on a sparse grid. I then construct the optimal choices from these solution using Algorithm 4, where I choose the same tolerance for all optimal choices, i.e., $\epsilon_u = \bar{\epsilon}$ for $u = 1, \ldots, k$. Finally, I compute the Euler error implied by these optimal choices. I vary four different choices of the refinement tolerance $\bar{\epsilon} \in \{5\text{e}{-}01, \ldots, 5\text{e}{-}04\}$. Table 4 depicts the resulting Euler errors for the experiment with base level $l = 3$. The results for the other base levels are similar. In no case the Euler errors are lower when spatially adaptively refined. For the four-dimensional problem, the Euler errors increase drastically, showing erratic optimization behaviour. This is in line with the increasingly frequent failure of the optimization routine with increasing dimension.

Table 4 Spatially adaptive refinement of the optimal choices

	Two-dimensional problem			Three-dimensional problem			Four-dimensional problem		
$\bar{\epsilon}$	L_1^{Euler}	L_2^{Euler}	L_∞^{Euler}	L_1^{Euler}	L_2^{Euler}	L_∞^{Euler}	L_1^{Euler}	L_2^{Euler}	L_∞^{Euler}
5e−01	6.18e−02	1.41e−01	4.97e−01	9.65e−02	1.62e−01	5.67e−01	1.86e−01	2.80e−01	1.50e+00
5e−02	6.18e−02	1.41e−01	4.97e−01	1.12e−01	1.80e−01	6.06e−01	3.72e−01	5.47e−01	6.53e+00
5e−03	1.09e−01	2.40e−01	8.95e−01	3.70e−01	5.09e−01	2.34e+00	4.93e+07	6.19e+09	1.25e+12
5e−04	8.31e−02	1.79e−01	7.01e−01	3.44e−01	4.80e−01	2.19e+00	3.48e+08	7.11e+04	1.07e+13

The solution to the optimization problem on base level $l = 3$ with refinement tolerance $\epsilon = 5e{-}04$ for the value function is starting point for generating the optimal choices. These are generated for varying $\bar{\epsilon}$. Reported are the Euler errors L^{Euler}

Nonetheless, spatially adaptive refinement of the optimal choices might prove useful, when a different set of basis functions is used. Algorithm 4 can be applied to various basis functions of the value function used within the optimization, as long as the underlying grid is constructed on a nested sequence of grids with one-dimensional growth 2^l+1 for level l. One example would be the B-splines basis [43].

6 Conclusion

In this paper, I develop a spatially adaptive sparse grid dynamic programming scheme with the goal to solve high-dimensional dynamic portfolio choice models.

While on the one hand sparse grids can break the curse of dimensionality inherent in the dynamic programming approach to solving dynamic portfolio choice models, on the other hand local basis functions allow for spatial adaptivity. The latter cannot only be employed to adapt the value function approximation for locally sharp behaviour in the course of the iterative solution procedure, it also relieves the researcher from choosing a thrifty discretization of the state space. Assuming a regularly spaced grid structure, one has to choose $3d$ parameters (the boundaries and the number of grid points) compared to only $2d+2$ parameters (the boundaries, the level, and the refinement constant) with spatially adaptive refinement to determine a grid. At worst, this has to be repeated for different time steps and different discrete states. In addition, spatially adaptive refinement can be used to improve the approximation accuracy of the optimal choices, the subject of interest to economists, that are inferred from the solution of the optimization problem. In this paper, I not only present algorithms that solve the optimization problem for dynamic portfolio choice models, but also propose a procedure how to construct optimal choices from a solution to the optimization problem on a sparse grid.

The results show that on the sparse grid and on the full grid comparable solutions to the numerical example problem can be obtained. The sparse grid value function converges in the pointwise sense to its full grid counterpart with an increasing base level of the sparse grid and a decreasing refinement tolerance. However, when the refinement tolerance is decreased too much, the Euler and pointwise errors increase again, similar to overfitting. As the Euler errors, as a measure of the quality of the optimal choices, remain rather high and increase with the dimensionality of the problem, it could be beneficial to use different basis functions than the linear basis functions in order to increase the approximation quality of the value function and its gradients. Another possibility is considering a different optimization routine. In this setup, spatially adaptive refinement of the optimal choices does not improve on the Euler errors, but might prove more useful, when a different set of basis functions or a different optimization routine is used. With a sparse grid, significant run time gains can be realized for three and more dimensions.

A likely remedy without giving up local adaptivity would be to use local basis functions that are globally smooth and continuously differentiable, such as B-splines [43]. However, surplus-based adaptive refinement is only sensibly defined for linear basis functions and fitting the approximation using conventional B-spline basis functions involves solving a linear system of equations, which can take considerable amount of time when there are many grid points.

With globally smooth basis functions or a different optimization routine, my approach breaks the curse of dimensionality for a large class of dynamic portfolio choice models.

Acknowledgements I thank Julian Valentin for many fruitful discussions on the proposed numerical schemes as well as his support in implementing the MEX file interface to SG++ and parts of the proposed algorithms in MATLAB. I thank Yannick Dillschneider for discussing and helping me to develop the Euler error measure as well for his feedback to draft versions of this paper. I thank two anonymous referees for their rigorous reviews that helped to improve the numerical schemes and the analysis of the results considerably. Valuable feedback was provided by Johannes Brumm, Andreas Hubener, Kenneth Judd, Dirk Pflüger, and Miroslav Stoyanov. I thank the initiative High Performance Computing in Hessen for granting me computing time at the LOEWE-CSC and Lichtenberg Cluster. Finally, I thank Raimond Maurer for supporting this research in every way possible.

Appendix: Euler Equation Errors

To derive the Euler equation errors (25) and (26), I reformulate optimization problem (18)–(24) in terms of the stock investment $s_t^i = w_t x_t^i + \delta_t^{i+} - \delta_t^{i-}$, $i = 1, \ldots, d$:

$$j_t(w_t, \boldsymbol{x}_t) = \max_{b_t, \boldsymbol{s}_t, \boldsymbol{\delta}_t^+, \boldsymbol{\delta}_t^-} \underbrace{u(c_t) + \rho \pi_t \mathscr{E}_t \left[j_{t+1}(w_{t+1}, \boldsymbol{x}_{t+1}) \right]}_{\tilde{j}_t(b_t, \boldsymbol{s}_t, \boldsymbol{\delta}_t^+, \boldsymbol{\delta}_t^-, w_t, \boldsymbol{x}_t)} .$$

with consumption $c_t = w_t - \sum_{i=1}^d s_t^i - b_t - \tau \sum_{i=1}^d (\delta_t^{i-} + \delta_t^{i-})$. This is subject to the constraints $s_t^i \geq 0$, $\delta_t^{i+} \geq 0$, $\delta_t^{i-} \in \left[0, s_t^i\right]$, $i = 1, \ldots, d$ as well as $c_t \geq c_{\min}$, $b_t \geq 0$, and $\sum_{i=1}^d x_t^i \leq 1$. The state dynamics thus become:

$$w_{t+1} = \sum_{i=1}^{d} s_t^i r_{t+1}^i + b_t r_f + \alpha_R ,$$

$$x_{t+1}^i = \frac{s_t^i r_{t+1}^i}{w_{t+1}} , \quad i = 1, \ldots, d .$$

Following not yet published notes on gradient-based optimization and Euler equation errors for discrete time dynamic portfolio choice written together with Yannick

Dillschneider, I set up the Lagrangian

$$\mathcal{L}_t = \tilde{j}_t - \mu_{c_t}(c_t - c_{min}) - \mu_{b_t} b_t$$
$$- \sum_{i=1}^{d} \left(\mu_{s_t^i} s_t^i + \mu_{\delta_t^{i+}} \delta_t^{i+} + \mu_{\delta_t^{i-}\geq} \delta_t^{i-} + \mu_{\delta_t^{i-}\leq} \left(s_t^i - \delta_t^{i-} \right) \right),$$

where μ_X is the value of the respective Kuhn-Tucker multiplier for choice X and I translated the box constraints for δ_t^{i-}, $i = 1, \ldots, d$, into two greater-equal constraints, each denoted by $\mu_{\delta_t^{i-}\geq}$ for the lower and $\mu_{\delta_t^{i-}\leq}$ for the upper bound. At the optimum, the first order conditions for the Lagrangian with respect to the stocks s_t^i and the bond b_t must hold:

$$\frac{\partial \mathcal{L}_t}{\partial s_t^i} : \frac{\partial \tilde{j}_t}{\partial s_t^i} + \mu_{c_t} - \mu_{s_t^i} - \mu_{\delta_t^{i-}\leq} = 0, \quad i = 1, \ldots, d,$$
$$\frac{\partial \mathcal{L}_t}{\partial b_t} : \frac{\partial \tilde{j}_t}{\partial b_t} + \mu_{c_t} - \mu_{b_t} = 0.$$

Neglecting occasional binding constraints (i.e., I consider only optimal choices where $\mu_X = 0$ for any X) and expanding the partial derivatives of $\tilde{j}_t$ with respect to stocks s_t^i and the bond b_t, the first order conditions become

$$-c_t^{-\gamma} + \rho \pi_t \mathcal{E}_t \left[\left(\frac{\partial j_{t+1}}{\partial w_{t+1}} \frac{\partial w_{t+1}}{\partial s_t^i} + \sum_{j=1}^{d} \frac{\partial j_{t+1}}{\partial x_{t+1}^j} \frac{\partial x_{t+1}^j}{\partial s_t^i} \right) \right] = 0, \quad i = 1, \ldots, d,$$
$$-c_t^{-\gamma} + \rho \pi_t \mathcal{E}_t \left[\left(\frac{\partial j_{t+1}}{\partial w_{t+1}} \frac{\partial w_{t+1}}{\partial b_t} + \sum_{i=1}^{d} \frac{\partial j_{t+1}}{\partial x_{t+1}^i} \frac{\partial x_{t+1}^i}{\partial b_t} \right) \right] = 0.$$

I then apply the envelope theorem [36]

$$\frac{\partial j_t}{\partial w_t} = \frac{\partial \tilde{j}_t}{\partial w_t} \left(b_t^*, \boldsymbol{s}_t^*, \boldsymbol{\delta}_t^{+*}, \boldsymbol{\delta}_t^{-*}, w_t, \boldsymbol{x}_t \right),$$
$$\frac{\partial j_t}{\partial x_t^i} = \frac{\partial \tilde{j}_t}{\partial x_t^i} \left(b_t^*, \boldsymbol{s}_t^*, \boldsymbol{\delta}_t^{+*}, \boldsymbol{\delta}_t^{-*}, w_t, \boldsymbol{x}_t \right), \quad i = 1, \ldots, d.$$

evaluated at the optimal choices $(b_t^*, \boldsymbol{s}_t^*, \boldsymbol{\delta}_t^{+*}, \boldsymbol{\delta}_t^{-*})$. Because u does not directly depend on the states, its partial derivatives with respect to the states are zero. Hence,

$$\frac{\partial \tilde{j}_t}{\partial w_t} = \rho \pi_t \mathcal{E}_t \left[\left(\frac{\partial j_{t+1}}{\partial w_{t+1}} \frac{\partial w_{t+1}}{\partial w_t} + \sum_{i=1}^{d} \frac{\partial j_{t+1}}{\partial x_{t+1}^i} \frac{\partial x_{t+1}^i}{\partial w_t} \right) \right],$$

$$\frac{\partial \tilde{j}_t}{\partial x_t^i} = \rho\pi_t \mathscr{E}_t \left[\left(\frac{\partial j_{t+1}}{\partial w_{t+1}} \frac{\partial w_{t+1}}{\partial x_t^i} + \sum_{j=1}^{d} \frac{\partial j_{t+1}}{\partial x_{t+1}^j} \frac{\partial x_{t+1}^j}{\partial x_t^i} \right) \right], \quad i = 1, \ldots, d .$$

Since the derivatives are $\partial w_{t+1}/\partial s_t^i = r_{t+1}^i$, $i = 1, \ldots, d$, and $\partial w_{t+1}/\partial b_t = r_f$, it remains to show that:

$$\frac{\partial \tilde{j}_t}{\partial w_t} + \sum_{i=1}^{d} \frac{\partial \tilde{j}_t}{\partial x_t^i} \frac{x_t^i}{w_t} = c_t^{-\gamma} . \tag{28}$$

To show Eq. (28) holds involves lengthy algebraic computations. The results are the basic pricing equations of asset pricing [13]

$$\rho\pi_t \mathscr{E}_t \left[\frac{c_{t+1}^{-\gamma}}{c_t^{-\gamma}} r_{t+1}^i \right] = 1 , \quad i = 1, \ldots, d ,$$

$$\rho\pi_t \mathscr{E}_t \left[\frac{c_{t+1}^{-\gamma}}{c_t^{-\gamma}} r_f \right] = 1 ,$$

which serve as my error measure.

References

1. R. Bellman, The theory of dynamic programming. Technical report, The RAND Corporation, 1954
2. O. Bokanowski, J. Garcke, M. Griebel, I. Klompmaker, An adaptive sparse grid semi-lagrangian scheme for first order Hamilton-Jacobi Bellman equations. J. Sci. Comput. **55**(3), 575–605 (2013)
3. J. Brumm, M. Grill, Computing equilibria in dynamic models with occasionally binding constraints. J. Econ. Dyn. Control. **38**, 142–160 (2014)
4. J. Brumm, S. Scheidegger, Using adaptive sparse grids to solve high-dimensional dynamic models. Econometrica **85**(5), 1575–1612 (2017)
5. H.-J. Bungartz, Finite elements of higher order on sparse grids. PhD thesis, Technische Universität München, 1998
6. H.-J. Bungartz, M. Griebel, Sparse grids. Acta Numer. **13**, 147–269 (2004)
7. Y. Cai, Dynamic programming and its application in economics and finance. PhD thesis, Stanford University, 2009
8. Y. Cai, K.L. Judd, Stable and efficient computational methods for dynamic programming. J. Eur. Econ. Assoc. **8**(2-3), 626–634 (2010)
9. Y. Cai, K.L. Judd, G. Thain, S.J. Wright, Solving dynamic programming problems on a computational grid. Comput. Econ. **45**(2), 261–284 (2015)
10. J. Chai, W. Horneff, R. Maurer, O.S. Mitchell, Optimal portfolio choice over the life cycle with flexible work, endogenous retirement, and lifetime payouts. Rev. Financ. **15**(4), 875–907 (2011)
11. J.F. Cocco, Portfolio choice in the presence of housing. Rev. Financ. Stud. **18**(2), 535–567 (2005)

12. J.F. Cocco, F.J. Gomes, P.J. Maenhout, Consumption and portfolio choice over the life cycle. Rev. Financ. Stud. **18**(2), 491–533 (2005)
13. J.H. Cochrane, *Asset Pricing* (Princeton University Press, Princeton, 2009)
14. C. Feuersänger, M. Griebel, Principal manifold learning by sparse grids. Computing **85**(4), 267–299 (2009)
15. E. French, The effects of health, wealth, and wages on labour supply and retirement behaviour. Rev. Econ. Stud. **72**(2), 395–427 (2005)
16. J. Garcke, Sparse grids in a nutshell, in *Sparse Grids and Applications*, ed. by J. Garcke, M. Griebel (Springer, Berlin, 2012), pp. 57–80
17. J. Garcke, A. Kröner, Suboptimal feedback control of PDEs by solving HJB equations on adaptive sparse grids. J. Sci. Comput. **70**(1), 1–28 (2017)
18. P.E. Gill, W. Murray, M.A. Saunders, SNOPT: an SQP algorithm for large-scale constrained optimization. SIAM Rev. **47**(1), 99–131 (2005)
19. M. Griebel, A parallelizable and vectorizable multi-level algorithm on sparse grids, in *Parallel Algorithms for Partial Differential Equations*, ed. by W. Hackbusch (Vieweg, Braunschweig, 1991), pp. 94–199
20. M. Griebel, Adaptive sparse grid multilevel methods for elliptic PDEs based on finite differences. Computing **61**(2), 151–179 (1998)
21. F. Heiss, V. Winschel, Likelihood approximation by numerical integration on sparse grids. J. Econ. **144**(1), 62–80 (2008)
22. W.J. Horneff, R. Maurer, M.Z. Stamos, Life-cycle asset allocation with annuity markets. J. Econ. Dyn. Control. **32**(11), 3590–3612 (2008)
23. W.J. Horneff, R. Maurer, O.S. Mitchell, M.Z. Stamos, Variable payout annuities and dynamic portfolio choice in retirement. J. Pension Econ. Financ. **9**(2), 163–183 (2010)
24. V. Horneff, R. Maurer, O.S. Mitchell, R. Rogalla, Optimal life cycle portfolio choice with variable annuities offering liquidity and investment downside protection. Insur. Math. Econ. **63**, 91–107 (2015)
25. V. Horneff, R. Maurer, P. Schober, Efficient parallel solution methods for dynamic portfolio choice models in discrete time. Available at SSRN 2665031 (2016)
26. A. Hubener, R. Maurer, O.S. Mitchell, How family status and social security claiming options shape optimal life cycle portfolios. Rev. Financ. Stud. **29**(4), 937–978 (2015)
27. K.L. Judd, Projection methods for solving aggregate growth models. J. Econ. Theory **58**(2), 410–452 (1992)
28. K.L. Judd, *Numerical Methods in Economics* (MIT Press, Cambridge, 1998)
29. K.L. Judd, L. Maliar, S. Maliar, R. Valero, Smolyak method for solving dynamic economic models: Lagrange interpolation, anisotropic grid and adaptive domain. J. Econ. Dyn. Control. **44**, 92–123 (2014)
30. H.H. Kim, R. Maurer, O.S. Mitchell, Time is money: rational life cycle inertia and the delegation of investment management. J. Financ. Econ. **121**(2), 427–447 (2016)
31. A. Klimke, B. Wohlmuth, Algorithm 847: spinterp: piecewise multilinear hierarchical sparse grid interpolation in MATLAB. ACM Trans. Math. Softw. (TOMS) **31**(4), 561–579 (2005)
32. D. Krueger, F. Kubler, Computing equilibrium in OLG models with production. J. Econ. Dyn. Control. **28**(7), 1411–1436 (2004)
33. D.A. Love, The effects of marital status and children on savings and portfolio choice. Rev. Financ. Stud. **23**(1), 385–432 (2010)
34. B.A. Malin, D. Krueger, F. Kubler, Solving the multi-country real business cycle model using a Smolyak-collocation method. J. Econ. Dyn. Control. **35**(2), 229–239 (2011)
35. R.C. Merton, Lifetime portfolio selection under uncertainty: the continuous-time case. Rev. Econ. Stat. **51**, 247–257 (1969)
36. M.J. Osborne, *Mathematical Methods for Economic Theory: A Tutorial* (University of Toronto, Toronto, 2007)
37. D. Pflüger, Spatially adaptive sparse grids for high-dimensional problems. PhD thesis, Technische Universität München, 2010

38. D. Pflüger, Spatially adaptive refinement, in *Sparse Grids and Applications*, ed. by J. Garcke, M. Griebel (Springer, Berlin, 2012), pp. 243–262
39. J. Rust, Dynamic programming, in *The New Palgrave Dictionary of Economics*, ed. by S.N. Durlauf, L.E. Blume (Palgrave Macmillan, London, 2008)
40. P.A. Samuelson, Lifetime portfolio selection by dynamic stochastic programming. Rev. Econ. Stat. **51**, 239–246 (1969)
41. S.A. Smolyak, Quadrature and interpolation formulas for tensor products of certain classes of functions. Dokl. Akad. Nauk SSSR **4**, 123 (1963)
42. M. Stoyanov, User manual: TASMANIAN sparse grids v3.1. Technical report, Oak Ridge National Laboratory, 2016
43. J. Valentin, D. Pflüger, Hierarchical gradient-based optimization with B-splines on sparse grids, in *Sparse Grids and Applications - Stuttgart 2014*, ed. by J. Garcke, D. Pflüger (Springer, Berlin, 2016), pp. 315–336
44. V. Winschel, M. Krätzig, Solving, estimating, and selecting nonlinear dynamic models without the curse of dimensionality. Econometrica **78**(2), 803–821 (2010)
45. C. Zenger, Sparse grids, in *Parallel Algorithms for Partial Differential Equations*, ed. by W. Hackbusch (Vieweg, Braunschweig, 1991), pp. 241–251

Adaptive Sparse Grid Construction in a Context of Local Anisotropy and Multiple Hierarchical Parents

Miroslav Stoyanov

Abstract We consider general strategy for hierarchical multidimensional interpolation based on sparse grids, where the interpolation nodes and locally supported basis functions are constructed from tensors of a one dimensional hierarchical rule. We consider four different hierarchies that are tailored towards general functions, high or low order polynomial approximation, or functions that satisfy homogeneous boundary conditions. The main advantage of the locally supported basis is the ability to choose a set of functions based on the observed behavior of the target function. We present an alternative to the classical surplus refinement techniques, where we exploit local anisotropy and refine using functions with not strictly decreasing support. The more flexible refinement strategy improves stability and reduces the total number of expensive simulations, resulting in significant computational saving. We demonstrate the advantages of the different hierarchies and refinement techniques by application to series of simple functions as well as a system of ordinary differential equations given by the Kermack-McKendrick SIR model.

1 Introduction

Computer simulations of complex physical phenomena are an invaluable tool for advancing modern science and engineering. A very common approach is to model the behavior of a system of interest via a set of deterministic differential equations, where the result depends on a number of input parameters, e.g., heat conductivity, infection rate, temperature, etc. The equations are then discretized and solved with standard techniques of numerical analysis, often requiring computationally heavy software and a powerful supercomputer. However, the questions of interest to science are seldom answerable by a single deterministic simulation, since, in practice, the inputs could vary over a wide range due to changing or uncertain

M. Stoyanov (✉)
Oak Ridge National Laboratory, Oak Ridge, TN, USA
e-mail: stoyanovmk@ornl.gov

J. Garcke et al. (eds.), *Sparse Grids and Applications – Miami 2016*,
Lecture Notes in Computational Science and Engineering 123,
https://doi.org/10.1007/978-3-319-75426-0_8

operation conditions. Applications that rely on statistical analysis and optimal design require a large number of simulations with different values of the input parameters, which is often times prohibitively expensive if a simulation needs hundreds of CPU hours. Computationally cheap surrogate models are needed to alleviate the computational burden, and the methods for creating such models are fundamental for advancing modern science.

In this work, we present a general strategy for multidimensional interpolation based on hierarchical sparse grids using basis functions with local support. Of specific considerations are strategies for achieving maximum accuracy with fewest number of interpolation nodes, which is desirable since each node corresponds to an evaluation of the target function which in turn requires a computationally expensive simulation. We present several one dimensional hierarchies and multidimensional adaptive refinement techniques that aim at achieving optimal convergence when applied to different classes of models.

Interpolation is one of the most widely used techniques for creating surrogates, due to their non-intrusive nature and high accuracy per number of deterministic simulations. Intrusive methods, such as Galerkin projection [1, 13] and reduced basis [4, 40, 42], have limited applicability since they are often restricted to a specific set of equations and always require the complete rewriting of the simulation software. The non-intrusive methods rely on series of deterministic simulation and thus can be easily coupled with existing complex solvers. Statistical sampling methods, such as Monte Carlo (MC) [16, 18], are attractive when facing a problem with huge number of inputs; however, the convergence rate of those methods is slow and in cases where the inputs are few, e.g., ≤ 10, polynomial based approximation yields better results. Optimization techniques using ℓ^1 and ℓ^2 minimization work well with noisy, large and random data [7, 12, 14, 25]; however, in both cases, the number of simulations needed to construct the surrogate exceeds the cardinality of the polynomial space where the final approximation resides. Deterministic simulations are noise free and hence interpolation methods that use one sample per basis function are preferable.

Sparse grids is the most common technique for constructing multidimensional interpolation rules by using tensors of one dimensional nodes and basis functions [2, 3, 5, 6, 10, 11, 20–22, 26, 30, 31, 33]. The sparse grid construction offer faster convergence than simple tensor techniques due to the better error balance, which helps alleviate the *curse of dimensionality* associated with a large number of model inputs. In the context of high order polynomial approximation of analytic function, several strategies have been presented that aim at optimal or quasi-optimal convergence [3, 6, 29, 32, 41, 43]. However, the globally supported polynomials are not well suited for approximating functions with local sharp behavior, hence piece-wise polynomial basis with decreasing support is preferable. The hallmark work in the field [6] presents an algorithm for adaptive hierarchical construction of interpolant that has been used in many applications [7, 19, 28, 34–36]. However, the algorithm is prone to stagnation in some cases and the isotropic nature of the refinement strategy can yield an excess number of nodes.

We present several hierarchies of one dimensional rules that construct approximations with low or high polynomial order or homogeneous boundary conditions. We introduce a novel construction of piece-wise constant approximation that does not rely on assumptions about left and right continuity. Following the standard sparse grids method, we extend the one dimensional hierarchy to a multidimensional context using tensors. The hierarchical structure translates to the high dimensions and allows for adaptive refinement of the interpolant based on hierarchical coefficients. The standard refinement strategy uses either strictly parent-to-children refinement which can result in stagnation, or includes entire hierarchical ancestry which leads to excessive model evaluations. We present a more flexible approach that considers only the intermediate parents, thus improving stability and minimizing the number of nodes. We also present a method for detecting local anisotropy, which can yield significant improvements even when dealing with a globally isotropic model.

The rest of the paper is organized as follows: in Sect. 2 we present the general sparse grids construction and the four one dimensional hierarchies; in Sect. 3 we present the different refinement strategies; in Sect. 4 we demonstrate the advantages of the different techniques with several simple functions as well as Kermack-McKendrick SIR model.

2 Multidimensional Hierarchical Interpolation Strategy

In this section, our objective is to construct an interpolatory approximation to a target function that achieves desired tolerance with fewest possible evaluations, i.e., number of samples. Let $\Gamma \subset \mathbb{R}^d$ indicate a d-dimensional box $\Gamma = \otimes_{k=1}^{d} [a_k, b_k]$ and $f : \Gamma \to \mathbb{R}$ indicate a continuous function defined over Γ, i.e., $f(\boldsymbol{x}) \in C^0(\Gamma)$. By linear translation $\boldsymbol{\xi} \to \boldsymbol{x}$, where $x_k = 0.5(b_k - a_k)\xi_k + 0.5(b_k + a_k)$, every finite interval can be mapped to $[-1, 1]$, hence, w.l.o.g., we assume $a_k = -1$ and $b_k = 1$ for all k. Let $\{\boldsymbol{x}_j\}_{j \in X} \subset \Gamma$ indicate a pre-defined set of interpolation nodes and an associated set of basis functions $\{\phi_j(\boldsymbol{x})\}_{j \in X}$, then an interpolation rule is defined as

$$G_X[f](\boldsymbol{x}) = \sum_{j \in X} s_j \phi_j(\boldsymbol{x}),$$

where the coefficients s_j are chosen so that $G_X[f](\boldsymbol{x}_j) = f(\boldsymbol{x}_j)$ for all $j \in X$. The point sets that we use in this paper are always unisolvent for the set of functions, i.e., the coefficients s_j exist for any function $f(\boldsymbol{x}) \in C^0(\Gamma)$. Furthermore, we assume that the computational cost of finding s_j is dominated by the evaluations $f(\boldsymbol{x}_j)$, e.g., $f(\boldsymbol{x})$ is associated with an expensive simulation [15, 42]. Thus, we seek an algorithm for constructing a suitable $G_X[f](\boldsymbol{x})$ that approximates $f(\boldsymbol{x})$ to desired tolerance and requires as few nodes $\boldsymbol{x}_j$, as possible.

We begin by introducing a one dimensional hierarchy of interpolation nodes and basis functions, then we extend the hierarchy to a multidimensional context using tensors. We use the resulting hierarchy in an algorithm for adaptive interpolation similar to [6, 7, 19, 28, 34, 35].

2.1 Multidimensional Hierarchy of Nodes and Functions

In this section, we describe three widely used hierarchical rules with different polynomial order and boundary conditions [5, 6, 34], in Sect. 2.3 we present an alternative piece-wise constant hierarchy. Let $\mathbb{N}$ denote the set of natural numbers including zero and define the sequence of one dimensional nodes $\{x_j\}_{j\in\mathbb{N}} \subset [-1, 1]$ by

$$x_0 = 0, \quad x_{1,2} = \pm 1, \quad \text{for } j > 2, \quad x_j = (2j-1) \times 2^{-\lfloor \log_2(j-1) \rfloor} - 3, \tag{1}$$

where $\lfloor z \rfloor = \max\{j \in \mathbb{N} : j \leq z\}$ is the floor function. The first nine nodes of the sequence are given in Fig. 1. Note that for any $L \in \mathbb{N}$ the nodes, $\{x_j\}_{j \leq 2^L+1}$, are equidistant in $[-1, 1]$, thus we partition the nodes into *levels*, where the level of x_j is $L_j = \min\{L \in \mathbb{N} : j \leq 2^L + 1\}$. The nodes form a hierarchical three structure, where each node is associated with a *parent* and two *offspring* (children) nodes, i.e., the parent is the closest node on the previous level and we define the parent set P_j

$$P_j = \{i \in \mathbb{N} : x_i \text{ is parent of } x_j\} = \left\{ \operatorname*{argmin}_{i \leq 2^{L_j - 1}+1} |x_i - x_j| \right\},$$

where we take $P_0 = \emptyset$. The simple one dimensional hierarchy in Fig. 1 associates one parent with each point; however, alternative hierarchies allow for multiple parents, e.g., see Fig. 3, and we adopt the set notation for consistency across all cases. The children sets are defined by the reciprocal relation to the parents, i.e.,

$$O_j = \{i \in \mathbb{N} : x_i \text{ is offspring of } x_j\} = \{i \in \mathbb{N} : j \in P_i\}.$$

For the example in Fig. 1, P_j and O_j are explicitly defined as

$$\begin{aligned} P_0 = \emptyset,\ P_1 = \{0\},\ P_2 = \{0\},\ P_3 = \{1\}, \text{ for } j > 3\ P_j = \left\{ \left\lfloor \frac{j+1}{2} \right\rfloor \right\}, \\ O_0 = \{1, 2\},\ O_1 = \{3\},\ O_2 = \{4\}, \text{ for } j > 2\ O_j = \{2j-1, 2j\}. \end{aligned} \tag{2}$$

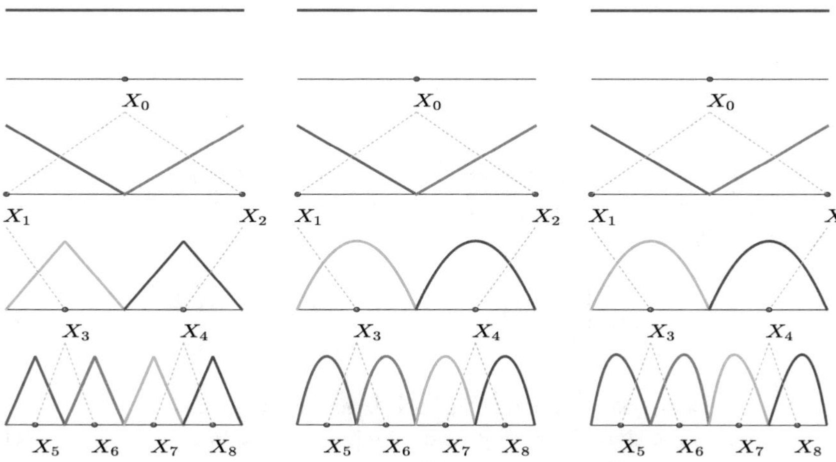

Fig. 1 Linear (**left**), quadratic (**middle**), and cubic (**right**) nodes and basis functions; the dashed lines correspond to parent-offspring relation. Note that in each case level 0 is a constant and level 1 is linear, only at level 2 the hierarchy provide enough ancestors for a quadratic approximation

The interpolation nodes x_j are associated with basis functions $\phi_j(x) : [-1, 1] \to \mathbb{R}$ and each function is piece-wise polynomial with support in $[x_j - \Delta x_j, x_j + \Delta x_j]$. Here, Δx_j is the distance to the closest parent, i.e., $\Delta x_j = \min\{|x_i - x_j| : i \in P_j\}$, and $\Delta x_0 = 1$, i.e., the support of $\phi_0(x)$ is $[-1, 1]$. Explicitly, for $j > 1$, $\Delta x_j = 2^{-\lfloor \log_2(j-1) \rfloor}$.

The polynomial order of each $\phi_j(x)$ depends on the total number of ancestors of x_j, where the *ancestor* set $A_j \subset \mathbb{N}$ is defined as the smallest set satisfying

$$P_j \subset A_j, \qquad \text{and } P_i \subset A_j \text{ for all } i \in A_j.$$

The basis functions of maximal order are defined as

$$\phi_j(x) = \begin{cases} \prod_{i \in A_j} \frac{x - x_i}{x_j - x_i}, & x \in [x_j - \Delta x_j, x_j + \Delta x_j], \\ 0, & x \notin [x_j - \Delta x_j, x_j + \Delta x_j]. \end{cases} \tag{3}$$

Thus, $\phi_0(x)$ is a constant, $\phi_1(x)$ and $\phi_2(x)$ are linear, $\phi_3(x)$ and $\phi_4(x)$ are quadratic, etc. Alternatively, the order of the polynomials can be restricted to a specific $p > 1$ by selecting only the nearest p ancestors F_j^p and defining

$$\phi_j^p(x) = \begin{cases} \prod_{i \in F_j^p} \frac{x - x_i}{x_j - x_i}, & x \in [x_j - \Delta x_j, x_j + \Delta x_j], \\ 0, & x \notin [x_j - \Delta x_j, x_j + \Delta x_j]. \end{cases} \tag{4}$$

In particular,

$$F_j^p = \operatorname*{argmin}_{F \subset A_j, \#F = p} \sum_{i \in F} |x_i - x_j|, \tag{5}$$

where $\#F$ indicates the number of elements in F and for notational convenience we take $F_j^p = A_j$ whenever $\#A \leq p$. As an exception to the above, the linear basis functions are defined as

$$\phi_j^1(x) = \begin{cases} 1 - \frac{|x - x_j|}{\Delta x_j}, & x \in [x_j - \Delta x_j, x_j + \Delta x_j], \\ 0, & x \notin [x_j - \Delta x_j, x_j + \Delta x_j]. \end{cases} \tag{6}$$

Examples of linear, quadratic and cubic basis functions are given in Fig. 1.

Variations of the above hierarchy have been presented in literature, e.g., [7, 34]. In the numerical experiments in this paper, we consider two examples where we either assume homogeneous boundary or aim at higher order approximation. In some applications, we have a priori knowledge that $f(\boldsymbol{x})$ satisfies $f(\boldsymbol{x}) = 0$ for $\boldsymbol{x} \in \partial \Gamma$, thus, we consider a hierarchical rules that excludes the boundary

$$x_j = (2j+3) \times 2^{-\lfloor \log_2(j+1) \rfloor} - 3, \qquad P_j = \left\{ \left\lfloor \frac{j-1}{2} \right\rfloor \right\}, \qquad Q_j = \{2j+1, 2j+2\},$$

with $x_0 = 0$ and $P_0 = \emptyset$. The definition of the level also changes to $L_j = \lfloor \log_2(j+1) \rfloor$. Figure 2 shows the first seven nodes and functions for linear, quadratic and cubic basis. The linear basis functions are constructed identical to (6) with $\Delta x_j = 2^{-\lfloor \log_2(j+1) \rfloor}$, while the higher order basis uses (4) with the end points -1 and 1 added to the ancestry sets A_j in (3) and (5). Note that compared to the hierarchy described in Fig. 1, the homogeneous boundary case allows for higher order basis to be used at a lower level. In Sect. 4 we demonstrate that incorporating boundary information directly into the basis gives better approximation that using the standard hierarchy and simply setting $f(x_1) = f(x_2) = 0$.

Interpolation using high-order polynomials with global support and an equidistant set of nodes results in a very large penalty, i.e., Lebesgue constant, see (8) in Sect. 2.2. However, high-order basis can be used only for the first two levels and still obtain stable interpolant. Using the nodes and levels described in (1) and Fig. 1, we set $\phi_1(x) = 0.5x(x-1)$, and $\phi_2(x) = 0.5x(x+1)$, and augment the sets $P_3 = P_4 = \{1, 2\}$ and $O_1 = O_2 = \{3, 4\}$. For every other j, we define $\phi_j(x)$, P_j and O_j identical to (3), (4), (2). Modifying P_3 and P_4 changes the ancestry sets A_j and F_j^p, which results in higher order basis functions. Figure 3 shows the first nine nodes and corresponding basis. In Sect. 4 we demonstrate that increasing the polynomial order gives better approximation when $f(x)$ is smooth, despite the higher penalty due to the global support of $\phi_1(x)$ and $\phi_2(x)$.

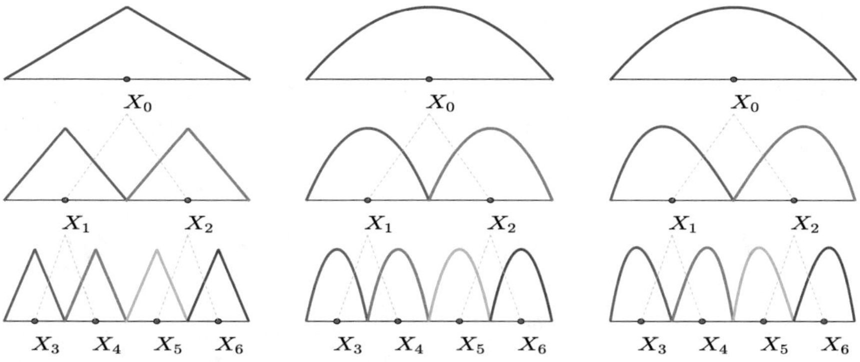

Fig. 2 Linear (**left**), quadratic (**middle**), and cubic (**right**) nodes and basis functions assuming zero boundary conditions. The dashed lines correspond to parent-offspring relation. Note that in this case, at level $l = 0, 1, 2, \ldots$, the homogeneous assumption allows the use of polynomials of order $p = l + 1$

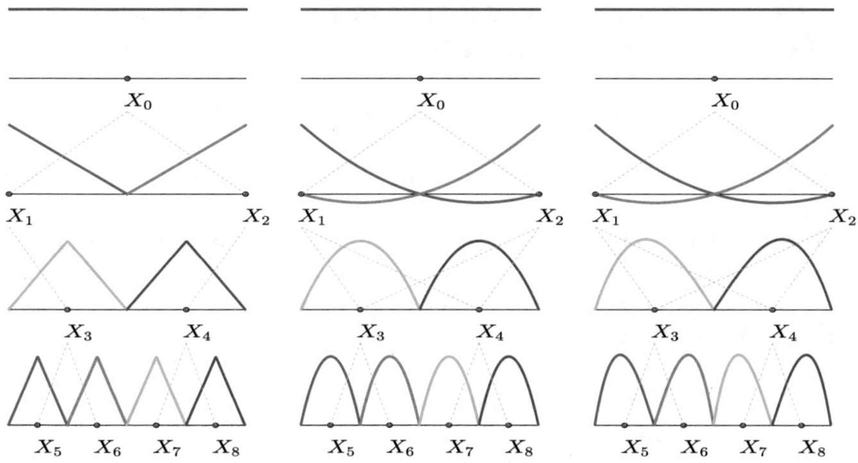

Fig. 3 Linear (**left**), quadratic (**middle**), and cubic (**right**) nodes and basis functions assuming zero boundary conditions. The dashed lines correspond to parent-offspring relation, when using quadratic or higher basis nodes 3 and 4 have two parents each due to the global support of $\phi_1(x)$ and $\phi_2(x)$. Note that in this case, at level $l = 0, 1, 2, \ldots$, basis of order $p = l + 1$ is allowed

2.2 Multidimensional Interpolation

Using standard sparse grids techniques [6, 19, 33, 34], we extend the one dimensional hierarchy to a d-dimensional context ($d > 1$) using tensors and multi-index notation:

$$\boldsymbol{j} = (j_1, j_2, \dots, j_d) \in \mathbb{N}^d, \ \boldsymbol{x_j} = (x_{j_1}, x_{j_2}, \dots, x_{j_d}) \in \Gamma, \ \phi_{\boldsymbol{j}}(\boldsymbol{x}) = \prod_{k=1}^{d} \phi_{j_k}(x^{(k)}),$$

where $\boldsymbol{x} = (x^{(1)}, x^{(2)}, \dots, x^{(d)}) \in \mathbb{R}^d$ and we use the notation $x^{(k)}$ to indicate the component of the vector $\boldsymbol{x}$ and differentiate from the index or position in the hierarchy $\boldsymbol{j}$. A node $\boldsymbol{x_j}$ has parents and children for each direction k, i.e.,

$$P_{\boldsymbol{j},k} = \{\boldsymbol{i} \in \mathbb{N}^d : \boldsymbol{i}_{-k} = \boldsymbol{j}_{-k}, \text{ and } i_k \in P_{j_k}\}, \ O_{\boldsymbol{j},k} = \{\boldsymbol{i} \in \mathbb{N}^d : \boldsymbol{i}_{-k} = \boldsymbol{j}_{-k}, \text{ and } i_k \in O_{j_k}\},$$

where $\boldsymbol{j}_{-k}$ indicates the multi-index resulting from the removal of the k-th component of $\boldsymbol{j}$. The multidimensional ancestry sets $A_{\boldsymbol{j}}$ are defined in a similar fashion including all parents in all directions, i.e., for all $k = 1, 2, \dots, d$,

$$P_{\boldsymbol{j},k} \subset A_{\boldsymbol{j}}, \qquad \text{and } P_{\boldsymbol{i},k} \subset A_{\boldsymbol{j}} \text{ for all } \boldsymbol{i} \in A_{\boldsymbol{j}}.$$

Finally, the multidimensional *level* of a node is the sum of the one dimensional levels.

The multidimensional hierarchy provide convenient way of defining interpolation rules; since the support of every $\phi_{\boldsymbol{j}}(x)$ is disjointed from every function on the same level, and since $\phi_{\boldsymbol{j}}(\boldsymbol{x_i}) = 0$ for all $\boldsymbol{i} \in A_{\boldsymbol{j}}$, it follows that for any set of basis functions X, the corresponding set of nodes is unisolvent. Specifically,

$$G_X[f](\boldsymbol{x}) = \sum_{\boldsymbol{j} \in X} s_{\boldsymbol{j}} \phi_{\boldsymbol{j}}(\boldsymbol{x}), \quad \text{where } s_{\boldsymbol{j}} = f(\boldsymbol{x_j}) - \sum_{\boldsymbol{i} \in A_{\boldsymbol{j}} \bigcap X} s_{\boldsymbol{i}} \phi_{\boldsymbol{i}}(\boldsymbol{x_j}). \tag{7}$$

The coefficients $s_{\boldsymbol{j}}$ can be computed one level at a time, i.e., the equations have a (usually) sparse lower triangular matrix form. Fast algorithms for computing $s_{\boldsymbol{j}}$ have been proposed [8], but those make assumptions that $P_{\boldsymbol{i},k} \subset X$ for all $\boldsymbol{i} \in X$ and all $k = 1, 2, \dots, d$; if $P_{\boldsymbol{i},k} \not\subset X$ expanding the index set would require additional model simulations, which is usually orders of magnitude more expensive than solving the sparse triangular system. The general formula (7) allows us to avoid extraneous evaluations of $f(\boldsymbol{x})$.

Consider the error in approximation of $f(\boldsymbol{x})$; from (7) we have that $s_{\boldsymbol{j}}$ are bounded by some constant times $\|f(\boldsymbol{x})\|_\infty$, therefor $G_X : \mathbb{C}^0(\Gamma) \to \mathbb{C}^0(\Gamma)$ is a bounded linear operator with norm depending on X and $\phi_{\boldsymbol{j}}(\boldsymbol{x})$. From the disjoint support of the basis $G_X[\phi_j] = \phi_j$ for all $\boldsymbol{j} \in X$, and therefore, for any $g \in span\{\phi_j\}$

$$f(\boldsymbol{x})-G_X[f](\boldsymbol{x}) = f(\boldsymbol{x})-g(\boldsymbol{x})+g(\boldsymbol{x})-G_X[f](\boldsymbol{x}) = f(\boldsymbol{x})-g(\boldsymbol{x})+G_X[f-g](\boldsymbol{x}).$$

Taking the inf over all $g(\boldsymbol{x})$, we obtain the $L^{\infty}(\Gamma)$ error bound

$$\|f(\boldsymbol{x}) - G_X[f](\boldsymbol{x})\|_{\infty} \leq (1 + \|G_X\|_{\infty}) \inf_{g \in span\{\phi_j\}} \|f(\boldsymbol{x}) - g(\boldsymbol{x})\|_{\infty}, \tag{8}$$

where $\|G_X\|_{\infty}$ denotes the operator norm and is often called the Lebesgue constant of the interpolant. The $L^2(\Gamma)$ error can be bounded

$$\|f(\boldsymbol{x}) - G_X[f](\boldsymbol{x})\|_2 \leq |\Gamma|^{1/2} \|f(\boldsymbol{x}) - G_X[f](\boldsymbol{x})\|_{\infty},$$

where $|\Gamma|$ indicates the d-dimensional volume of Γ. In the context of interpolation and surrogate modeling, applications most often seek an approximation that is accurate for every point in the domain, hence, the error is commonly measured in the $L^{\infty}(\Gamma)$ norm. However, continuous functions are dense in $C^0(\Gamma)$ but not in $L^{\infty}(\Gamma)$ and a discontinuous $f(\boldsymbol{x})$ cannot be approximated by any set of continuous basis functions. A discontinuous basis suffers from similar limitation, unless the discontinuities in $\phi_j(\boldsymbol{x})$ can be aligned to the discontinuity in $f(\boldsymbol{x})$, which is seldom feasible. Thus, when $f(\boldsymbol{x})$ is discontinuous, the approximation error should be measured in the more appropriate way, which usually depends on the specific application.

For detail error analysis and convergence results for different classes of functions, see [6]. In our context, we are interested in constructing an interpolant that reaches desired tolerance $\epsilon > 0$ with fewest nodes, hence, we need an easily computable error indicator for every possible node. Observe that for any $\boldsymbol{j} \notin X$

$$|s_j| = |f(\boldsymbol{x}_j) - G_X[f](\boldsymbol{x}_j)| \leq \|f(\boldsymbol{x}) - G_X[f](\boldsymbol{x})\|_{\infty}, \tag{9}$$

thus, $|s_j| \leq \epsilon$ is a necessary, albeit not a sufficient condition for accuracy. Nevertheless, s_j is a good indicator for the local approximation error. Furthermore, disjoint basis supports guarantees that s_j are the same for all choices of X such that $A_j \subset X$, in which case s_j is also called the hierarchical surplus associated with $\boldsymbol{x}_j$ and $\phi_j(\boldsymbol{x})$. In Sect. 3 we present several algorithms that rely on s_j to construct an accurate approximation with minimal number of function evaluations, but first we present an alternative hierarchy of nodes and piece-wise constant basis.

2.3 Piece-Wise Constant Hierarchy

Piece-wise constant approximation is desired in some applications, e.g., when approximating models with very sharp or discontinuous response. Here we present a new hierarchy of piece-wise constant basis functions, which we derive from a series of non-hierarchical zero-order interpolants.

Suppose that $I[f](x)$ is an interpolant of $f(x) : [-1, 1] \to \mathbb{R}$ on a set of nodes $\{x_i\}_{i=0}^N \subset [-1, 1]$ with piece-wise constant basis $\phi_i(x)$, then, for any $x \in [-1, 1]$, $I[f](x) = f(x_i)$ for some i. Consider the relation between the distribution of the nodes, the support of the basis, and the approximation error. If $f(x)$ is absolutely continuous, then $|f(x) - f(x_i)| \leq C|x - x_i|$, for some constant C; and for a general $f(x) \in L^2([-1, 1])$, by definition of the Lebesgue integral [37]:

$$\|f(x) - I[f](x)\|_2 \to 0, \quad \text{as } N \to \infty \quad \text{and} \quad \max_{x\in[-1,1]} |x - x_i| \to 0,$$

where for each x in the max we take the corresponding x_i from $I[f](x) = f(x_i)$. Using this estimate, the interpolant that is generally optimal (i.e., not considering specific structure of $f(x)$), is the one that minimizes $|x - x_i|$. Specifically, we seek an interpolant that minimizes the following two conditions:

1. $I[f](x) = f(x_i)$, where x_i is the node nearest to x;
2. The distribution of the nodes $\{x_i\}_{i=0}^N$ is such that it minimizes the distance to the nearest node, i.e.,

$$\{x_0, x_1, \dots, x_N\} = \operatorname*{argmin}_{x_0,x_1,\dots,x_N\in[-1,1]} \max_{x\in[-1,1]} \min_{0\leq i\leq N} |x_i - x|. \tag{10}$$

Condition 1 implies that x_i is in the middle of the support of $\phi_i(x)$ and condition 2 implies that the nodes are uniformly distributed and strictly in the interior of $[-1, 1]$. Note that condition (10) is specific to zero order approximation, for example, when using the nodal (non-hierarchical) linear basis functions associated with the basis in Figs. 1 and 3 (commonly known as "hat functions"), the approximation is constructed from the nearest two nodes, hence the nodes are equidistant and include the boundary.

The hallmark sparse grid paper by Bungartz and Griebel [6] presents a piece-wise constant hierarchy; however, that example violates both of our optimality conditions. Furthermore, the hierarchical surpluses depend on a choice of left/right continuity of the basis, which is ambiguous and therefore undesirable.

The two conditions presented here give us unambiguous way to construct piece-wise constant approximation with N nodes; however, we desire a hierarchy of nodes and functions, thus we need to choose the number of nodes on each level as well as the parents-offspring relation sets. The specific construction that we propose is illustrated in Fig. 4. At level 0 we use one node, i.e., $x_0 = 0$. The next level requires at least two more nodes, otherwise we cannot satisfy the equidistant property, i.e., adding a single node to the left or right of 0 will break the symmetry; thus, we

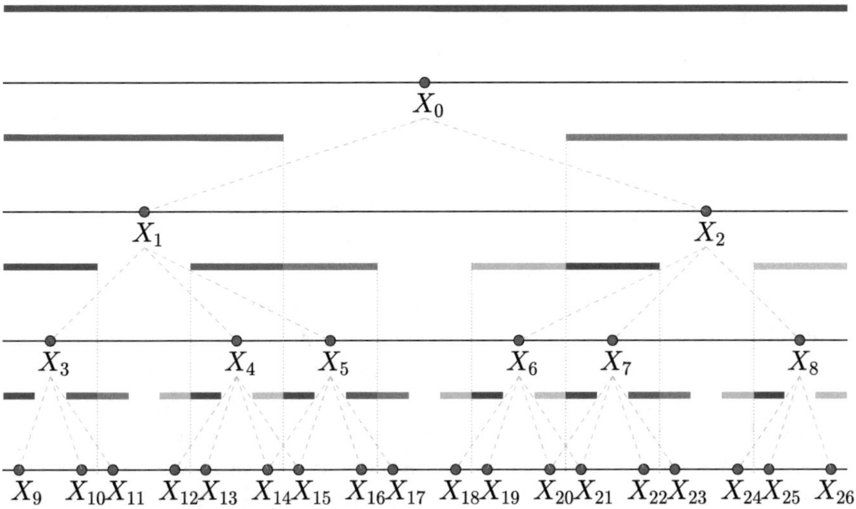

Fig. 4 One dimensional hierarchy of nodes and piece-wise constant basis functions. Note that in this hierarchy interior nodes have multiple parents and up to four children

have $x_1 = -2/3$ and $x_2 = 2/3$. Going to the third level, we have a choice of the number of nodes to add between x_1 and x_0. If we add a single node half-way (i.e., at $1/3$) then satisfying the equidistant property would imply adding a node at -1 which would in turn violate the optimality condition (10). Furthermore, $1/3$ is right on the edge of the support of $\phi_1(x)$ and the surplus of node $1/3$ would depend on whether $\phi_1(x)$ is left continuous. Therefore, we elect to add two nodes between x_1 and x_0, namely $x_4 = -4/9$ and $x_5 = -2/9$, which leads to $x_3 = -8/9$; x_6, x_7 and x_8 follow by symmetry. Using the same logic, on the next level, we add two nodes between any two existing nodes and one node near each boundary (but not on the boundary). The number of nodes triples on each level and the explicit formula is

$$x_j = \begin{cases} -2.0 + (3j+2)\Delta x_j, & \text{when } j \text{ is even,} \\ -2.0 + (3j+1)\Delta x_j, & \text{when } j \text{ is odd,} \end{cases}$$

where $\Delta x_j = 3^{-\lceil \log_3(j+1) \rceil}$ and $\lceil z \rceil = \min\{i \in \mathbb{N} : i \geq z\}$ is the ceiling function. The nodes on the third level, i.e., $x_3, \ldots, x_8$, are not equidistant; however, the optimality condition (10) is satisfied when we combine all previous levels, i.e., nodes $x_0, \ldots, x_8$ are equidistant.

Each $\phi_j(x)$ is supported on the interior of $(x_j - \Delta x_j, x_j + \Delta x_j)$. The mid-point $x_j + \Delta x_j$ is at an equal distance from both x_j and x_{j+1} and following the error estimate, there is no unambiguous way to associate the value with either $f(x_j)$ or $f(x_{j+1})$; however, left/right continuity does not affect the hierarchical coefficients and the choice makes little difference in practice. For example, we assume that the boundary functions $\phi_j(x)$ for $j \in \{3^{\mathbb{N}}, 3^{\mathbb{N}} - 1\}$ are always supported on a closed interval and all other functions are right continuous.

The complete hierarchy requires a parent-offspring relation, see Fig. 4. Clearly, $P_0 = \emptyset$ and $P_1 = P_2 = \{0\}$. Then, following the principle that the parent is the closest node on the previous level, we have $P_3 = P_4 = P_5 = \{1\}$ and $P_6 = P_7 = P_8 = \{2\}$. It is important to note that the support of $\phi_5(x)$ falls outside of the support of the parent $\phi_1(x)$, which is required so that the descendants of x_5 are dense in the interval $(-1, x_0)$, i.e., the closure of $\{x_i : 1 \in A_i\}$ contains $(-1, 0)$. In other approximation schemes, e.g., projection and finite element method, the (projection) coefficient of a basis function contains information about the behavior of $f(x)$ over the full support; but, in our interpolation context, the surpluses contain information localized at the nodes. For example, if s_1 is large, that implies $f(x)$ exhibits sharp behavior (e.g., jump or large gradient) somewhere between x_1 and x_0, and it is incorrect to assume that the transition has occurred within the support of $\phi_1(x)$. Hence, the descendants of a node must cover the entire interval between the closest nodes on the current and previous levels. This is always satisfied for the hierarchies in Figs. 1, 2 and 3 since each $\phi_i(x)$ has ancestors at the edge of the support; for our piece-wise constant hierarchy, some descendants must lie outside of the support of the parent.

Analogously, a large surplus for x_5 indicates a transition between the nearest nodes x_4 and x_0; the transition of $f(x)$ can still occur within the support of $\phi_4(x)$ even if the surplus s_4 is small. Therefore, the descendants of x_5 must include x_{14}, i.e., $5 \in P_{14}$. Furthermore, $4 \in P_{14}$ due to the overlap in support between $\phi_4(x)$ and $\phi_{14}(x)$. A quirk of this hierarchy is the fact that two of every six nodes must share two parents. While this presents a challenge in a one dimensional context, the multidimensional hierarchy always results in multiple parents, hence, this is not of specific consideration when $d > 1$. The parents and offspring are given explicitly by:

$$P_j = \begin{cases} \{\lfloor j/3 \rfloor, \lfloor j/3 \rfloor + 1\}, & \text{when } j \notin \{3^{\mathbb{N}}, 3^{\mathbb{N}} - 1\} \text{ and } \frac{j-2}{6} \in \mathbb{N} \text{ or } \frac{j-3}{6} \in \mathbb{N}, \\ \{\lfloor j/3 \rfloor\}, & \text{otherwise}, \end{cases}$$

$$O_j = \begin{cases} \{3j, 3j+1, 3j+2\}, & \text{when } j \in \{3^{\mathbb{N}}, 3^{\mathbb{N}} - 1\}, \\ \{3j, 3j+1, 3j+2, 3j+3\}, & \text{when } j \notin \{3^{\mathbb{N}}, 3^{\mathbb{N}} - 1\} \text{ and } j \text{ is even}, \\ \{3j-1, 3j, 3j+1, 3j+2\}, & \text{when } j \notin \{3^{\mathbb{N}}, 3^{\mathbb{N}} - 1\} \text{ and } j \text{ is odd}, \end{cases}$$

where $\{3^{\mathbb{N}}, 3^{\mathbb{N}} - 1\}$ is the set of boundary indexes, $\{1, 2, 3, 8, 9, 26, 27, 80, 81, \ldots\}$, and on the first level $O_0 = \{1, 2\}$ and $P_0 = \emptyset$.

3 Adaptive Interpolation

The four hierarchical rules described in the previous section give flexibility in the interpolation scheme so we can choose a family of basis functions suitable for approximating a specific $f(\boldsymbol{x})$; however, we also need to select a suitable multi-index set X. Standard sparse grid construction uses level sets X_l defined recursively

$$X_0 = \{\mathbf{0}\}, \qquad X_l = X_{l-1} \bigcup \left(\bigcup_{\boldsymbol{j} \in X_{l-1}} \bigcup_{k=1}^{d} O_{\boldsymbol{j},k} \right), \tag{11}$$

and it has been shown that sparse grids interpolants produce convergent approximation for large classes of functions [6, 17, 36, 38]. However, such construction does not take into account the specific $f(\boldsymbol{x})$ and the nodes are distributed uniformly and isotropically, i.e., the children for every direction are considered for all nodes. A more advanced interpolation technique constructs X_l in manner that adapts to the specific target function.

Using the necessary condition (9) and assuming that the coefficients $s_{\boldsymbol{j}}$ decay monotonically, i.e., $|s_{\boldsymbol{j}}| \leq |s_{\boldsymbol{i}}|$ for $\boldsymbol{j} \in O_{\boldsymbol{i}}$, we seek the minimal set X such that $|s_{\boldsymbol{j}}| \geq \epsilon$. In general, the larger the index set X, the more accurate the interpolant becomes; however, we want to avoid nodes with small $|s_{\boldsymbol{j}}|$, since (9) implies that those have negligible contribution to the accuracy and each node requires expensive evaluations of $f(\boldsymbol{x})$. Thus, we are willing to sacrifice *some* accuracy for the sake of avoiding nodes with small surpluses.

Algorithm 1 outlines a general adaptive approach that aims at achieving a desired error tolerance ϵ by constructing a set X using $s_{\boldsymbol{j}}$ as a local error estimator. Specifically, X is the final set in a finite sequence $\{X_l\}$, where $X_{l+1} = X_l \bigcup R_l$ for some refinement set R_l. The classical approach for selecting R_l first considers the set of indexes of coefficients larger than ϵ, i.e.,

$$B = \{\boldsymbol{j} \in X_l : |s_{\boldsymbol{j}}| > \epsilon\}, \quad \text{then} \quad R_l = \bigcup_{\boldsymbol{j} \in B_l} \left(\bigcup_{k=1}^{d} O_{\boldsymbol{j},k} \right). \tag{12}$$

The classical refinement strategy has been successfully applied to many problem in science and engineering [7, 27, 28, 36, 42, 45], unfortunately, this approximation scheme is not always stable. Condition (9) is not sufficient and the hierarchical surpluses do not always decay monotonically, especially for the first few levels of the hierarchy. The nodes in the multidimensional hierarchy from a directed acyclic graph with edges corresponding to the parent-offspring relation, and most nodes are associated with multiple acyclic paths leading back to the root $\boldsymbol{x_0}$. Following (12), each node $\boldsymbol{j} \in X_l$ has at least one path going back to $\boldsymbol{x_0}$, but some paths may include nodes not present in X. Following Algorithm 1, a path is terminated when the hierarchical coefficient of one of the nodes falls below ϵ, thus, if the coefficients do not decay monotonically, a path may be terminated before it reaches an $\boldsymbol{x}_j$ with large coefficient, which leads on an issue with missing parents. This can be the case if $f(\boldsymbol{x})$ exhibits sharp localized behavior, the coarse levels of the grid will not to put nodes close to the sharp region and the nodes on the coarse levels will have smaller surpluses. The commonly used workaround to this problems is to select a large initial set X_0, e.g., a full grid for some pre-defined level selected according (11), or recursively add all parents of all nodes, ensuring there are no missing paths in the graph. Selecting large X_l can improve stability and avoid stagnation; however, this also implies adding nodes with small surpluses, which leads to excessive computational burden.

Algorithm 1: Spatially adaptive sparse grids

Data: Given a hierarchy rule, $f(\boldsymbol{x})$ and desired tolerance ϵ
Result: An interpolant $G_X[f](\boldsymbol{x})$ that aims to achieve tolerance ϵ
Let $l = 0$, $X_0 = \{\boldsymbol{0}\}$ and construct $G_{X_0}[f](\boldsymbol{x}) = f(\boldsymbol{x_0})\phi_{\boldsymbol{0}}(\boldsymbol{x})$;
repeat
  Select R_l using one of the methods (12), (14), (16), or (17);
  Let $X_{l+1} = X_l \bigcup R_l$;
  Construct $G_{X_{l+1}}[f](\boldsymbol{x})$ by evaluating $f(\boldsymbol{x}_j)$ for all $\boldsymbol{j} \in R_l \setminus X_l$;
  Let $l = l + 1$;
until $R_{l-1} = \emptyset$;
Finally, return $X = X_l$

For example, consider the function

$$f(\boldsymbol{x}) = \frac{1}{1 + \exp\left(16 - 40\sqrt{x^{(1)}x^{(1)} + x^{(2)}x^{(2)}}\right)}, \qquad x^{(1,2)} \in [-1, 1]. \tag{13}$$

We apply Algorithm 1 with the standard hierarchy (1), cubic ($p = 3$) basis functions described in (4), and classical refinement strategy (12) with $\epsilon = 10^{-4}$ and X_0 containing 705 interpolation nodes. After 13 iterations, the algorithm stagnates, i.e., for each l the algorithm introduces 160 new nodes around four symmetric points defined by $x^{(1)} = \pm 0.265625$ and $x^{(2)} = \pm 0.265625$. The left plots in Fig. 5 show the nodes clustering towards $(x^{(1)}, x^{(2)}) = (0.265625, 0.265625)$, but never reaching that node. According to the hierarchy, $x_{104} = 0.265625$, but most ancestors of $\boldsymbol{j} = (104, 104)$ are located away from the region where $f(\boldsymbol{x})$ exhibits sharp behavior and therefore have small coefficients, while the cluster nodes fall within the sharp region but cannot resolve the local dynamics without the parent.

We propose an alternative refinement strategy that considers both parents and children of nodes with large coefficients. In addition to B_l, for each node, we identify the set of "orphan" directions, i.e.,

$$T_j = \{k \in \{1, 2, \ldots, d\} : P_{j,k} \not\subset X_l\}.$$

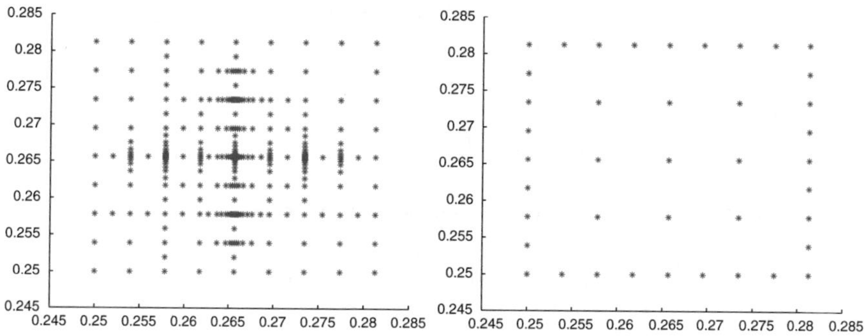

Fig. 5 The nodes near $(x_1, x_2) = (0.265625, 0.265625)$ resulting from Algorithm 1 applied to (13). **Left**: using classical adaptivity (12) the method clusters nodes near the critical point, but fails to converge. **Right**: the hierarchy adaptive scheme (14) needs but a single node to resolve $f(\boldsymbol{x})$ in the region

Then, the refinement strategy favors the parents in the orphan directions, i.e., we add the children only if all the parents are present. The parental aware refinement set is

$$R_l = \bigcup_{\boldsymbol{j}\in B_l} \left(\left(\bigcup_{k\in T_j} P_{\boldsymbol{j},k} \right) \bigcup \left(\bigcup_{k\notin T_j} O_{\boldsymbol{j},k} \right) \right). \tag{14}$$

Note that we do not recursively include all ancestors of $\boldsymbol{j} \in B_l$ as many of those may lie far away from $\boldsymbol{x}_{\boldsymbol{j}}$ and may indeed have small coefficients. Large $s_{\boldsymbol{j}}$ indicate sharp *local* behavior of $f(\boldsymbol{x})$ and we only consider the intermediate patents, since those include the closest ancestors and are most likely to also fall in the transition region. It is possible that the algorithm would include all ancestors over several iterations, but this will happen only if the majority of the ancestors have large coefficients.

Applying the hierarchy selective strategy (14) to function (13), the algorithm converges with final approximation using 9833 nodes. In contrast, if we impose the restriction that all parents have to be included in each step, there is no stagnation but the algorithm terminates with a grid requiring 59,209 nodes. Using a full level sparse grid that includes $\boldsymbol{j} = (104,104)$ would require 147,457 nodes. Both classical solutions are at best impractical and often times infeasible when evaluations of $f(\boldsymbol{x})$ are computationally expensive.

Observe that the ancestry aware refinement (14) treats the d-directions differently for each node, which is in contrast to the isotropic classical refinement (12). In many applications, variability in the d inputs to $f(\boldsymbol{x})$ does not have equal influence on the response and anisotropic approximation strategies have been considered [3, 7, 25, 26, 30, 41]; however, those were usually applied in a context of approximation with globally supported Lagrange polynomials, or data mining problem using large number of random samples. In this work, we propose a direction adaptive strategy that refines in a subset of the d directions. As a directional error estimator, we propose the coefficient of an interpolant constructed from the nodes aligned in that direction. Specifically, for each $\boldsymbol{j} \in X$, let $W_{\boldsymbol{j},k}$ correspond to the nodes aligned with $\boldsymbol{x}_{\boldsymbol{j}}$ in the k-th direction

$$W_{\boldsymbol{j},k} = \{\boldsymbol{i} \in X : \boldsymbol{i}_{-k} = \boldsymbol{j}_{-k}\}.$$

Then, we construct the interpolant associated with $W_{\boldsymbol{j},k}$, i.e.,

$$G_{W_{\boldsymbol{j},k}}[f](\boldsymbol{x}) = \sum_{\boldsymbol{i} \in W_{\boldsymbol{j},k}} c_{\boldsymbol{i},k} \phi_{\boldsymbol{i}}(\boldsymbol{x}), \tag{15}$$

where we put k in the subscript of $c_{\boldsymbol{i},k}$ to indicate explicitly the direction that we are considering. For each node, (15) is a one dimensional hierarchical interpolant with much fewer nodes and the coefficients can be computed with the same negligible complexity as (7). Furthermore, the nodes $W_{\boldsymbol{j},k}$ are aligned in a single direction and thus $c_{\boldsymbol{j},k}$ is a not only local but also directional error indicator. The refinement set R_l is constructed using only directions where $c_{\boldsymbol{j},k}$ is large, i.e., let

$$C_{\boldsymbol{j}} = \left\{k \in \{1, 2, \ldots, d\} : |c_{\boldsymbol{j},k}| > \epsilon\right\}$$

then, the direction selective refinement set is

$$R_l = \bigcup_{\boldsymbol{j} \in B_l} \left(\bigcup_{k \in C_{\boldsymbol{j}}} O_{\boldsymbol{j},k} \right) \tag{16}$$

Combining (14) and (16) yields the fully adaptive selection set

$$R_l = \bigcup_{\boldsymbol{j} \in B_l} \left(\left(\bigcup_{k \in C_{\boldsymbol{j}} \cap T_{\boldsymbol{j}}} P_{\boldsymbol{j},k} \right) \bigcup \left(\bigcup_{k \in C_{\boldsymbol{j}} \setminus T_{\boldsymbol{j}}} O_{\boldsymbol{j},k} \right) \right) \tag{17}$$

The function presented in (13) is globally isotropic, i.e., it is invariant under rotation around the origin; however, $f(\boldsymbol{x})$ is anisotropic on a local scale. We apply the fully adaptive strategy (17) to the function and observe that Algorithm 1 terminates after only 8085 evaluations of $f(\boldsymbol{x})$. The nearly 18% improvement is significant in many applications, e.g., [15].

Remark in the above example we consider the number of interpolation nodes; however, the final accuracy of the interpolant is also important. Reducing the number of nodes and refining in only select directions, i.e., using strategy (16), increases the likelihood of encountering missing parents and stagnation. Furthermore, fewer interpolation nodes, i.e., using either (16) or (17), generally results in larger error. However, in all our examples, the neglected nodes have small coefficients and the final approximation is still $O(\epsilon)$, hence a direction reduction strategy is a viable option in reducing the total number of interpolation nodes.

4 Examples

In this section, we present numerical results from applying the sparse grids interpolation techniques to several functions, as well as the output to a system of differential equations. The sparse grid construction is done using the Oak Ridge National Laboratory Toolkit for Adaptive Stochastic Modeling and Non-Intrusive ApproximatioN (TASMANIAN) [39].

The examples presented here focus on two dimensional problems in order to apply brute-force error estimates. In examples where $f(\boldsymbol{x}) \in C^0(\Gamma)$, we measure the $L^\infty(\Gamma)$ error by evaluating $f(\boldsymbol{x})$ and $G_X[f](x)$ on a dense grid with thousands of points. Forming a dense grid in higher dimensions would be prohibitive even if $f(\boldsymbol{x})$ is simple to evaluate. In the example where $f(\boldsymbol{x})$ is a discontinuous disk, we consider how well the integral of $G_X[f](x)$ approximates the area of the disk.

4.1 Influence of the Type of Hierarchy

Consider the order of approximation of different hierarchies when applied a smooth function. Specifically, consider

$$f(\boldsymbol{x}) = \exp\left(x^{(1)}x^{(1)} + x^{(2)}x^{(2)}\right), \qquad \left(x^{(1)}, x^{(2)}\right) \in [-1, 1] \otimes [-1, 1]. \quad (18)$$

Figure 6 shows a comparison between different orders of approximation, as expected increasing p results in much lower error. In addition, using the semi-local hierarchy described in Fig. 3 allows for even higher accuracy, which can lead to significant savings when dealing with complex $f(\boldsymbol{x})$. We estimate the error from 10,000 samples on a dense Cartesian grid.

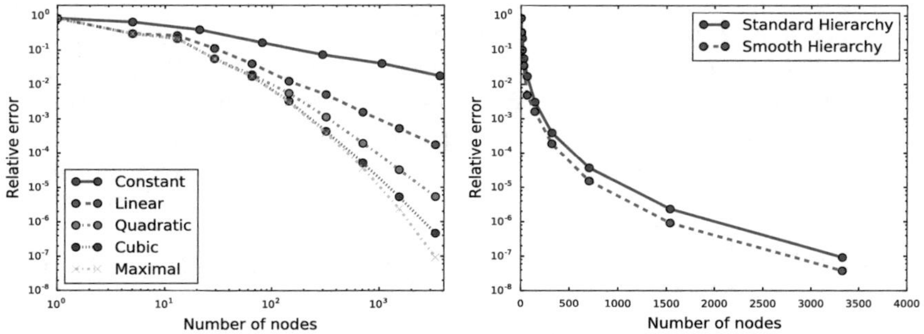

Fig. 6 Comparison of different hierarchical rules applied to function (18). **Left**: standard hierarchy, in Fig. 1, with functions of increasing order $p = 0, 1, 2, 3$ and maximum available based on the set of ancestors. **Right**: standard hierarchy, given in Fig. 1, and the higher order hierarchy, given in Fig. 3, both using maximal p

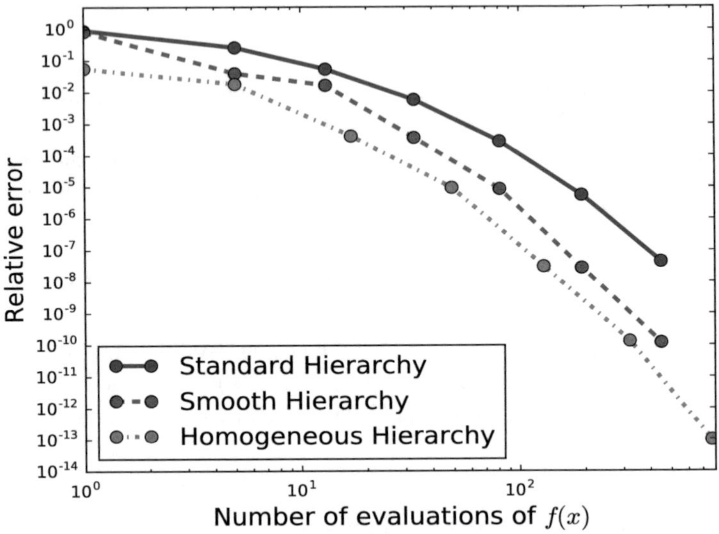

Fig. 7 Comparison of different hierarchical rules applied to function (19), the error is estimated from a dense grid with 10,000 samples. The abscissa counts total function evaluations in the interior of the domain, nodes along the boundary are not counted

Smoothness of $f(\boldsymbol{x})$ is only one of the many useful properties that hierarchal approaches can exploit, in many applications we have prior knowledge of $f(\boldsymbol{x})$ around the boundary of Γ. Consider the function with homogeneous boundary

$$f(\boldsymbol{x}) = \cos\left(\frac{\pi}{2}x^{(1)}\right)(x^{(2)}x^{(2)} - 1), \qquad \left(x^{(1)}, x^{(2)}\right) \in [-1, 1]\otimes[-1, 1]. \quad (19)$$

We test the standard hierarchy, semi-local hierarchy with globally supported $\phi_1(\boldsymbol{x})$ and $\phi_2(\boldsymbol{x})$, and the hierarchy presented in Fig. 2, all using maximal order of polynomial approximation. Figure 7 shows the comparison between the three hierarchies and we clearly see the advantage of incorporating the boundary conditions directly into $f(\boldsymbol{x})$. Note that in error plot, we only consider the required evaluations of $f(\boldsymbol{x})$ in the interior of the domain, i.e., nodes placed on the boundary are not counted.

We also consider the advantages of a piece-wise constant approximation of a discontinuous function, specifically we consider the indicator functions of a disk

$$f(\boldsymbol{x}) = \begin{cases} 1, & \sqrt{x^{(1)} + x^{(2)}} \leq \frac{1}{2}, \\ 0, & \text{otherwise}, \end{cases} \qquad \left(x^{(1)}, x^{(2)}\right) \in [-1, 1] \otimes [-1, 1]. \quad (20)$$

No hierarchy with continuous basis functions could approximate (20) in the $L^{\infty}(\Gamma)$. Neither, is such approximation feasible to the discontinuous hierarchy presented in Fig. 4, since the support of the basis does not align with the edges of the disk. We use an alternative measure of accuracy, specifically, we look at the integral of the

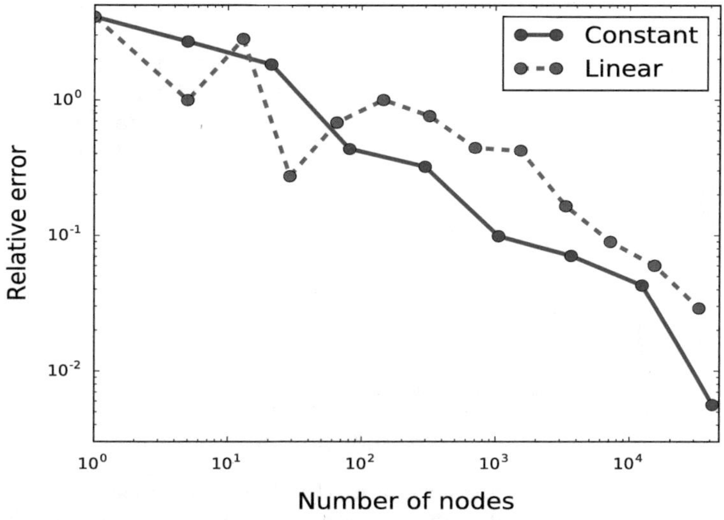

Fig. 8 Comparison between hierarchies with piece-wise constant and linear basis functions when approximating the integral of the disk indicator function (20)

sparse grid interpolant and compare to the actual area of the disk, which is $\frac{\pi}{4}$. This problem is similar to the one considered in [44], with the main difference that we employ a full sparse grid instead of a hyper-spherical coordinate transformation. Figure 8 shows the result form the comparison; while the linear hierarchy does produce better approximation in some cases when the grid is course (i.e., error is >10%), the linear basis has very erratic behavior. In contrast, the piece-wise constant hierarchy converges monotonically (i.e., has stable behavior) and gives much better approximation when the level increases.

4.2 *Influence of the Refinement Method*

In Sect. 3 we gave an example of a refinement strategy that fails due to missing parents. In this section, we consider the advantages of the anisotropic refinement schemes (16) and (17). For the first example, we consider the globally isotropic function presented in (18). Using global adaptive techniques, i.e., [30, 41], there is no way to distinguish between the $x^{(1)}$ and $x^{(2)}$ direction; however, the function is still locally anisotropic. We use a sequence of values of ϵ

$$\{5 \times 10^{-2}, 10^{-2}, 5 \times 10^{-3}, 10^{-3}, 5 \times 10^{-4}, 10^{-4}, 5 \times 10^{-5}, 10^{-5}, 5 \times 10^{-6}, 10^{-6}\}$$

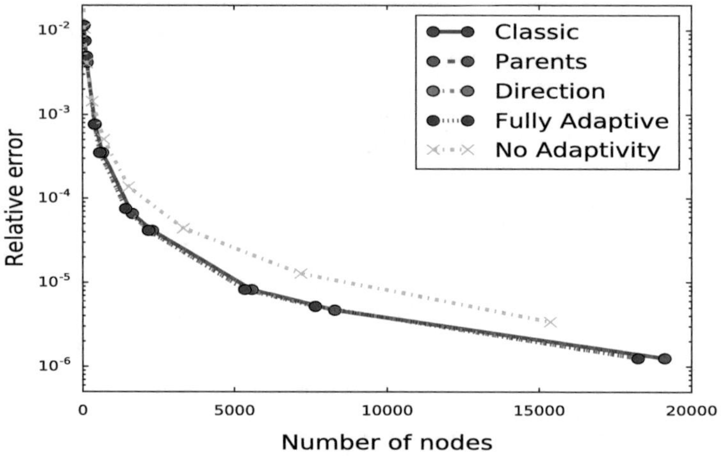

Fig. 9 Comparison of applying Algorithm 1 to the globally isotropic function (18) with different refinement strategies. The locally anisotropic schemes (16) and (17) result in noticeable 4–5% improvement, albeit hard to see on the plot

and we execute Algorithm 1 with each one starting from the same initial state where the sum of levels of all nodes is no more than 3. We compare the final error in the sparse grids approximation, estimated from a dense Cartesian grid of 10,000 samples, and the final number of interpolation nodes. Figure 9 shows the result of applying the four refinement strategies and, for reference purposes, the sparse grids constructed without adaptivity. Every adaptive method overtakes the "blind" unrefined construction, and the best performance is given by the two anisotropic refinement strategies. The semi-log plot highlights the 4–5% improvement, which is small but significant for many scientific and engineering applications, thus local anisotropy can be exploited even for globally isotropic problems. In this particular example, the parental refinement has no effect on the approximation.

We also consider a function with weak globally anisotropic behavior, specifically

$$f(\boldsymbol{x}) = \exp\left(-x^{(1)}x^{(1)}\right)\cos\left(\frac{\pi}{2}x^{(2)}\right), \qquad \left(x^{(1)}, x^{(2)}\right) \in [-1,1]\otimes[-1,1]. \tag{21}$$

The anisotropy comes from the different expressions associated with the two components $x^{(1)}$ and $x^{(2)}$. We repeat the test performed to (18) but using the new function. Figure 10 shows the result of applying the four refinement strategies and the unrefined grid for reference purposes. The anisotropy is more strongly present in this example, resulting in overall 9–10% savings over the classical refinement strategy.

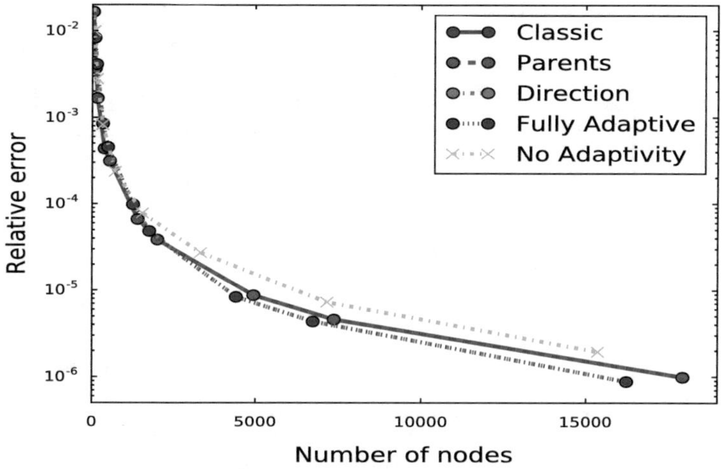

Fig. 10 Comparison of applying Algorithm 1 to the anisotropic function (21) with different refinement strategies. The locally anisotropic schemes (16) and (17) result in noticeable 9–10% improvement

4.3 Application to Kermack-McKendrick SIR Model

The Kermack-McKendrick SIR model [23, 24] for the spread of an infectious disease is given by the system of ordinary differential equations

$$\begin{aligned} \tfrac{d}{dt}F_I(t) &= \beta F_I(t)S(t) - \gamma F_I(t), & F_I(0) &= 0.05, \\ \tfrac{d}{dt}F_S(t) &= -\beta F_I(t)S(t), & F_S(0) &= 0.95, \\ \tfrac{d}{dt}F_R(t) &= \gamma F_I(t), & F_R(0) &= 0.00, \end{aligned} \tag{22}$$

where $t > 0$, and β and γ are parameters that we vary in the range $[0.1, 0.3]$. In this context, $F_I(t)$, $F_S(t)$ and $F_R(t)$ correspond to the infected susceptible and removed portion of the population, and β and γ are the infection and recovery rates respectively. We are interested in the value of the removed variable at $T = 30$, i.e.,

$$f(\beta, \gamma) = F_R(30), \tag{23}$$

where $F_R(t)$ satisfies (22). Using a linear transformation we translate the interval $[0.1, 0.3]$ to the canonical $[-1, 1]$. For the purpose of the numerical simulation, we integrate (22) with Runge-Kutta 4-th order method [9] with time step $\Delta t = 10^{-3}$, the high accuracy of the 4-th order method eliminates the consideration of the numerical error when constructing the interpolant.

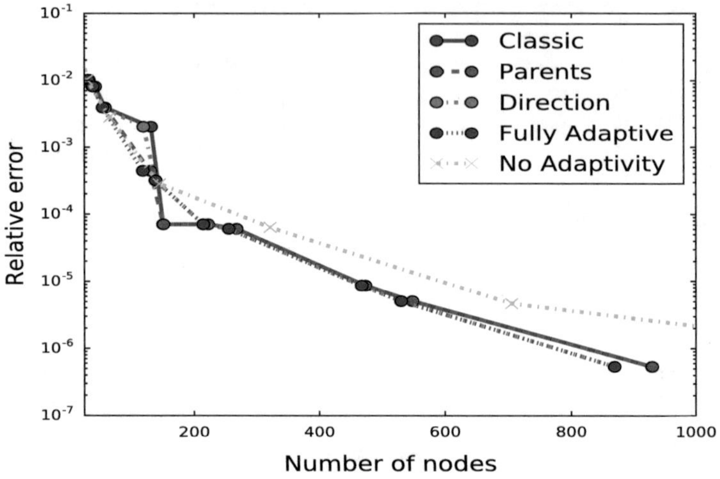

Fig. 11 Comparison between refinement schemes applied to the system (22). All refinement schemes provide significant improvement in accuracy as compared to the unrefined grid

Figure 11 shows the result of applying Algorithm 1 and the four refinement schemes to the SIR model. We vary the refinement tolerance from 5×10^{-2} to 10^{-6} and compare the accuracy of the constructed grids. As expected, all refinement strategies outperform the non-adaptive approximation scheme. Comparing the different selection strategies for R_l, the anisotropic refinement of the direction (16) and fully adaptive (17) schemes outperforms the isotropic approaches; however, this trend is not universally true for all values of the refinement tolerance. As for the parent-to-child refinement, including the parents in to the fully adaptive scheme results in more monotonic behavior and therefore improved stability. While it is not possible to guarantee that the interpolant constructed using the fully adaptive scheme will always outperform the other approaches, it is generally the best first choice when selecting a refinement strategy.

5 Conclusions

In this work, we presented several schemes for constructing an interpolant to a multidimensional functions defined over a hypercube domain. The multidimensional construction uses tensors of one dimensional nodes and basis functions that form a hierarchy, which allows for adaptive interpolation that aims to maximize accuracy while minimizing the number of interpolation nodes. We present several hierarchical rules, including a novel piece-wise constant approach; as well as several strategies for adaptive refinement including locally anisotropic approach based on one dimensional surplus coefficients. In several numerical examples, we demonstrate some advantages and drawbacks of the different methods.

Acknowledgements This research was supported by the Exascale Computing Project (17-SC-20-SC), a collaborative effort of the U.S. Department of Energy Office of Science and the National Nuclear Security Administration; the U.S. Defense Advanced Research Projects Agency, Defense Sciences Office under contract and award numbers HR0011619523 and 1868-A017-15; and by the Laboratory Directed Research and Development program at the Oak Ridge National Laboratory, which is operated by UT-Battelle, LLC., for the U.S. Department of Energy under Contract DE-AC05-00OR22725.

References

1. I. Babuška, R. Tempone, G.E. Zouraris, Galerkin finite element approximations of stochastic elliptic partial differential equations. SIAM J. Numer. Anal. **42**(2), 800–825 (2004)
2. J. Bäck, F. Nobile, L. Tamellini, R. Tempone, Stochastic spectral galerkin and collocation methods for PDEs with random coefficients: a numerical comparison, in *Spectral and High Order Methods for Partial Differential Equations* (Springer, Berlin, 2011), pp. 43–62
3. J. Bäck, F. Nobile, L. Tamellini, R. Tempone, Convergence of quasi-optimal stochastic Galerkin methods for a class of PDEs with random coefficients. Comput. Math. Appl. **67**(4), 732–751 (2014)
4. P. Binev, A. Cohen, W. Dahmen, R. DeVore, G. Petrova, P. Wojtaszczyk, Convergence rates for greedy algorithms in reduced basis methods. SIAM J. Math. Anal. **43**(3), 1457–1472 (2011)
5. H.-J. Bungartz, *Finite Elements of Higher Order on Sparse Grids* (Shaker, Maastricht, 1998)
6. H.-J. Bungartz, M. Griebel, Sparse grids. Acta Numer. **13**, 147–269 (2004)
7. H.-J. Bungartz, D. Pflüger, S. Zimmer, Adaptive sparse grid techniques for data mining, in *Modeling, Simulation and Optimization of Complex Processes* (Springer, Berlin, 2008), pp. 121–130
8. G. Buse, D. Pfluger, A. Murarasu, R. Jacob, A non-static data layout enhancing parallelism and vectorization in sparse grid algorithms, in *11th International Symposium on Parallel and Distributed Computing (ISPDC), 2012* (IEEE, Washington, 2012), pp. 195–202
9. J.C. Butcher, *Numerical Methods for Ordinary Differential Equations* (Wiley, Hoboken, 2016)
10. M.A. Chkifa, On the Lebesgue constant of Leja sequences for the complex unit disk and of their real projection. J. Approx. Theory **166**, 176–200 (2013)
11. A. Chkifa, A. Cohen, C. Schwab, High-dimensional adaptive sparse polynomial interpolation and applications to parametric PDEs. Found. Comput. Math. **14**(4), 601–633 (2014)
12. A. Chkifa, N. Dexter, H. Tran, C.G. Webster, Polynomial approximation via compressed sensing of high-dimensional functions on lower sets (2016). ArXiv e-prints
13. N.C. Dexter, C.G. Webster, G. Zhang, Explicit cost bounds of stochastic Galerkin approximations for parameterized PDEs with random coefficients. Comput. Math. Appl. **71**(11), 2231–2256 (2016)
14. A. Doostan, H. Owhadi, A non-adapted sparse approximation of PDEs with stochastic inputs. J. Comput. Phys. **230**(8), 3015–3034 (2011)
15. C.E. Finney, M.K. Stoyanov, S. Pannala, C.S. Daw, R.M. Wagner, K.D. Edwards, C.G. Webster, J.B. Green, Application of high performance computing for simulating the unstable dynamics of dilute spark-ignited combustion, in *International Conference on Theory and Application in Nonlinear Dynamics (ICAND 2012)* (Springer, Cham, 2014), pp. 259–270
16. G. Fishman, *Monte Carlo: Concepts, Algorithms, and Applications*. Springer Series in Operations Research (Springer, New York, 1996)
17. J. Garcke, Sparse grids in a nutshell, in *Sparse Grids and Applications* (Springer, Cham, 2012), pp. 57–80

18. M.B. Giles, Multilevel monte carlo methods. Acta Numer. **24**, 259 (2015)
19. J.D. Jakeman, S.G. Roberts, Local and dimension adaptive stochastic collocation for uncertainty quantification, in *Sparse Grids and Applications* (Springer, Berlin, 2012), pp. 181–203
20. J.D. Jakeman, R. Archibald, D. Xiu, Characterization of discontinuities in high-dimensional stochastic problems on adaptive sparse grids. J. Comput. Phys. **230**(10), 3977–3997 (2011)
21. J.D. Jakeman, A. Narayan, D. Xiu, Minimal multi-element stochastic collocation for uncertainty quantification of discontinuous functions. J. Comput. Phys. **242**, 790–808 (2013)
22. P. Jantsch, C.G. Webster, G. Zhang, On the Lebesgue constant of weighted Leja points for Lagrange interpolation on unbounded domains (2016). arXiv preprint arXiv:1606.07093
23. D.S. Jones, M. Plank, B.D. Sleeman, *Differential Equations and Mathematical Biology* (CRC Press, Boca Raton, 2009)
24. W.O. Kermack, A.G. McKendrick, A contribution to the mathematical theory of epidemics, in *Proceedings of the Royal Society of London A: Mathematical, Physical and Engineering Sciences*, vol. 115 (The Royal Society, London, 1927), pp. 700–721
25. V. Khakhutskyy, *Sparse Grids for Big Data: Exploiting Parsimony for Large-Scale Learning*, Ph.D. thesis, Universität München, 2016
26. V. Khakhutskyy, M. Hegland, Spatially-dimension-adaptive sparse grids for online learning, in *Sparse Grids and Applications-Stuttgart 2014* (Springer, Cham, 2016), pp. 133–162
27. A. Klimke, B. Wohlmuth, Algorithm 847: Spinterp: piecewise multilinear hierarchical sparse grid interpolation in MATLAB. ACM Trans. Math. Softw. (TOMS) **31**(4), 561–579 (2005)
28. X. Ma, N. Zabaras, An adaptive hierarchical sparse grid collocation algorithm for the solution of stochastic differential equations. J. Comput. Phys. **228**(8), 3084–3113 (2009)
29. A. Narayan, J.D. Jakeman, Adaptive Leja sparse grid constructions for stochastic collocation and high-dimensional approximation. SIAM J. Sci. Comput. **36**(6), A2952–A2983 (2014)
30. F. Nobile, R. Tempone, C.G. Webster, An anisotropic sparse grid stochastic collocation method for partial differential equations with random input data. SIAM J. Numer. Anal. **46**(5), 2411–2442 (2008)
31. F. Nobile, R. Tempone, C.G. Webster, A sparse grid stochastic collocation method for partial differential equations with random input data. SIAM J. Numer. Anal. **46**(5), 2309–2345 (2008)
32. F. Nobile, L. Tamellini, R. Tempone, Convergence of quasi-optimal sparse-grid approximation of hilbert-space-valued functions: application to random elliptic PDEs. Numer. Math. **134**(2), 343–388 (2016)
33. E. Novak, K. Ritter, Simple cubature formulas with high polynomial exactness. Constr. Approx. **15**(4), 499–522 (1999)
34. D. Pflüger, *Spatially Adaptive Sparse Grids for High-Dimensional Problems* (Verlag Dr. Hut, München, 2010)
35. D. Pflüger, Spatially adaptive refinement, in *Sparse Grids and Applications* (Springer, Berlin, 2012), pp. 243–262
36. D. Pflüger, B. Peherstorfer, H.-J. Bungartz, Spatially adaptive sparse grids for high-dimensional data-driven problems. J. Complex. **26**(5), 508–522 (2010)
37. H. Royden, *Real Analysis*, vol. 1 (Macmillan, New York, 1968), p. 963
38. M. Stoyanov, Hierarchy-direction selective approach for locally adaptive sparse grids. Technical Report ORNL/TM-2013/384, Oak Ridge National Laboratory, Oak Ridge, 2013
39. M. Stoyanov, User manual: Tasmanian sparse grids. Technical Report ORNL/TM-2015/596, Oak Ridge National Laboratory, Oak Ridge, 2015
40. M. Stoyanov, C.G. Webster, A gradient-based sampling approach for dimension reduction of partial differential equations with stochastic coefficients. Int. J. Uncertain. Quantif. **5**(1), 49–72 (2015)
41. M.K. Stoyanov, C.G. Webster, A dynamically adaptive sparse grids method for quasi-optimal interpolation of multidimensional functions. Comput. Math. Appl. **71**(11), 2449–2465 (2016)
42. M. Stoyanov, P. Seleson, C. Webster, Predicting fracture patterns in simulations of brittle materials under variable load and material strength, in *19th AIAA Non-Deterministic Approaches Conference* (2017), p. 1326

43. G. Zhang, M. Gunzburger, Error analysis of a stochastic collocation method for parabolic partial differential equations with random input data. SIAM J. Numer. Anal. **50**(4), 1922–1940 (2012)
44. G. Zhang, C. Webster, M. Gunzburger, J. Burkardt, A hyperspherical adaptive sparse-grid method for high-dimensional discontinuity detection. SIAM J. Numer. Anal. **53**(3), 1508–1536 (2015)
45. G. Zhang, C.G. Webster, M. Gunzburger, J. Burkardt, Hyperspherical sparse approximation techniques for high-dimensional discontinuity detection. SIAM Rev. **58**(3), 517–551 (2016)

Smolyak's Algorithm: A Powerful Black Box for the Acceleration of Scientific Computations

Raúl Tempone and Sören Wolfers

Abstract We provide a general discussion of Smolyak's algorithm for the acceleration of scientific computations. The algorithm first appeared in Smolyak's work on multidimensional integration and interpolation. Since then, it has been generalized in multiple directions and has been associated with the keywords: sparse grids, hyperbolic cross approximation, combination technique, and multilevel methods. Variants of Smolyak's algorithm have been employed in the computation of high-dimensional integrals in finance, chemistry, and physics, in the numerical solution of partial and stochastic differential equations, and in uncertainty quantification. Motivated by this broad and ever-increasing range of applications, we describe a general framework that summarizes fundamental results and assumptions in a concise application-independent manner.

1 Introduction

We study Smolyak's algorithm for the convergence acceleration of general numerical approximation methods

$$\mathscr{A} \colon \mathbb{N}^n := \{0, 1, \ldots, \}^n \to Y,$$

which map discretization parameters $\boldsymbol{k} = (\imath_1, \ldots, k_n) \in \mathbb{N}^n$ to outputs $\mathscr{A}(\boldsymbol{k})$ in a Banach space Y.

For instance, a straightforward way to approximate the integral of a function $f \colon [0, 1]^n \to \mathbb{R}$ is to employ tensor-type quadrature formulas, which evaluate f at the nodes of a regular grid within $[0, 1]^n$. This gives rise to an approximation method where k_j determines the grid resolution in direction of the jth coordinate axis,

R. Tempone · S. Wolfers (✉)
King Abdullah University of Science and Technology (KAUST), Thuwal, Kingdom of Saudi Arabia
e-mail: soeren.wolfers@kaust.edu.sa

J. Garcke et al. (eds.), *Sparse Grids and Applications – Miami 2016*,
Lecture Notes in Computational Science and Engineering 123,
https://doi.org/10.1007/978-3-319-75426-0_9

$j \in \{1, \dots, n\}$. Smolyak himself derived and studied his algorithm in this setting, where it leads to evaluations in the nodes of *sparse grids* [24, 26]. Another example, which emphasizes the distinctness of sparse grids and the general version of Smolyak's algorithm considered in this work, is integration of a univariate function $f: \mathbb{R} \to \mathbb{R}$ that is not compactly supported but exhibits sufficient decay at infinity. In this case, k_1 could as before determine the resolution of regularly spaced quadrature nodes and k_2 could be used to determine a truncated quadrature domain. Smolyak's algorithm then leads to quadrature nodes whose density is high near the origin and decreases at infinity, as intuition would dictate.

To motivate Smolyak's algorithm, assume that the approximation method $\mathscr{A}$ converges to a limit $\mathscr{A}_\infty \in Y$ at the rate

$$\|\mathscr{A}(\boldsymbol{k}) - \mathscr{A}_\infty\|_Y \leq K_1 \sum_{j=1}^{n} k_j^{-\beta_j} \quad \forall \boldsymbol{k} \in \mathbb{N}^n \tag{1}$$

and requires the work

$$\mathrm{Work}(\mathscr{A}(\boldsymbol{k})) = K_2 \prod_{j=1}^{n} k_j^{\gamma_j} \quad \forall \boldsymbol{k} \in \mathbb{N}^n \tag{2}$$

for some $K_1 > 0$, $K_2 > 0$ and $\beta_j > 0$, $\gamma_j > 0$, $j \in \{1, \dots, n\}$. An approximation of $\mathscr{A}_\infty$ with accuracy $\epsilon > 0$ can then be obtained with the choice

$$k_j := -\Big(\frac{\epsilon}{nK_1}\Big)^{-1/\beta_j}, \quad j \in \{1, \dots, n\}, \tag{3}$$

which requires the work

$$C(n, K_1, K_2, \gamma_1, \dots, \gamma_n, \beta_1, \dots, \beta_n)\epsilon^{-(\gamma_1/\beta_1 + \dots + \gamma_n/\beta_n)}. \tag{4}$$

Here and in the remainder of this work we denote by $C(\dots)$ generic constants that depend only on the quantities in parentheses but may change their value from line to line and from equation to equation.

The appearance of the sum $\gamma_1/\beta_1 + \dots + \gamma_n/\beta_n$ in the exponent above is commonly referred to as the *curse of dimensionality*. Among other things, we will show (see Example 1) that if the bound in Eq. (1) holds in a slightly stronger sense, then Smolyak's algorithm can replace this dreaded sum by $\max_{j=1}^{n} \gamma_j/\beta_j$, which means that it yields convergence rates that are, up to possible logarithmic factors, independent of the number of discretization parameters. In the general form presented here, Smolyak's algorithm forms linear combinations of the values $\mathscr{A}(\boldsymbol{k})$, $\boldsymbol{k} \in \mathbb{N}^n$, based on

1. an infinite decomposition of $\mathscr{A}_\infty$ and
2. a knapsack approach to truncate this decomposition.

Since the decomposition is independent of the particular choice of $\mathscr{A}$ and the truncation relies on easily verifiable assumptions on the decay and work of the decomposition terms, Smolyak's algorithm is a powerful black box for the non-intrusive acceleration of scientific computations. In the roughly 50 years since its first description, applications in various fields of scientific computation have been described; see, for example, the extensive survey article [2]. The goal of this work is to summarize previous results in a common framework and thereby encourage further research and exploration of novel applications. While some of the material presented here may be folklore knowledge in the sparse grids community, we are not aware of any published sources that present this material in a generally applicable fashion.

The remainder of this work is structured as follows. In Sect. 2, we introduce the infinite decomposition of $\mathscr{A}_\infty$ that is at the core of Smolyak's algorithm. In Sect. 3, we introduce spaces of approximation methods $\mathscr{A}\colon \mathbb{N}^n \to Y$ that allow for efficient solutions of the resulting truncation problem. In Sect. 4, we derive explicit convergence rates for Smolyak's algorithm in common examples of such spaces. Finally, in Sect. 5, we discuss how various previous results can be deduced within the framework presented here.

2 Decomposition

Smolyak's algorithm is based on a decomposition of $\mathscr{A}_\infty$ that is maybe most simply presented in the continuous setting. Here, Fubini's theorem and the fundamental theorem of calculus show that any function $f\colon \mathbb{R}^n_{\geq} := [0,\infty)^n \to Y$ with $f \equiv 0$ on $\partial\mathbb{R}^n_{\geq}$ satisfies

$$f(x) = \int_{\prod_{j=1}^n [0,x_j]} \partial_1 \dots \partial_n f(s)\, ds \quad \forall x \in \mathbb{R}^n_{\geq}, \tag{5}$$

questions of integrability and differentiability aside. Moreover, if f converges to a limit $f_\infty \in Y$ as $\min_{j=1}^n x_j \to \infty$, then

$$f_\infty = \lim_{\min_{j=1}^n x_j \to \infty} \int_{\prod_{j=1}^n [0,x_j]} \partial_1 \dots \partial_n f(s)\, ds = \int_{\mathbb{R}^n_{\geq}} \partial_{\mathrm{mix}} f(s)\, ds, \tag{6}$$

where we introduced the shorthand ∂_{mix} for the mixed derivative $\partial_1 \dots \partial_n$. The crucial observation is now that an approximation of f_∞ can be achieved not only by rectangular truncation of the integral in Eq. (6), which according to Eq. (5) is equivalent to a simple evaluation of f at a single point, but also by truncation to more complicated domains. These domains should ideally correspond to large values of $\partial_{\mathrm{mix}} f$ in order to minimize the truncation error, but also have to take into consideration the associated computational work.

To transfer the decomposition in Eq. (6) to the discrete setting, we denote by $Y^{\mathbb{N}^n} := \{\mathscr{A} : \mathbb{N}^n \to Y\}$ the space of all functions from $\mathbb{N}^n$ into the Banach space Y. Next, we define the discrete unidirectional difference and sum operators

$$\Delta_j : Y^{\mathbb{N}^n} \to Y^{\mathbb{N}^n}$$

$$(\Delta_j \mathscr{A})(\boldsymbol{k}) := \begin{cases} \mathscr{A}(k_1, \ldots, k_n) - \mathscr{A}(k_1, \ldots, k_{j-1}, k_j - 1, k_{j+1}, \ldots, k_n) & \text{if } k_j > 0, \\ \mathscr{A}(k_1, \ldots, k_n) & \text{else,} \end{cases}$$

$$\Sigma_j := \Delta_j^{-1} : Y^{\mathbb{N}^n} \to Y^{\mathbb{N}^n}$$

$$(\Sigma_j \mathscr{A})(\boldsymbol{k}) := \sum_{s=0}^{k_j} \mathscr{A}(k_1, \ldots, k_{j-1}, s, k_{j+1}, \ldots, k_n),$$

Finally, we introduce their compositions, the mixed difference operator

$$\Delta_{\text{mix}} := \Delta_1 \circ \cdots \circ \Delta_n : Y^{\mathbb{N}^n} \to Y^{\mathbb{N}^n},$$

and the rectangular sum operator

$$\Sigma_R := \Sigma_1 \circ \cdots \circ \Sigma_n : Y^{\mathbb{N}^n} \to Y^{\mathbb{N}^n},$$

which replace the mixed derivative and integral operators that map $f : \mathbb{R}^n \to Y$ to $f \mapsto \partial_{\text{mix}} f$ and $x \mapsto \int_{\prod_{j=1}^n [0,x_j]} f(s)\, ds$, respectively.

The discrete analogue of Eq. (5) is now a matter of simple algebra.

Proposition 1

(i) We have $\Sigma_R = \Delta_{mix}^{-1}$*, that is*

$$\mathscr{A}(\boldsymbol{k}) = \sum_{s_1=0}^{k_1} \cdots \sum_{s_n=0}^{k_n} \Delta_{mix} \mathscr{A}(s_1, \ldots, s_n) \quad \forall \boldsymbol{k} \in \mathbb{N}^n.$$

(ii) We have $\Delta_{mix} = \sum_{\mathbf{e} \in \{0,1\}^n} (-1)^{|\mathbf{e}|_1} S_{\mathbf{e}}$*, where* $S_{\mathbf{e}}$ *is the shift operator defined by*

$$(S_{\mathbf{e}} \mathscr{A})(\boldsymbol{k}) := \begin{cases} \mathscr{A}(\boldsymbol{k} - \mathbf{e}), & \text{if } \boldsymbol{k} - \mathbf{e} \in \mathbb{N}^n \\ 0 & \text{else} \end{cases}.$$

Proof Part (i) follows directly from the commutativity of the operators $\{\Sigma_j\}_{j=1}^n$. Part (ii) follows from plugging the representation $\Delta_j = \text{Id} - S_{\mathbf{e}_j}$, where $\mathbf{e}_j$ is the jth standard basis vector in $\mathbb{N}^n$, into the definition $\Delta_{\text{mix}} = \Delta_1 \circ \cdots \circ \Delta_n$, and subsequent expansion.

Part (i) of the previous proposition shows that, ignoring questions of convergence, discrete functions $\mathscr{A}\colon \mathbb{N}^n \to Y$ with limit $\mathscr{A}_\infty$ satisfy

$$\mathscr{A}_\infty = \sum_{\boldsymbol{k}\in\mathbb{N}^n} \Delta_{\text{mix}}\mathscr{A}(\boldsymbol{k}) \tag{7}$$

in analogy to Eq. (6). In the next section, we define spaces of discrete functions for which this sum converges absolutely and can be efficiently truncated. We conclude this section by the observation that a necessary condition for the sum in Eq. (7) to converge absolutely is that the unidirectional limits $\mathscr{A}(k_1,\dots,\infty,\dots,k_n) := \lim_{k_j\to\infty}\mathscr{A}(k_1,\dots,k_j,\dots,k_n)$ exist. Indeed, by part (i) of the previous proposition, these limits correspond to summation of $\Delta_{\text{mix}}\mathscr{A}$ over hyperrectangles that are growing in direction of the jth coordinate axis and fixed in all other directions. For instance, in the context of time-dependent partial differential equations this implies stability requirements for the underlying numerical solver, prohibiting explicit time-stepping schemes that diverge when the space-discretization is refined while the time-discretization is fixed.

3 Truncation

For any index set $\mathscr{I} \subset \mathbb{N}^n$, we may define Smolyak's algorithm as the approximation of $\mathscr{A}_\infty$ that is obtained by truncation of the infinite decomposition in Eq. (7) to $\mathscr{I}$,

$$\mathscr{S}_{\mathscr{I}}(\mathscr{A}) := \sum_{\boldsymbol{k}\in\mathscr{I}} \Delta_{\text{mix}}\mathscr{A}(\boldsymbol{k}). \tag{8}$$

By definition of $\Delta_{\text{mix}}\mathscr{A}$, the approximation $\mathscr{S}_{\mathscr{I}}(\mathscr{A})$ is a linear combination of the values $\mathscr{A}(\boldsymbol{k})$, $\boldsymbol{k} \in \mathbb{N}^n$ (see Sect. 3.2 for explicit coefficients). This is the reason for the name *combination technique* that was given to approximations of this form in the context of the numerical approximation of partial differential equations [11]. When one talks about *the* Smolyak algorithm, or *the* combination technique, a particular truncation is usually implied. The general idea here is to include those indices for which the ratio between contribution (measured in the norm of Y) and required work of the corresponding decomposition term is large. To formalize this idea, we require decay of the norms of the decomposition terms and bounds on the work required for their evaluation. To express the former, we define for strictly decreasing functions $e_j\colon \mathbb{N} \to \mathbb{R}_> := (0,\infty)$, $j \in \{1,\dots,n\}$ the spaces

$$\mathscr{E}_{(e_j)_{j=1}^n}(Y) := \Big\{\mathscr{A}\colon \mathbb{N}^n \to Y : \exists K_1 > 0\ \forall \boldsymbol{k} \in \mathbb{N}^n\ \|\Delta_{\text{mix}}\mathscr{A}(\boldsymbol{k})\|_Y \le K_1 \prod_{j=1}^n e_j(k_j)\Big\}.$$

Proposition 2

(i) If

$$\sum_{\boldsymbol{k}\in\mathbb{N}^n}\prod_{j=1}^{n} e_j(k_j) < \infty, \tag{9}$$

then any $\mathscr{A} \in \mathscr{E}_{(e_j)_{j=1}^n}(Y)$ *has a limit* $\mathscr{A}_\infty := \lim_{\min_{j=1}^n k_j\to\infty}\mathscr{A}(\boldsymbol{k})$. *Furthermore, the decomposition in Eq. (7) holds and converges absolutely.*

(ii) The spaces $\mathscr{E}_{(e_j)_{j=1}^n}(Y)$ *are linear subspaces of* $Y^{\mathbb{N}^n}$.

*(iii) (*Error expansions*) Assume that the ratios* $e_j(k)/e_j(k+1)$ *are uniformly bounded above for* $k \in \mathbb{N}$ *and* $j \in \{1,\dots,n\}$. *For* $\boldsymbol{k} \in \mathbb{N}^n$ *and* $J \subset \{1,\dots,n\}$ *let* $\boldsymbol{k}_J := (k_j)_{j\in J} \in \mathbb{N}^{|J|}$. *If the approximation error can be written as*

$$\mathscr{A}(\boldsymbol{k}) - \mathscr{A}_\infty = \sum_{\varnothing\neq J\subset\{1,\dots,n\}} \mathscr{A}_J(\boldsymbol{k}_J) \quad \forall \boldsymbol{k}\in\mathbb{N}^n$$

with functions $\mathscr{A}_J : \mathbb{N}^{|J|} \to Y$, $J \subset \{1,\dots,n\}$ *that satisfy*

$$\|\mathscr{A}_J(\boldsymbol{k}_J)\|_Y \le \prod_{j\in J} e_j(k_j)$$

then

$$\mathscr{A} \in \mathscr{E}_{(e_j)_{j=1}^n}(Y).$$

*(iv) (*Multilinearity *[20]) Assume* $(Y_i)_{i=1}^m$ *and* Y *are Banach spaces and* $\mathscr{M}\colon \prod_{i=1}^m Y_i \to Y$ *is a continuous multilinear map. If*

$$\mathscr{A}_i \in \mathscr{E}_{(e_j)_{j=n_1+\dots+n_{i-1}+1}^{n_1+\dots+n_i}}(Y_i) \quad \forall i \in \{1,\dots,m\},$$

then

$$\mathscr{M}(\mathscr{A}_1,\dots,\mathscr{A}_m) \in \mathscr{E}_{(e_j)_{j=1}^n}(Y),$$

where $n := n_1 + \dots + n_m$ *and*

$$\mathscr{M}(\mathscr{A}_1,\dots,\mathscr{A}_m)(\boldsymbol{k}) := \mathscr{M}(\mathscr{A}_1(\boldsymbol{k}_1),\dots,\mathscr{A}_m(\boldsymbol{k}_m)) \quad \forall \boldsymbol{k} = (\boldsymbol{k}_1,\dots,\boldsymbol{k}_m) \in \mathbb{N}^n.$$

Proof Since Y is a Banach space, the assumption in part (i) shows that for any $\mathscr{A} \in \mathscr{E}_{(e_j)_{j=1}^n}(Y)$ the infinite sum in Eq. (7) converges absolutely to some limit $\bar{\mathscr{A}}$. Since rectangular truncations of this sum yield point values $\mathscr{A}(\boldsymbol{k})$, $\boldsymbol{k} \in \mathbb{N}^n$ by part

(i) of Proposition 1, the limit $\mathscr{A}_\infty := \lim_{\min_{j=1}^n k_j \to \infty} \mathscr{A}(\boldsymbol{k})$ exists and equals $\bar{\mathscr{A}}$. Part (ii) follows from the triangle inequality.

For part (iii), observe that by part (ii) it suffices to show $\mathscr{A}_J \in \mathscr{E}_{(e_j)_{j=1}^n}(Y)$ for all $J \subset \{1, \dots, n\}$, where we consider $\mathscr{A}_J$ as functions on $\mathbb{N}^n$ depending only on the parameters indexed by J. Since $\Delta_{\text{mix}} = \Delta^J_{\text{mix}} \circ \Delta^{J^C}_{\text{mix}}$, where Δ^J_{mix} denotes the mixed difference operator acting on the parameters in J, we then obtain

$$\Delta_{\text{mix}} \mathscr{A}_J(\boldsymbol{k}) = \begin{cases} \Delta^J_{\text{mix}} \mathscr{A}_J(\boldsymbol{k}_J) & \text{if } \forall j \in J^C : \boldsymbol{k}_j = 0 \\ 0 \quad \text{else.} \end{cases}$$

Hence, it suffices to consider $J = \{1, \dots, n\}$. In this case, the assumption $\|\mathscr{A}_J(\boldsymbol{k}_J)\|_Y \leq C \prod_{j \in J} e_j(k_j)$ is equivalent to $\Delta^{-1}_{\text{mix}} \mathscr{A}_J \in \mathscr{E}_{(e_j)_{j=1}^n}(Y)$. Thus, it remains to show that Δ_{mix} preserves $\mathscr{E}_{(e_j)_{j=1}^n}(Y)$. This holds by part (ii) of this proposition together with part (ii) of Proposition 1 and the fact that shift operators preserve $\mathscr{E}_{(e_j)_{j=1}^n}(Y)$, which itself follows from the assumption that the functions $e_j(\cdot)/e_j(\cdot + 1)$ are uniformly bounded.

Finally, for part (iv) observe that by multilinearity of $\mathscr{M}$ we have

$$\Delta_{\text{mix}} \mathscr{M}(\mathscr{A}_1, \dots, \mathscr{A}_m) = \mathscr{M}(\Delta^{(1)}_{\text{mix}} \mathscr{A}_1, \dots, \Delta^{(m)}_{\text{mix}} \mathscr{A}_m),$$

where the mixed difference operator on the left hand side acts on $n = n_1 + \dots + n_m$ coordinates, whereas those on the right hand side only act on the n_i coordinates of $\mathscr{A}_i$. By continuity of $\mathscr{M}$ we have

$$\|\mathscr{M}(\Delta^{(1)}_{\text{mix}} \mathscr{A}_1, \dots, \Delta^{(m)}_{\text{mix}} \mathscr{A}_m)(\boldsymbol{k})\|_Y \leq C \prod_{i=1}^m \|\Delta^{(i)}_{\text{mix}} \mathscr{A}_i(\boldsymbol{k}_i)\|_{Y_i},$$

for some $C > 0$, from which the claim follows.

Parts (iii) and (iv) of the previous proposition provide sufficient conditions to verify $\mathscr{A} \in \mathscr{E}_{(e_j)_{j=1}^n}(Y)$ without analyzing mixed differences directly.

Example 1

(i) After an exponential reparametrization, the assumptions in Eqs. (1) and (2) become

$$\|\mathscr{A}(\boldsymbol{k}) - \mathscr{A}_\infty\|_Y \leq K_1 \sum_{j=1}^n \exp(-\beta_j k_j) \quad \forall \boldsymbol{k} \in \mathbb{N}^n$$

and

$$\text{Work}(\mathscr{A}(\boldsymbol{k})) = K_2 \prod_{j=1}^{n} \exp(\gamma_j k_j) \quad \forall \boldsymbol{k} \in \mathbb{N}^n,$$

respectively. If we slightly strengthen the first and assume that

$$\mathscr{A}(\boldsymbol{k}) - \mathscr{A}_\infty = \sum_{j=1}^{n} \mathscr{A}_j(k_j) \quad \forall \boldsymbol{k} \in \mathbb{N}^n$$

with functions $\mathscr{A}_j$ that satisfy

$$\|\mathscr{A}_j(k_j)\|_Y \leq C \exp(-\beta_j k_j), \quad \forall k_j \in \mathbb{N}$$

for some $C > 0$ and $\beta_j > 0$, $j \in \{1, \ldots, n\}$, then

$$\mathscr{A} \in \mathscr{E}_{(e_j)_{j=1}^n} \quad \text{with } e_j(k_j) := \exp(-\beta_j k_j),$$

by part (iii) of Proposition 2. Theorem 1 below then shows that Smolyak's algorithm applied to $\mathscr{A}$ requires only the work $\epsilon^{-\max_{j=1}^n\{\gamma_j/\beta_j\}}$, up to possible logarithmic factors, to achieve the accuracy $\epsilon > 0$.

(ii) Assume we want to approximate the integral of a function $f : [0, 1] \to \mathbb{R}$ but are only able to evaluate approximations f_{k_2}, $k_2 \in \mathbb{N}$ of f with increasing cost as $k_2 \to \infty$. Given a sequence S_{k_1}, $k_1 \in \mathbb{N}$ of linear quadrature formulas, the straightforward approach would be to fix sufficiently large values of k_1 and k_2 and then approximate the integral of f_{k_2} with the quadrature formula S_{k_1}. Formally, this can be written as

$$\mathscr{A}(k_1, k_2) := S_{k_1} f_{k_2}.$$

To show decay of the mixed differences $\Delta_{\text{mix}}\mathscr{A}$, observe that the application of quadrature formulas to functions is linear in both arguments, which means that we may write

$$\mathscr{A}(k_1, k_2) = \mathscr{M}(S_{k_1}, f_{k_2}) = \mathscr{M}(\mathscr{A}_1(k_1), \mathscr{A}_2(k_2))$$

where $\mathscr{A}_1(k_1) := S_{k_1}$, $\mathscr{A}_2(k_2) := f_{k_2}$, and $\mathscr{M}$ is the application of linear functionals to functions on [0, 1]. Assume, for example, that the functions f_{k_2} converge to f in some Banach space B of functions on [0, 1] as $k_2 \to \infty$, and that the quadrature formulas S_{k_1} converge to the integral operator $\int$ in the continuous dual space B^* as $k_1 \to \infty$. The decay of the mixed differences $\Delta_{\text{mix}}\mathscr{A}(k_1, k_2)$ then follows from part (iv) of Proposition 2, since $\mathscr{M}$ is a continuous bilinear map from $B^* \times B$ to $\mathbb{R}$. We will see in Sect. 5.3 below that the application of Smolyak's algorithm in this example yields so called *multilevel quadrature formulas*. This connection between Smolyak's algorithm and multilevel formulas was observed in [14].

(iii) Assume that we are given approximation methods $\mathscr{A}_j : \mathbb{N} \to Y_j$, $j \in \{1, \dots n\}$ that converge at the rates $\|\mathscr{A}_j(k_j) - \mathscr{A}_{\infty,j}\|_{Y_j} \leq e_j(k_j)$ to limits $\mathscr{A}_{\infty,j} \in Y_j$, where $e_j : \mathbb{N} \to \mathbb{R}_>$ are strictly decreasing functions. Define the tensor product algorithm

$$\mathscr{A} : \mathbb{N}^n \to Y := Y_1 \otimes \cdots \otimes Y_n,$$
$$\mathscr{A}(\boldsymbol{k}) := \mathscr{A}_1(k_1) \otimes \cdots \otimes \mathscr{A}_n(k_n).$$

If the algebraic tensor product Y is equipped with a norm that satisfies $\|y_1 \otimes \cdots \otimes y_n\|_Y \leq \|y_1\|_{Y_1} \dots \|y_n\|_{Y_n}$, then $\mathscr{A} \in \mathscr{E}_{(e_j)_{j=1}^n}(Y)$. Indeed, $\mathscr{A}_j \in \mathscr{E}_{e_j}(Y_j)$ by part (iii) of Proposition 2, thus $\mathscr{A} \in \mathscr{E}_{(e_j)_{j=1}^n}(Y)$ by part (iv) of the same proposition.

Similar to the product type decay assumption on the norms $\|\Delta_{\text{mix}}\mathscr{A}(\boldsymbol{k})\|_Y$, which we expressed in the spaces $\mathscr{E}_{(e_j)_{j=1}^n}$, we assume in the remainder that

$$\text{Work}(\Delta_{\text{mix}}\mathscr{A}(\boldsymbol{k})) \leq K_2 \prod_{j=1}^{n} w_j(k_j) \quad \forall \boldsymbol{k} \in \mathbb{N}^n \tag{10}$$

for some $K_2 > 0$ and increasing functions $w_j : \mathbb{N} \to \mathbb{R}_>$. By part (ii) of Proposition 1, such a bound follows from the same bound on the evaluations $\mathscr{A}(\boldsymbol{k})$ themselves.

3.1 Knapsack Problem

The goal of this subsection is to describe quasi-optimal truncations of the decomposition in Eq. (7) for functions $\mathscr{A} \in \mathscr{E}_{(e_j)_{j=1}^n}(Y)$ that satisfy Eq. (10). Given a work budget $W > 0$, a quasi-optimal index set solves the knapsack problem

$$\begin{aligned} \max_{\mathscr{I} \subset \mathbb{N}^n} \quad & |\mathscr{I}|_e := K_1 \sum_{\boldsymbol{k} \in \mathscr{I}} \prod_{j=1}^{n} e_j(k_j) \\ \text{subject to} \quad & |\mathscr{I}|_w := K_2 \sum_{\boldsymbol{k} \in \mathscr{I}} \prod_{j=1}^{n} w_j(k_j) \leq W. \end{aligned} \tag{11}$$

The term that is maximized here is motivated by

$$\|\mathcal{S}_{\mathcal{I}}(\mathcal{A})-\mathcal{A}_\infty\|_Y=\|\sum_{\boldsymbol{k}\in\mathcal{I}^c}\Delta_{\text{mix}}\mathcal{A}(\boldsymbol{k})\|_Y\approx\sum_{\boldsymbol{k}\in\mathcal{I}^c}\|\Delta_{\text{mix}}\mathcal{A}(\boldsymbol{k})\|_Y\approx|\mathcal{I}^c|_e$$

Proposition 3 below shows that for any $W>0$ the knapsack problem has an optimal value. However, finding corresponding optimal sets is NP-hard [19, Section 1.3]. As a practical alternative one can use Dantzig's approximation algorithm [19, Section 2.2.1], which selects indices for which the ratio between contribution and work is above some threshold $\delta(W)>0$,

$$\mathcal{I}_W:=\{\boldsymbol{k}\in\mathbb{N}^n:\prod_{j=1}^n e_j(k_j)/w_j(k_j)>\delta(W)\},\qquad(12)$$

where $\delta(W)$ is chosen minimally such that $|\mathcal{I}_W|_w\leq W$.

Proposition 3

(i) The knapsack problem in Eq. (11) *has a (not necessarily unique) solution, in the sense that a maximal value of* $|\mathcal{I}|_e$ *is attained. We denote this maximal value by* $E^*(W)$.

(ii) Any set $\mathcal{I}^*$ *for which* $|\mathcal{I}|_e=E^*(W)$ *is finite and downward closed: If* $\boldsymbol{k}\in\mathcal{I}^*$ *and* $\tilde{\boldsymbol{k}}\in\mathbb{N}^n$ *satisfies* $\tilde{\boldsymbol{k}}\leq\boldsymbol{k}$ *componentwise, then* $\tilde{\boldsymbol{k}}\in\mathcal{I}^*$. *The same holds for the set* $\mathcal{I}_W$ *from Eq.* (12).

(iii) The set $\mathcal{I}_W$ *from Eq.* (12) *satisfies*

$$|\mathcal{I}_W|_e\geq\frac{|\mathcal{I}_W|_w}{W}E^*(W).$$

This means that if $\mathcal{I}_W$ *uses all of the available work budget,* $|\mathcal{I}_W|_w=W$, *then it is a solution to the knapsack problem. In particular, Dantzig's solutions are optimal for the work* $|\mathcal{I}_W|_w$ *they require, but not necessarily for the work* W *they were designed for.*

Proof There is an upper bound N on the cardinality of admissible sets in Eq. (11) since the functions w_j are increasing and strictly positive. Furthermore, replacing an element $\boldsymbol{k}$ of an admissible set by $\tilde{\boldsymbol{k}}$ with $\tilde{\boldsymbol{k}}\leq\boldsymbol{k}$ decreases $|\cdot|_w$ and increases $|\cdot|_e$. This proves parts (i) and (ii), as there are only finitely many downward closed sets of cardinality less than N (for example, all such sets are subsets of $\{0,\ldots,N-1\}^n$). Part (iii) follows directly from the inequality $|\mathcal{I}_W|_e/|\mathcal{I}_W|_w\geq|\mathcal{I}^*|_e/|\mathcal{I}^*|_w$, where $\mathcal{I}^*$ is a set that attains the maximal value $E^*(W)$.

Even in cases where no bounding functions e_j and w_j are available, parts (ii) and (iii) of the previous proposition serve as motivation for adaptive algorithms that progressively build a downward closed set $\mathcal{I}$ by adding at each step a multi-index that maximizes a gain-to-work estimate [6, 15].

3.2 Combination Rule

Part (ii) of Proposition 1 provides a way to express the approximations $\mathcal{S}_{\mathcal{I}}(\mathcal{A})$ in a succinct way as linear combinations of different values of $\mathcal{A}$. This yields the *combination rule*, which in its general form says that

$$\mathcal{S}_{\mathcal{I}}(\mathcal{A}) = \sum_{\boldsymbol{k}\in\mathcal{I}} c_{\boldsymbol{k}}\mathcal{A}(\boldsymbol{k})$$

with

$$c_{\boldsymbol{k}} = \sum_{e\in\{0,1\}^n:\boldsymbol{k}+e\in\mathcal{I}} (-1)^{|e|_1} \tag{13}$$

for any downward closed set $\mathcal{I}$. It is noteworthy that $c_{\boldsymbol{k}} = 0$ for all $\boldsymbol{k}$ with $\boldsymbol{k} + (1,\dots,1) \in \mathcal{I}$, because for such $\boldsymbol{k}$ the sum in Eq. (13) is simply the expansion of $(1-1)^n$.

When $\mathcal{I}$ is a standard simplex, $\mathcal{I} = \{\boldsymbol{k} \in \mathbb{N}^n : |\boldsymbol{k}|_1 \leq L\}$, the following explicit formula holds [25]:

$$c_{\boldsymbol{k}} = \begin{cases} (-1)^{L-|\boldsymbol{k}|_1}\binom{n-1}{L-|\boldsymbol{k}|_1} & \text{if } L-n+1 \leq |\boldsymbol{k}|_1 \leq L \\ 0 \quad \text{else.} \end{cases}$$

4 Convergence Analysis

4.1 Finite-Dimensional Case

We consider an approximation method $\mathcal{A} \in \mathcal{E}_{(e_j)_{j=1}^n}(Y)$ with

$$e_j(k_j) = K_{j,1}\exp(-\beta_j k_j)(k_j+1)^{s_j} \quad \forall j \in \{1,\dots,n\} \tag{14}$$

and assume that

$$\text{Work}(\mathcal{A}(\boldsymbol{k})) \leq \prod_{j=1}^{n} K_{j,2}\exp(\gamma_j k_j)(k_j+1)^{t_j} \quad \forall \boldsymbol{k} \in \mathbb{N}^n \tag{15}$$

with $K_{j,1} > 0$, $K_{j,2} > 0$, $\beta_j > 0$, $\gamma_j > 0$, $s_j \geq 0$, $t_j \geq 0$. The required calculations with $\boldsymbol{s} \equiv \boldsymbol{t} \equiv 0$ were previously done in various specific contexts, see for example

[13]. According to Proposition 3, quasi-optimal index sets are given by

$$\begin{aligned}\mathcal{I}_\delta := &\left\{\boldsymbol{k}\in\mathbb{N}^n : \prod_{j=1}^n \frac{K_{j,1}\exp(-\beta_j k_j)(k_j+1)^{s_j}}{K_{j,2}\exp(\gamma_j k_j)(k_j+1)^{t_j}} > \delta\right\}\\ &= \left\{\boldsymbol{k}\in\mathbb{N}^n : \frac{K_1}{K_2}\exp(-(\boldsymbol{\beta}+\boldsymbol{\gamma})\cdot\boldsymbol{k})\prod_{j=1}^n (k_j+1)^{s_j-t_j} > \delta\right\}\end{aligned}$$

for $\delta > 0$, where $K_1 := \prod_{j=1}^n K_{j,1}$, $K_2 := \prod_{j=1}^n K_{j,2}$, and $\boldsymbol{\beta} := (\beta_1,\dots,\beta_n)$, $\boldsymbol{\gamma} := (\gamma_1,\dots,\gamma_n)$. For the analysis in this section, we use the slightly simplified sets

$$\mathcal{I}_L := \left\{\boldsymbol{k}\in\mathbb{N}^n : \exp((\boldsymbol{\beta}+\boldsymbol{\gamma})\cdot\boldsymbol{k}) \le \exp(L)\right\} = \left\{\boldsymbol{k}\in\mathbb{N}^n : (\boldsymbol{\beta}+\boldsymbol{\gamma})\cdot\boldsymbol{k} \le L\right\},$$

with $L\to\infty$, where, by abuse notation, we distinguish the two families of sets by the subscript letter.

The work required by $\mathcal{S}_L(\mathcal{A}) := \mathcal{S}_{\mathcal{I}_L}(\mathcal{A})$ satisfies

$$\mathrm{Work}(\mathcal{S}_L(\mathcal{A})) \le \sum_{\boldsymbol{k}\in\mathcal{I}_L}\prod_{j=1}^n K_{j,2}\exp(\gamma_j k_j)(k_j+1)^{t_j} = K_2 \sum_{(\boldsymbol{\beta}+\boldsymbol{\gamma})\cdot\boldsymbol{k}\le L} \exp(\boldsymbol{\gamma}\cdot\boldsymbol{k})(\boldsymbol{k}+\boldsymbol{1})^{\boldsymbol{t}}$$

with $(\boldsymbol{k}+\boldsymbol{1})^{\boldsymbol{t}} := \prod_{j=1}^n (k_j+1)^{t_j}$. Similarly, the approximation error satisfies

$$\|\mathcal{S}_L(\mathcal{A})-\mathcal{A}_\infty\|_Y \le \sum_{\boldsymbol{k}\in\mathcal{I}_L^c}\prod_{j=1}^n K_{j,1}\exp(-\beta_j k_j)k_j^{s_j} = K_1 \sum_{(\boldsymbol{\beta}+\boldsymbol{\gamma})\cdot\boldsymbol{k}> L} \exp(-\boldsymbol{\beta}\cdot\boldsymbol{k})(\boldsymbol{k}+\boldsymbol{1})^{\boldsymbol{s}}. \tag{16}$$

The exponential sums appearing in the work and residual bounds above are estimated in the appendix of this work, with the results

$$\mathrm{Work}(\mathcal{S}_L(\mathcal{A})) \le K_2 C(\boldsymbol{\gamma},\boldsymbol{t},n)\exp(\frac{\rho}{1+\rho}L)(L+1)^{n^*-1+t^*} \tag{17}$$

and

$$\|\mathcal{S}_L(\mathcal{A})-\mathcal{A}_\infty\|_Y \le K_1 C(\boldsymbol{\beta},\boldsymbol{s},n)\exp(-\frac{1}{1+\rho}L)(L+1)^{n^*-1+s^*}, \tag{18}$$

where $\rho := \max_{j=1}^n \gamma_j/\beta_j$, $J := \{j\in\{1,\dots,n\} : \gamma_j/\beta_j = \rho\}$, $n^* := |J|$, $s^* := \sum_{j\in J} s_j$, $t^* := \sum_{j\in J} t_j$. We may now formulate the main result of this section by rewriting the bound in Eq. (17) in terms of the right-hand side of Eq. (18).

Theorem 1 *Under the previously stated assumptions on* $\mathscr{A}$ *and for small enough* $\epsilon > 0$, *we may choose* $L > 0$ *such that*

$$\|\mathscr{S}_L(\mathscr{A}) - \mathscr{A}_\infty\|_Y \leq \epsilon$$

and

$$Work(\mathscr{S}_L(\mathscr{A})) \leq K_1^\rho K_2 C(\boldsymbol{\beta}, \boldsymbol{\gamma}, \boldsymbol{s}, \boldsymbol{t}, n)\epsilon^{-\rho}|\log \epsilon|^{(n^*-1)(1+\rho)+\rho s^*+t^*}. \qquad \square$$

This means that we have eliminated the sum in the exponent of the bound in Eq. (4), as announced in Sect. 1. The additional logarithmic factors in Theorem 1 vanish if the worst ratio of work and convergence exponents, ρ, is attained only for a single index $j_{\max} \in \{1, \ldots, n\}$ and if $t_{j_{\max}} = s_{j_{\max}} = 0$.

Remark 1 If $\boldsymbol{\gamma} \equiv 0$ and $\boldsymbol{\beta} \equiv 0$, that is when both work and residual depend algebraically on all parameters, then an exponential reparametrization, $\exp(\tilde{\boldsymbol{k}}) := \boldsymbol{k}$, takes us back to the situation considered above. The preimage of $\mathscr{I}_L = \{\tilde{\boldsymbol{k}} : (\boldsymbol{s} + \boldsymbol{t}) \cdot \tilde{\boldsymbol{k}} \leq L\}$ under this reparametrization is $\{\boldsymbol{k} : \prod_{j=1}^n k_j^{s_j+t_j} \leq \exp(L)\}$, whence the name *hyperbolic cross approximation* [5].

Remark 2 When the terms $\Delta_{\text{mix}}\mathscr{A}(\boldsymbol{k})$, $\boldsymbol{k} \in \mathbb{N}^n$ are orthogonal to each other, we may substitute the Pythagorean theorem for the triangle inequality in Eq. (16). As a result, the exponent of the logarithmic factor in Theorem 1 reduces to $(n^* - 1)(1 + \rho/2) + \rho s^* + t^*$.

4.2 Infinite-Dimensional Case

The theory of the previous sections can be extended to the case $n = \infty$. In this case the decomposition in Eq. (7) becomes

$$\mathscr{A}_\infty = \sum_{\boldsymbol{k} \in \mathbb{N}_c^\infty} \Delta_{\text{mix}}\mathscr{A}(\boldsymbol{k}), \tag{19}$$

where $\mathbb{N}_c^\infty$ are the sequences with finite support, and $\Delta_{\text{mix}}\mathscr{A}(\boldsymbol{k})$ is defined as $\Delta_1 \circ \cdots \circ \Delta_{n_{\max}}\mathscr{A}(\boldsymbol{k})$, where $n_{\max}$ is a bound on the support of $\boldsymbol{k}$. In particular, every term in Eq. (19) is a linear combination of values of $\mathscr{A}$ with only finitely many nonzero discretization parameters.

We consider the case $\mathscr{A} \in \mathscr{E}_{(e_j)_{j=1}^n}(Y)$ for

$$e_j(k_j) := K_{j,1} \exp(-\beta_j k_j)(k_j + 1)^s \quad \forall j \geq 1$$

and $s \geq 0$, $K_1 := \prod_{j=1}^\infty K_{j,1} < \infty$ $s \geq 0$, and we assume constant computational work for the evaluation of the mixed differences $\Delta_{\text{mix}}\mathscr{A}(\boldsymbol{k})$, i.e. $w_j \equiv C$ in Eq. (10)

for all $j \geq 1$. Similarly to the finite-dimensional case, we consider sets

$$\mathscr{I}_L := \{\boldsymbol{k} \in \mathbb{N}_c^\infty : \sum_{j=1}^{\infty} \beta_j k_j \leq L\}$$

and the associated Smolyak algorithm

$$\mathscr{S}_L(\mathscr{A}) := \sum_{\mathscr{I}_L} \Delta_{\text{mix}} \mathscr{A}(\boldsymbol{k}).$$

The following theorem is composed of results from [10] on interpolation and integration of analytic functions; the calculations there transfer directly to the general setting.

Theorem 2 *Let $L > 0$ and define $N := |\mathscr{I}_L| = \mathit{Work}(\mathscr{S}_L(\mathscr{A}))$.*

(i) Assume $s = 0$.

- *[10, Theorem 3.2] If there exists $\beta_0 > 1$ such that $M := M(\beta_0, (\beta_j)_{j=1}^\infty) := \sum_{j=1}^\infty \frac{1}{\exp(\beta_j/\beta_0)-1} < \infty$, then*

$$\|\mathscr{S}_L(\mathscr{A}) - \mathscr{A}_\infty\|_Y \leq \frac{K_1}{\beta_0} \exp(\beta_0 M) N^{-(\beta_0-1)},$$

which implies

$$\mathit{Work}(\mathscr{S}_L(\mathscr{A})) \leq C(K, \beta_0, M)\epsilon^{-1/(\beta_0-1)}$$

for $\epsilon := \frac{K_1}{\beta_0} \exp(\beta_0 M) N^{-(\beta_0-1)}$.

- *[10, Theorem 3.4] If $\beta_j \geq \beta_0 j$ for $\beta_0 > 0$, $j \geq 1$, then*

$$\|\mathscr{S}_L(\mathscr{A}) - \mathscr{A}_\infty\|_Y \leq \frac{2}{\beta_0 \sqrt{\log N}} N^{1+\frac{1}{4}\beta_0 - \frac{3}{8}\beta_0 (\log N)^{1/2}}.$$

(ii) Assume $s > 0$.

- *[10, Corollary 4.2 (i)] If there exist $\beta_0 > 1$ and $\delta > 0$ such that $M(\beta_0, ((1-\delta)\beta_j)_{j=1}^\infty) < \infty$, then*

$$\|\mathscr{S}_L(\mathscr{A}) - \mathscr{A}_\infty\|_Y \leq C(K_1, \delta, \beta_0, M, (\beta_j)_{j\in\mathbb{N}}, s) N^{-(\beta_0-1)},$$

which implies

$$\mathit{Work}(\mathscr{S}_L(\mathscr{A})) \leq C(K_1, \delta, \beta_0, M, (\beta_j)_{j\in\mathbb{N}}, s)\epsilon^{-1/(\beta_0-1)}$$

for $\epsilon := C(K_1, \delta, \beta_0, M, (\beta_j)_{j\in\mathbb{N}}, s)N^{-(\beta_0-1)}$.

- *[10, Corollary 4.2 (ii)] If* $\beta_j \geq \beta_0 j$ *for* $\beta_0 > 0$, *then for every* $\hat{\beta}_0 < \beta_0$ *we have*

$$\|\mathscr{S}_L(\mathscr{A}) - \mathscr{A}_\infty\|_Y \leq \frac{C(\hat{\beta}_0, M, b)}{\sqrt{\log N}} N^{1+\frac{\hat{\beta}_0}{4}-\frac{3}{8}\hat{\beta}_0(\log N)^{1/2}}. \qquad \square$$

Remark 3 For alternative approaches to infinite-dimensional problems, which allow even for exponential type work bounds, $w_j(k_j) = K_{j,2}\exp(\gamma_j k_j)$, consider for example [4, 22, 23].

5 Applications

5.1 High-Dimensional Interpolation and Integration

Smolyak introduced the algorithm that now bears his name in [24] to obtain efficient high-dimensional integration and interpolation formulas from univariate building blocks. For example, assume we are given univariate interpolation formulas S_k, $k \in \mathbb{N}$ for functions in a Sobolev space $H^\beta([0, 1])$ that are based on evaluations in 2^k points in $[0, 1]$ and converge at the rate

$$\|S_k - \mathrm{Id}\|_{H^\beta([0,1])\to H^\alpha([0,1])} \leq C2^{-k(\beta-\alpha)}$$

for some $0 \leq \alpha < \beta$. A straightforward high-dimensional interpolation formula is then the corresponding tensor product formula

$$\bigotimes_{j=1}^n S_{k_j} : H^\beta([0, 1])^{\otimes n} =: H^\beta_{\mathrm{mix}}([0, 1]^n) \to H^\alpha([0, 1])^{\otimes n} =: H^\alpha_{\mathrm{mix}}([0, 1]^n)$$

for $(k_1, \ldots, k_n) \in \mathbb{N}^n$, where we consider both tensor product spaces to be completed with respect to the corresponding Hilbert space tensor norm [12]. This can be interpreted as a numerical approximation method with values in a space of linear operators,

$$\mathscr{A}(\boldsymbol{k}) := \bigotimes_{j=1}^n S_{k_j} \in \mathscr{L}(H^\beta_{\mathrm{mix}}([0, 1]^n), H^\alpha_{\mathrm{mix}}([0, 1]^n)) =: Y,$$

whose discretization parameters $\boldsymbol{k} = (k_1, \dots, k_n)$ determine the resolution of interpolation nodes in each direction $j \in \{1, \dots, n\}$.

If we associate as work with $\mathscr{A}(\boldsymbol{k})$ the number of required point evaluations,

$$\text{Work}(\mathscr{A}(\boldsymbol{k})) := \prod_{j=1}^{n} 2^{k_j},$$

then we are in the situation described in Sect. 4.1. Indeed, we have $\mathscr{A} \in \mathscr{E}_{(e_j)_{j=1}^n}(Y)$ with $e_j(k_j) := 2^{-k_j(\beta-\alpha)}$ by part (iii) of Example 1, since the operator norm of a tensor product operator between Hilbert space tensor products factorizes into the product of the operator norms of the constituent operators (see [12, Proposition 4.127] and [3, Section 26.7]).

In particular, the straightforward tensor product formulas $\mathscr{A}(k, \dots, k)$ require the work

$$\epsilon^{-n/(\beta-\alpha)}$$

to approximate the identity operator with accuracy $\epsilon > 0$ in the operator norm, whereas Smolyak's algorithm $\mathscr{S}_L(\mathscr{A})$ with an appropriate choice of $L = L(\epsilon)$ achieves the same accuracy with

$$\text{Work}(\mathscr{S}_L(\mathscr{A})) \lesssim \epsilon^{-1/(\beta-\alpha)} |\log \epsilon|^{(n-1)(1+1/(\beta-\alpha))},$$

according to Theorem 1. Here and in the following, we denote by $\lesssim$ estimates that hold up to factors that are independent of ϵ. As a linear combination of tensor product operators, Smolyak's algorithm $\mathscr{S}_L(\mathscr{A})$ is a linear interpolation formula based on evaluations in the union of certain tensor grids. These unions are commonly known as *sparse grids* [2, 7, 26].

Remark 4 Interpolation of functions in general Banach spaces, with convergence measured in different general Banach spaces can be treated in the same manner. However, more care has to be taken with the tensor products. Once the algebraic tensor products of the function spaces are equipped with *reasonable cross norms* [12] and completed, it has to be verified that the operator norm of linear operators between the tensor product spaces factorizes. Unlike for Hilbert spaces, this is not always true for general Banach spaces. However, it is true whenever the codomain is equipped with the *injective tensor norm*, or when the domain is equipped with the *projective tensor norm* [12, Sections 4.2.9 and 4.2.12]. For example, the L^∞-norm (and the similar C^k-norms) is an injective tensor norm on the product of L^∞-spaces, while the L^1-norm is a projective norm on the tensor product of L^1-spaces.

5.2 Monte Carlo Path Simulation

Consider a stochastic differential equation (SDE)

$$\begin{cases} dS(t) = a(t, S(t))dt + b(t, S(t))dW(t) & 0 \leq t \leq T \\ S(0) = S_0 \in \mathbb{R}^d, \end{cases} \tag{20}$$

with a Wiener process $W(t)$ and sufficiently regular coefficients $a, b \colon [0, T] \times \mathbb{R}^d \to \mathbb{R}$. A common goal in the numerical approximation of such SDE is to compute expectations of the form

$$E[Q(S(T))],$$

where $Q \colon \mathbb{R}^d \to \mathbb{R}$ is a Lipschitz-continuous quantity of interest of the final state $S(T)$. To approach this problem numerically, we first define random variables $S_N(t), 0 \leq t \leq T$ as the forward Euler approximations of Eq. (20) with $N \geq 1$ time steps. Next, we approximate the expectations $E[Q(S_N(T))]$ by Monte Carlo sampling using $M \geq 1$ independent samples $S_N^1(T), \ldots, S_N^M(T)$ that are computed using independent realizations of the Wiener process. Together, this gives rise to the numerical approximation

$$\mathscr{A}(M, N) := \frac{1}{M} \sum_{i=1}^{M} Q(S_N^i(T)).$$

For fixed values of M and N this is a random variable that satisfies

$$\begin{aligned} E[(\mathscr{A}(M, N) - E[Q(S(T))])^2] &= (E[\mathscr{A}(M, N)] - E[Q(S(T))])^2 \\ &\quad + \mathrm{Var}[\mathscr{A}(M, N)] \\ &= (E[Q(S_N(T))] - E[Q(S(T))])^2 \\ &\quad + M^{-1} \mathrm{Var}[Q(S_N(T))] \\ &\lesssim N^{-2} + M^{-1}, \end{aligned}$$

where the last inequality holds by the weak rate of convergence of the Euler method [17, Section 14.1] and by its L^2-boundedness as $N \to \infty$. This shows that the random variables $\mathscr{A}(M, N)$ converge to the limit $\mathscr{A}_\infty = E[Q(S(T))]$, which itself is just a deterministic real number, in the sense of probabilistic mean square convergence as $M, N \to \infty$. To achieve a mean square error or order $\epsilon^2 > 0$, this straightforward approximation requires the simulation of $M \approx \epsilon^{-2}$ sample paths of Eq. (20), each with $N \approx \epsilon^{-1}$ time steps, which incurs the total work

$$\mathrm{Work}(\mathscr{A}(M, N)) = MN \approx \epsilon^{-3}.$$

Smolyak's algorithm allows us to achieve the same accuracy with the reduced work ϵ^{-2} of usual Monte Carlo integration. To apply the results of Sect. 4.1, we consider the reparametrized algorithm $\mathscr{A}(k,l)$ with

$$M_k := M_0 \exp(2k/3),$$

$$N_l := N_0 \exp(2l/3),$$

for which the convergence and work parameters of Sect. 4.1 attain the values $\beta_j = 1/3, \gamma_j = 2/3$, and $s_j = t_j = 0$, $j \in \{1,2\}$. (Here and in the following we implicitly round up non-integer values, which increases the required work only by a constant factor.) Indeed, we may write

$$\mathscr{A}(k,l) = \mathscr{M}(\mathscr{A}_1(k), \mathscr{A}_2(l))),$$

where $\mathscr{A}_1(k)$, $k \in \mathbb{N}$ is the operator that maps random variables to an empirical average over M_k independent samples, $\mathscr{A}_2(l)$, $l \in \mathbb{N}$ is the random variable $Q(S_{N_l}(T))$, and $\mathscr{M}$ denotes the application of linear operators to random variables. Since $\mathscr{A}_1(k)$ converges in the operator norm to the expectation operator on the space of square integrable random variables at the usual Monte Carlo convergence rate $M_k^{-1/2}$ as $k \to \infty$, and $\mathscr{A}_2(l)$ converges to $Q(S(T))$ at the strong convergence rate $N_l^{-1/2}$ of the Euler method in the L^2-norm [17, Section 10.2] as $l \to \infty$, and since $\mathscr{M}$ is linear in both arguments, the claimed values of the convergence parameters β_j, $j \in \{1,2\}$ hold by part (iv) of Proposition 2.

Theorem 1 now shows that choosing $L = L(\epsilon)$ such that

$$E[(\mathscr{S}_L(\mathscr{A}) - E[Q(S(T))])^2] \leq \epsilon^2$$

incurs the work

$$\text{Work}(\mathscr{S}_L(\mathscr{A})) \lesssim \epsilon^{-2}|\log \epsilon|^{-3}. \tag{21}$$

To link this result to the keyword *multilevel approximation*, we observe that, thanks to our particular choice of parametrization, Smolyak's algorithm from Sect. 4.1 takes the simple form

$$\mathscr{S}_L(\mathscr{A}) = \sum_{k+l \leq L} \Delta_{\text{mix}} \mathscr{A}(k,l).$$

Since $\Delta_{\text{mix}} = \Delta_1 \circ \Delta_2$ and $\Delta_1 = \Sigma_1^{-1}$ we may further write

$$\begin{aligned}\mathscr{S}_L(\mathscr{A}) &= \sum_{l=0}^{L}\sum_{k=0}^{L-l} \Delta_{\text{mix}}\mathscr{A}(k,l)\\ &= \sum_{l=0}^{L} \Delta_2\mathscr{A}(L-l,l)\\ &= \frac{1}{M_L}\sum_{i=1}^{M_L} Q(S^i_{N_0}(T)) + \sum_{l=1}^{L}\frac{1}{M_{L-l}}\sum_{i=1}^{M_{L-l}}\Big(Q(S^i_{N_l}(T)) - Q(S^i_{N_{l-1}}(T))\Big),\end{aligned} \tag{22}$$

which reveals that Smolyak's algorithm employs a large number of samples from the coarse approximation $S_{N_0}(T)$, and subsequently improves on the resulting estimate of $E[Q(S(T))]$ by adding approximations of the expectations $E\left[Q(S_{N_l}(T)) - Q(S_{N_{l-1}}(T))\right]$, $l \in \{1,\ldots,L\}$ that are computed using less samples.

Equation (22) is a multilevel formula of the form analyzed in [8, 16]. Alternatively, this formula could also be deduced directly from the combination rule for triangles in Sect. 5.4. Compared to the analysis in [8], our presentation has two shortcomings: First, our analysis only exploits the strong rate of the discretization method used to approximate Eq. (20). In the situation considered above, this does not affect the results, but for more slowly converging schemes a faster weak convergence rate may be exploited to obtain improved convergence rates. Second, the bound in Eq. (21) is larger than that in [8] by the factor $|\log\epsilon|$. This factor can be removed by using independent samples for different values of l in Eq. (22), since we may then apply Remark 2.

5.3 *Multilevel Quadrature*

As in Example 1 of Sect. 3, assume that we want to approximate the integral $\int_{[0,1]} f(x)\ dx \in \mathbb{R}$ using evaluations of approximations $f_l\colon [0,1] \to \mathbb{R}$, $l \in \mathbb{N}$. This is similar to the setting of the previous subsection, but with random sampling replaced by deterministic quadrature.

As before, denote by S_k, $k \in \mathbb{N}$ a sequence of quadrature formulas based on evaluations in 2^k nodes. If we assume that point evaluations of f_l require the work $\exp(\gamma l)$ for some $\gamma > 0$, that

$$\|f_l - f\|_B \lesssim 2^{-\kappa l}$$

for some $\kappa > 0$ and a Banach space B of functions on [0, 1] and that

$$\|S_k - \int_{[0,1]} \cdot \, dx\|_{B^*} \lesssim \exp(-\beta k)$$

for some $\beta > 0$, then $\mathscr{A}(k, l) := S_k f_l$ satisfies

$$|S_k f_l - \int_{[0,1]} f(x) \, dx| \lesssim \exp(-\beta k) + \exp(-\kappa l).$$

Hence, an accuracy of order $\epsilon > 0$ can be achieved by setting

$$k := -\log(\epsilon)/\beta, \quad l := -\log_2(\epsilon)/\kappa,$$

which requires the work

$$2^k \exp(\gamma l) = \epsilon^{-1/\beta - \gamma/\kappa}.$$

We have already shown the decay of the mixed differences,

$$|\Delta_{\text{mix}} \mathscr{A}(k, l)| \lesssim \exp(-\beta k) 2^{-\kappa l},$$

in Example 1. Thus, Theorem 1 immediately shows that we can choose $L = L(\epsilon)$ such that Smolyak's algorithm satisfies

$$|\mathscr{S}_L(\mathscr{A}) - \int_{[0,1]} f(x) \, dx| \leq \epsilon,$$

with

$$\text{Work}(\mathscr{S}_L(\mathscr{A})) \lesssim \epsilon^{-\max\{1/\beta, \gamma/\kappa\}} |\log \epsilon|^r$$

for some $r = r(\beta, \gamma, \kappa) \geq 0$.

As in Sect. 5.2, we may rewrite Smolyak's algorithm $\mathscr{S}_L(\mathscr{A})$ in a multilevel form, which reveals that a Smolyak's algorithm employs a large number of evaluations of f_0, and subsequently improves on the resulting integral approximation by adding estimates of the integrals $\int_{[0,1]} f_l(x) - f_{l-1}(x) \, dx, l > 0$, that are computed using less quadrature nodes.

5.4 *Partial Differential Equations*

The original Smolyak algorithm inspired two approaches to the numerical solution of partial differential equations (PDEs). The *intrusive* approach is to solve discretizations of the PDE that are built on sparse grids. The *non-intrusive* approach, which we describe here, instead applies the general Smolyak algorithm to product type discretizations whose resolution in the jth direction is described by the parameter k_j [9, 26].

We discuss here how the non-intrusive approach can be analyzed using error expansions of finite difference approximations. For example, the work [11], which introduced the name combination technique, exploited the fact that for the Poisson equation with sufficiently smooth data on $[0,1]^2$, finite difference approximations $u_{k_1,k_2} \in L^\infty([0,1]^2)$ with meshwidths $h_j = 2^{-k_j}$ in the directions $j \in \{1,2\}$ satisfy

$$u - u_{k_1,k_2} = w_1(h_1) + w_2(h_2) + w_{1,2}(h_1, h_2), \tag{23}$$

where u is the exact solution and $w_1(h_1), w_2(h_2), w_{1,2}(h_1,h_2) \in L^\infty([0,1]^2)$ are error terms that converge to zero in L^∞ at the rates $\mathcal{O}(h_1^2)$, $\mathcal{O}(h_2^2)$, and $\mathcal{O}(h_1^2 h_2^2)$, respectively. Since the work required for the computation of $\mathcal{A}(k_1,k_2) := u_{k_1,k_2}$ usually satisfies

$$\text{Work}(\mathcal{A}(k_1,k_2)) \approx (h_1 h_2)^{-\gamma}$$

for some $\gamma \geq 1$ depending on the employed solver, an error bound of size $\epsilon > 0$ could be achieved with the straightforward choice $k_1 := k_2 := -(\log_2 \epsilon)/2$, which would require the work

$$\text{Work}(\mathcal{A}(k_1,k_2)) \approx \epsilon^{-\gamma}.$$

Since Eq. (23) in combination with part (iii) of Proposition 2 shows that $\mathcal{A} \in \mathcal{E}_{(e_j)_{j=1}^2}$ with $e_j(k) := 2^{-2k_j}$, we may deduce from Theorem 1 that Smolyak's algorithm applied to $\mathcal{A}$ requires only the work

$$\epsilon^{-\gamma/2} |\log \epsilon|^{1+\gamma/2}$$

to achieve the same accuracy. The advantage of Smolyak's algorithm becomes even more significant in higher dimensions. All that is required to generalize the analysis presented here to high-dimensional problems, as well as to different PDE and different discretization methods, are error expansions such as Eq. (23).

5.5 *Uncertainty Quantification*

A common goal in uncertainty quantification [1, 13, 18] is the approximation of response surfaces

$$\Gamma \ni \mathbf{y} \mapsto f(\mathbf{y}) := Q(u_{\mathbf{y}}) \in \mathbb{R}.$$

Here, $\boldsymbol{y} \in \Gamma \subset \mathbb{R}^m$ represents parameters in a PDE and $Q(u_{\boldsymbol{y}})$ is a real-valued quantity of interest of the corresponding solution $u_{\boldsymbol{y}}$. For example, a thoroughly studied problem is the parametric linear elliptic second order equation with coefficients $a\colon U \times \Gamma \to \mathbb{R}$,

$$\begin{cases} -\nabla_x \cdot (a(x,\boldsymbol{y})\nabla_x u_{\boldsymbol{y}}(x)) = g(x) & \text{in } U \subset \mathbb{R}^d \\ u_{\boldsymbol{y}}(x) = 0 & \text{on } \partial U, \end{cases}$$

whose solution for any fixed $\boldsymbol{y} \in \Gamma$ is a function $u_{\boldsymbol{y}}\colon U \to \mathbb{R}$.

Approximations of response surfaces may be used for optimization, for worst-case analysis, or to compute statistical quantities such as mean and variance in the case where Γ is equipped with a probability distribution. The non-intrusive approach to compute such approximations, which is known as *stochastic collocation* in the case where Γ is equipped with a probability distribution, is to compute the values of f for finitely many values of $\boldsymbol{y}$ and then interpolate. For example, if we assume for simplicity that $\Gamma = \prod_{j=1}^m [0,1]$, then we may use, as in Sect. 5.1, a sequence of interpolation operators $S_k\colon H^\beta([0,1]) \to H^\alpha([0,1])$ based on evaluations in $(y_{k,i})_{i=1}^{2^k} \subset [0,1]$. However, unlike in Sect. 5.1, we cannot compute values of f exactly but have to rely on a numerical PDE solver. If we assume that this solver has discretization parameters $\boldsymbol{l} = (l_1,\dots,l_d) \in \mathbb{N}^d$ and returns approximations $u_{\boldsymbol{y},\boldsymbol{l}}$ such that the functions

$$\begin{aligned} f_{\boldsymbol{l}}\colon \Gamma &\to \mathbb{R} \\ \boldsymbol{y} &\mapsto f_{\boldsymbol{l}}(\boldsymbol{y}) := Q(u_{\boldsymbol{y},\boldsymbol{l}}) \end{aligned}$$

are elements of $H^\beta_{\text{mix}}([0,1]^m)$, then we may define the numerical approximation method

$$\begin{aligned} \mathscr{A}\colon \mathbb{N}^m \times \mathbb{N}^d &\to H^\alpha_{\text{mix}}([0,1]^m) =: Y \\ \mathscr{A}(\boldsymbol{k},\boldsymbol{l}) &:= \Big(\bigotimes_{j=1}^m S_{k_j}\Big) f_{\boldsymbol{l}}, \end{aligned}$$

with $n := m + d$ discretization parameters.

At this point the reader should already be convinced that straightforward approximation is a bad idea. We therefore omit this part of the analysis, and directly move on to the application of Smolyak's algorithm. To do so, we need to identify functions $e_j\colon \mathbb{N} \to \mathbb{R}_>$ such that $\mathscr{A} \in \mathscr{E}_{(e_j)_{j=1}^n}(Y)$. For this purpose, we write $\mathscr{A}$ as

$$\mathscr{A}(\boldsymbol{k},\boldsymbol{l}) = \mathscr{M}(\mathscr{A}_1(\boldsymbol{k}), \mathscr{A}_2(\boldsymbol{l})),$$

where

$$\mathscr{A}_1(\boldsymbol{k}) := \bigotimes_{j=1}^{m} S_{k_j} \in \mathscr{L}\Big(H^{\beta}_{\mathrm{mix}}([0,1]^m); H^{\alpha}_{\mathrm{mix}}([0,1]^m)\Big) =: Y_1 \quad \forall \boldsymbol{k} \in \mathbb{N}^m$$

$$\mathscr{A}_2(\boldsymbol{l}) := f_{\boldsymbol{l}} \in H^{\beta}_{\mathrm{mix}}([0,1]^m) =: Y_2 \quad \forall \boldsymbol{l} \in \mathbb{N}^d$$

and

$$\mathscr{M}\colon Y_1 \times Y_2 \to Y$$

is the application of linear operators in Y_1 to functions in Y_2. Since $\mathscr{M}$ is continuous and multilinear, we may apply part (iv) of Proposition 2 to reduce our task to the study of $\mathscr{A}_1$ and $\mathscr{A}_2$. The first part can be done exactly as in Sect. 5.1. The second part can be done similarly to Sect. 5.4. However, we now have to verify not only that the approximations $u_{\boldsymbol{y},\boldsymbol{l}}$ converge to the exact solutions $u_{\boldsymbol{y}}$ for each fixed value of $\boldsymbol{y}$ as $\min_{j=1}^{d} l_j \to \infty$, but that this convergence holds in some uniform sense over the parameter space.

More specifically, let us denote by $\Delta^{(\boldsymbol{l})}_{\mathrm{mix}}$ the mixed difference operator with respect to the parameters $\boldsymbol{l}$ and let us assume that

$$\|\Delta^{(\boldsymbol{l})}_{\mathrm{mix}} f_{\boldsymbol{l}}\|_{H^{\beta}_{\mathrm{mix}}([0,1]^m)} \lesssim \prod_{j=1}^{d} \exp(-\kappa_j l_j) =: \prod_{j=1}^{d} e^{(2)}_j(l_j) \quad \forall \boldsymbol{l} \in \mathbb{N}^d.$$

For example, such bounds are proven in [13, 14]. If the interpolation operators satisfy as before

$$\|S_k - \mathrm{Id}\|_{H^{\beta}([0,1]) \to H^{\alpha}([0,1])} \lesssim 2^{-k(\beta-\alpha)} =: e^{(1)}(k) \quad \forall k \in \mathbb{N},$$

then the results of Sect. 5.1 together with part (iv) of Proposition 2 shows that

$$\mathscr{A} \in \mathscr{E}_{(e^{(1)})_{j=1}^{m} \cup (e^{(2)}_j)_{j=1}^{d}}(Y).$$

If we further assume that the work required by the PDE solver with discretization parameters $\boldsymbol{l}$ is bounded by $\exp(\boldsymbol{\gamma}^{(2)} \cdot \boldsymbol{l})$ for some $\boldsymbol{\gamma} \in \mathbb{R}^d_{>}$, then we may assign as total work to the algorithm $\mathscr{A}(\boldsymbol{k},\boldsymbol{l})$ the value

$$\mathrm{Work}(\mathscr{A}(\boldsymbol{k},\boldsymbol{l})) := 2^{|\boldsymbol{k}|_1} \exp(\boldsymbol{\gamma} \cdot \boldsymbol{l}),$$

which is the number of required samples, $2^{|\boldsymbol{k}|_1}$, times the bound on the work per sample, $\exp(\boldsymbol{\gamma} \cdot \boldsymbol{l})$. Thus, by Theorem 1, Smolyak's algorithm achieves the accuracy

$$\|\mathscr{S}_L(\mathscr{A}) - f\|_Y \lesssim \epsilon$$

with

$$\mathrm{Work}(\mathscr{S}_L(\mathscr{A})) \lesssim \epsilon^{-\rho}|\log \epsilon|^r,$$

where $\rho := \max\{1/(\beta - \alpha), \max\{\gamma_j/\kappa_j\}_{j=1}^d\}$ and $r \geq 0$ as in Sect. 4.1.

6 Conclusion

We showed how various existing efficient numerical methods for integration, Monte Carlo simulations, interpolation, the solution of partial differential equations, and uncertainty quantification can be derived from two common underlying principles: decomposition and efficient truncation. The analysis of these methods was divided into proving decay of mixed differences by means of Proposition 2 and then applying general bounds on exponential sums in form of Theorem 1.

Besides simplifying and streamlining the analysis of existing methods, we hope that the framework provided in this work encourages novel applications. Finally, we believe that the general version of Smolyak's algorithm presented here may be helpful in designing flexible and reusable software implementations that can be applied to future problems without modification.

Appendix

Lemma 1 *Let* $\gamma_j > 0$, $\beta_j > 0$, *and* $t_j > 0$ *for* $j \in \{1, \dots, n\}$. *Then*

$$\sum_{(\boldsymbol{\beta}+\boldsymbol{\gamma})\cdot \boldsymbol{k} \leq L} \exp(\boldsymbol{\gamma} \cdot \boldsymbol{k})(\boldsymbol{k}+\mathbf{1})^{\boldsymbol{t}} \leq C(\boldsymbol{\gamma}, \boldsymbol{t}, n) \exp(\mu L)(L+1)^{n^*-1+t^*},$$

where $\rho := \max_{j=1}^n \gamma_j/\beta_j$, $\mu := \frac{\rho}{1+\rho}$, $J := \{j \in \{1, \dots, n\} : \gamma_j/\beta_j = \rho\}$, $n^* := |J|$, $t^* := \sum_{j \in J} t_j$, *and* $(\boldsymbol{k}+\mathbf{1})^{\boldsymbol{t}} := \prod_{j=1}^n (k_j+1)^{t_j}$.

Proof First, we assume without loss of generality that the dimensions are ordered according to whether they belong to J or $J^c := \{1, \dots, n\} \setminus J$. To avoid cluttered notation we then separate dimensions by plus or minus signs in the subscripts; for example, we write $\boldsymbol{t} = (\boldsymbol{t}_J, \boldsymbol{t}_{J^c}) =: (\boldsymbol{t}_+, \boldsymbol{t}_-)$.

Next, we may replace the sum by an integral over $\{(\boldsymbol{\beta}+\boldsymbol{\gamma}) \cdot \boldsymbol{x} \leq L\}$. Indeed, by monotonicity we may do so if we replace L by $L + |\boldsymbol{\beta}+\boldsymbol{\gamma}|_1$, but looking at the final result we observe that a shift of L only affects the constant $C(\boldsymbol{\gamma}, \boldsymbol{t}, n)$.

Finally, using a change of variables $y_j := (\beta_j + \gamma_j)x_j$ and the shorthand $\boldsymbol{\mu} := \boldsymbol{\gamma}/(\boldsymbol{\beta}+\boldsymbol{\gamma})$ (with componentwise division) we obtain

$$\begin{aligned}
&\int_{(\boldsymbol{\beta}+\boldsymbol{\gamma})\cdot\boldsymbol{x}\le L} \exp(\boldsymbol{\gamma}\cdot\boldsymbol{x})(\boldsymbol{x}+\mathbf{1})^{\boldsymbol{t}}\,d\boldsymbol{x}\\
&\quad\le C\int_{|\boldsymbol{y}|_1\le L} \exp(\boldsymbol{\mu}\cdot\boldsymbol{y})(\boldsymbol{y}+\mathbf{1})^{\boldsymbol{t}}\,d\boldsymbol{y}\\
&\quad= C\int_{|\boldsymbol{y}_+|_1\le L} \exp(\boldsymbol{\mu}_+\cdot\boldsymbol{y}_+)(\boldsymbol{y}_+ +\mathbf{1})^{\boldsymbol{t}_+}\\
&\qquad\times\int_{|\boldsymbol{y}_-|_1\le L-|\boldsymbol{y}_+|_1} \exp(\boldsymbol{\mu}_-\cdot\boldsymbol{y}_-)(\boldsymbol{y}_- +\mathbf{1})^{\boldsymbol{t}_-}\,d\boldsymbol{y}_-\,d\boldsymbol{y}_+\\
&\quad\le C\int_{|\boldsymbol{y}_+|_1\le L} \exp(\mu|\boldsymbol{y}_+|_1)(\boldsymbol{y}_+ +\mathbf{1})^{\boldsymbol{t}_+}\\
&\qquad\times\int_{|\boldsymbol{y}_-|_1\le L-|\boldsymbol{y}_+|_1} \exp(\mu_-|\boldsymbol{y}_-|_1)(\boldsymbol{y}_- +\mathbf{1})^{\boldsymbol{t}_-}\,d\boldsymbol{y}_-\,d\boldsymbol{y}_+ = (\star),
\end{aligned}$$

where the last equality holds by definition of $\mu = \max\{\boldsymbol{\mu}_+\}$ and $\mu_- := \max\{\boldsymbol{\mu}_-\}$. We use the letter C here and in the following to denote quantities that depend only on $\boldsymbol{\gamma}$, $\boldsymbol{t}$ and n but may change value from line to line. Using $(\boldsymbol{y}_+ +\mathbf{1})^{\boldsymbol{t}_+} \le (|\boldsymbol{y}_+|_1 + 1)^{|\boldsymbol{t}_+|_1}$ and $(\boldsymbol{y}_- +\mathbf{1})^{\boldsymbol{t}_-} \le (|\boldsymbol{y}_-|_1 + 1)^{|\boldsymbol{t}_-|_1}$ and the linear change of variables $\boldsymbol{y}\mapsto (|\boldsymbol{y}|_1, y_2,\dots,y_n)$ in both integrals, we obtain

$$\begin{aligned}
(\star) &\le C\int_{|\boldsymbol{y}_+|_1\le L} \exp(\mu|\boldsymbol{y}_+|_1)(|\boldsymbol{y}_+|_1+1)^{|\boldsymbol{t}_+|_1}\int_{|\boldsymbol{y}_-|_1\le L-|\boldsymbol{y}_+|_1} \exp(\mu_-|\boldsymbol{y}_-|_1)(|\boldsymbol{y}_-|_1+1)^{|\boldsymbol{t}_-|_1}\,d\boldsymbol{y}_-\,d\boldsymbol{y}_+\\
&\le C\int_0^L \exp(\mu u)(u+1)^{|\boldsymbol{t}_+|_1}u^{|J|-1}\int_0^{L-u}\exp(\mu_- v)(v+1)^{|\boldsymbol{t}_-|_1}v^{|J^c|-1}\,dv\,du\\
&\le C(L+1)^{|\boldsymbol{t}_+|_1}L^{|J|-1}\int_0^L \exp(\mu u)((L-u)+1)^{|\boldsymbol{t}_-|_1}(L-u)^{|J^c|-1}\int_0^{L-u}\exp(\mu_- v)\,dv\,du\\
&\le C(L+1)^{|\boldsymbol{t}_+|_1+|J|-1}\int_0^L \exp(\mu u)(L-u+1)^{|\boldsymbol{t}_-|_1}(L-u)^{|J^c|-1}\exp(\mu_-(L-u))\,du\\
&= C(L+1)^{|\boldsymbol{t}_+|_1+|J|-1}\exp(\mu L)\int_0^L \exp(-(\mu-\mu_-)w)(w+1)^{|\boldsymbol{t}_-|_1}w^{|J^c|-1}\,dw\\
&\le C(L+1)^{|\boldsymbol{t}_+|_1+|J|-1}\exp(\mu L),
\end{aligned}$$

where we used supremum bounds for both integrals for the third inequality, the change of variables $w := L-u$ for the penultimate equality, and the fact that $\mu > \mu_-$ for the last inequality.

Lemma 2 *Let $\gamma_j > 0$, $\beta_j > 0$, and $s_j > 0$ for $j \in \{1, \dots, n\}$. Then*

$$\sum_{(\boldsymbol{\beta}+\boldsymbol{\gamma})\cdot \boldsymbol{k} > L} \exp(-\boldsymbol{\beta}\cdot\boldsymbol{k})(\boldsymbol{k}+\mathbf{1})^{\boldsymbol{s}} \le C(\boldsymbol{\beta}, \boldsymbol{s}, n)\exp(-\nu L)(L+1)^{n^*-1+s^*},$$

where $\rho := \max_{j=1}^n \gamma_j/\beta_j$, $\nu := \frac{1}{1+\rho}$, $J := \{j \in \{1,\dots,n\} : \gamma_j/\beta_j = \rho\}$, $n^* := |J|$, $s^* := \sum_{j\in J} t_j$, and $(\boldsymbol{k}+\mathbf{1})^{\boldsymbol{s}} := \prod_{j=1}^n (k_j+1)^{s_j}$.

Proof First, we assume without loss of generality that the dimensions are ordered according to whether they belong to J or J^c. To avoid cluttered notation we then separate dimensions by plus or minus signs in the subscripts; for example, we write $\boldsymbol{s} = (\boldsymbol{s}_J, \boldsymbol{s}_{J^c}) =: (\boldsymbol{s}_+, \boldsymbol{s}_-)$.

Next, we may replace the sum by an integral over $\{(\boldsymbol{\beta}+\boldsymbol{\gamma})\cdot\boldsymbol{x} > L\}$. Indeed, by monotonicity we may do so if we replace L by $L - |\boldsymbol{\beta}+\boldsymbol{\gamma}|_1$, but looking at the final result we observe that a shift of L only affects the constant $C(\boldsymbol{\beta}, \boldsymbol{s}, n)$.

Finally, using a change of variables $y_j := (\beta_j+\gamma_j)x_j$ and the shorthand $\boldsymbol{\nu} := \boldsymbol{\beta}/(\boldsymbol{\beta}+\boldsymbol{\gamma})$ (with componentwise division) we obtain

$$\begin{aligned}
&\int_{(\boldsymbol{\beta}+\boldsymbol{\gamma})\cdot\boldsymbol{x}>L} \exp(-\boldsymbol{\beta}\cdot\boldsymbol{x})(\boldsymbol{x}+\mathbf{1})^{\boldsymbol{s}}\, d\boldsymbol{x} \le C\int_{|\boldsymbol{y}|_1>L} \exp(-\boldsymbol{\nu}\cdot\boldsymbol{y})(\boldsymbol{y}+\mathbf{1})^{\boldsymbol{s}}\, d\boldsymbol{y}\\
&= C\int_{|\boldsymbol{y}_+|_1>L} \exp(-\boldsymbol{\nu}_+\cdot\boldsymbol{y}_+)(\boldsymbol{y}_+ +\mathbf{1})^{\boldsymbol{s}_+}\int_{|\boldsymbol{y}_-|_1>(L-|\boldsymbol{y}_+|_1)^+} \exp(-\boldsymbol{\nu}_-\cdot\boldsymbol{y}_-)(\boldsymbol{y}_- +\mathbf{1})^{\boldsymbol{s}_-} d\boldsymbol{y}_- d\boldsymbol{y}_+\\
&\le C\int_{|\boldsymbol{y}_+|_1>L} \exp(-\nu|\boldsymbol{y}_+|_1)(\boldsymbol{y}_+ +\mathbf{1})^{\boldsymbol{s}_+}\int_{|\boldsymbol{y}_-|_1>(L-|\boldsymbol{y}_+|_1)^+} \exp(-\nu_-|\boldsymbol{y}_-|_1)(\boldsymbol{y}_- +\mathbf{1})^{\boldsymbol{s}_-} d\boldsymbol{y}_- d\boldsymbol{y}_+\\
&=: (\star),
\end{aligned}$$

where the last equality holds by definition of $\nu = \max\{\boldsymbol{\nu}_+\}$ and $\nu_- := \max\{\boldsymbol{\nu}_-\}$. We use the letter C here and in the following to denote quantities that depend only on $\boldsymbol{\beta}$, $\boldsymbol{s}$ and n but may change value from line to line. Using $(\boldsymbol{y}_+ +\mathbf{1})^{\boldsymbol{s}_+} \le (|\boldsymbol{y}_+|_1 + 1)^{|\boldsymbol{s}_+|_1}$ and $(\boldsymbol{y}_- +\mathbf{1})^{\boldsymbol{s}_-} \le (|\boldsymbol{y}_-|_1 + 1)^{|\boldsymbol{s}_-|_1}$ and the linear change of variables $\boldsymbol{y} \mapsto (|\boldsymbol{y}|_1, y_2, \dots, y_n)$ in both integrals, we obtain

$$\begin{aligned}
(\star) \le& C\int_{|\boldsymbol{y}_+|_1>0} \exp(-\nu|\boldsymbol{y}_+|_1)(|\boldsymbol{y}_+|_1+1)^{|\boldsymbol{s}_+|_1}\int_{|\boldsymbol{y}_-|_1>(L-|\boldsymbol{y}_+|_1)^+} \exp(-\nu_-|\boldsymbol{y}_-|_1)(|\boldsymbol{y}_-|_1+1)^{|\boldsymbol{s}_-|_1} d\boldsymbol{y}_- d\boldsymbol{y}_+\\
\le& C\int_0^\infty \exp(-\nu u)(u+1)^{|\boldsymbol{s}_+|_1} u^{|J|-1}\int_{(L-u)^+}^\infty \exp(-\nu_- v)(v+1)^{|\boldsymbol{s}_-|_1} v^{|J^c|-1}\, dv\, du\\
=& C\int_0^L \exp(-\nu u)(u+1)^{|\boldsymbol{s}_+|_1+|J|-1}\int_{L-u}^\infty \exp(-\nu_- v)(v+1)^{|\boldsymbol{s}_-|_1+|J^c|-1}\, dv\, du\\
&+ C\int_L^\infty \exp(-\nu u)(u+1)^{|\boldsymbol{s}_+|_1+|J|-1}\int_0^\infty \exp(-\nu_- v)(v+1)^{|\boldsymbol{s}_-|_1+|J^c|-1}\, dv\, du\\
=&: (\star\star) + (\star\star\star).
\end{aligned}$$

To bound (⋆⋆), we estimate the inner integral using the inequality $\int_a^\infty \exp(-bv)(v+1)^c\,dv \leq C\exp(-ba)(a+1)^c$ [21, (8.11.2)], which is valid for all positive a, b, c:

$$\begin{aligned}
(\star\star) &\leq C\int_0^L \exp(-\nu u)(u+1)^{|s_+|_1+|J|-1}\exp(-\nu_-(L-u))(L-u+1)^{|s_-|_1+|J^c|-1}\,du \\
&\leq C(L+1)^{|s_+|_1+|J|-1}\int_0^L \exp(-\nu(L-w))\exp(-\nu_- w)(w+1)^{|s_-|_1+|J^c|-1}\,dw \\
&= C(L+1)^{|s_+|_1+|J|-1}\exp(-\nu L)\int_0^L \exp(-(\nu_- - \nu)w)(w+1)^{|s_-|_1+|J^c|-1}\,dw \\
&\leq C(L+1)^{|s_+|_1+|J|-1}\exp(-\nu L),
\end{aligned}$$

where we used a supremum bound and the change of variables $w := L - u$ for the second inequality, and the fact that $\nu_- > \nu$ for the last inequality. Finally, to bound (⋆ ⋆ ⋆), we observe that the inner integral is independent of L, and bound the outer integral in the same way we previously bounded the inner integral. This shows

$$(\star\star\star) \leq C\exp(-\nu L)(L+1)^{|s_+|_1+|J|-1}.$$

References

1. I. Babuška, R. Tempone, G.E. Zouraris, Galerkin finite element approximations of stochastic elliptic partial differential equations. SIAM J. Numer. Anal. **42**(2), 800–825 (2004)
2. H.-J. Bungartz, M. Griebel, Sparse grids. Acta Numer. **13**, 147–269 (2004)
3. A. Defant, K. Floret, *Tensor Norms and Operator Ideals* (Elsevier, Burlington, 1992)
4. D. Dũng, M. Griebel, Hyperbolic cross approximation in infinite dimensions. J. Complexity **33**, 55–88 (2016)
5. D. Dũng, V.N. Temlyakov, T. Ullrich, Hyperbolic cross approximation (2015). arXiv:1601.03978
6. J. Garcke, A dimension adaptive sparse grid combination technique for machine learning. ANZIAM J. **48**(C), C725–C740 (2007)
7. J. Garcke, Sparse grids in a nutshell, in *Sparse Grids and Applications* (Springer, Berlin, 2012), pp. 57–80
8. M.B. Giles, Multilevel monte carlo path simulation. Oper. Res. **56**(3), 607–617 (2008)
9. M. Griebel, H. Harbrecht, On the convergence of the combination technique, in *Sparse Grids and Applications* (Springer, Cham, 2014), pp. 55–74
10. M. Griebel, J. Oettershagen, On tensor product approximation of analytic functions. J. Approx. Theory **207**, 348–379 (2016)
11. M. Griebel, M. Schneider, C. Zenger, A combination technique for the solution of sparse grid problems, in *Iterative Methods in Linear Algebra*, ed. by P. de Groen, R. Beauwens. IMACS, (Elsevier, Amsterdam, 1992), pp. 263–281
12. W. Hackbusch, Tensor Spaces and Numerical Tensor Calculus (Springer, Berlin, 2012)
13. A.-L. Haji-Ali, F. Nobile, R. Tempone, Multi-index Monte Carlo: when sparsity meets sampling. Numer. Math. **132**(4), 767–806 (2016)
14. H. Harbrecht, M. Peters, M. Siebenmorgen, Multilevel accelerated quadrature for PDEs with log-normally distributed diffusion coefficient. SIAM/ASA J. Uncertain. Quantif. **4**(1), 520–551 (2016)

15. M. Hegland, Adaptive sparse grids. ANZIAM J. **44**(C), C335–C353 (2002)
16. S. Heinrich, Monte Carlo complexity of global solution of integral equations. J. Complexity **14**(2), 151–175 (1998)
17. P.E. Kloeden, E. Platen, *Numerical Solution of Stochastic Differential Equations* (Springer, Berlin, 1992)
18. O.P. Le Maître, O.M. Knio, *Spectral Methods for Uncertainty Quantification* (Springer, Berlin, 2010)
19. S. Martello, P. Toth, *Knapsack Problems: Algorithms and Computer Implementations* (Wiley, New York, 1990)
20. F. Nobile, R. Tempone, S. Wolfers, Sparse approximation of multilinear problems with applications to kernel-based methods in UQ. Numer. Math. **139**(1), 247–280 (2018)
21. F.W.J. Olver, A.B. Olde Daalhuis, D.W. Lozier, B.I. Schneider, R.F. Boisvert, C.W. Clark, B.R. Miller, B.V. Saunders (eds.), *NIST Digital Library of Mathematical Functions*. http://dlmf.nist.gov/. Release 1.0.13 of 2016-09-16
22. A. Papageorgiou, H. Woźniakowski, Tractability through increasing smoothness. J. Complexity **26**(5), 409–421 (2010)
23. I.H. Sloan, H. Woźniakowski, Tractability of multivariate integration for weighted Korobov classes. J. Complexity **17**(4), 697–721 (2001)
24. S.A. Smolyak, Quadrature and interpolation formulas for tensor products of certain classes of functions. Soviet Math. Dokl. **4**, 240–243 (1963)
25. G.W. Wasilkowski, H. Woźniakowski, Explicit cost bounds of algorithms for multivariate tensor product problems. J. Complexity **11**(1), 1–56 (1995)
26. C. Zenger, Sparse grids, in *Parallel Algorithms for Partial Differential Equations. Proceedings of the Sixth GAMM-Seminar*, ed. by W. Hackbusch (Vieweg, Braunschweig, 1991)

Fundamental Splines on Sparse Grids and Their Application to Gradient-Based Optimization

Julian Valentin and Dirk Pflüger

Abstract Most types of hierarchical basis functions for sparse grids are not continuously differentiable. This can lead to problems, for example, when using gradient-based optimization methods on sparse grid functions. B-splines represent an interesting alternative to conventional basis types since they have displayed promising results for regression and optimization problems. However, their overlapping support impedes the task of hierarchization (computing the interpolant), as, in general, the solution of a linear system is required. To cope with this problem, we propose three general basis transformations. They leave the spanned function space on dimensionally adaptive sparse grids or full grids unchanged, but result in triangular linear systems. One of the transformations, when applied to the B-spline basis, yields the well-known fundamental splines. We suggest a modification of the resulting sparse grid basis to enable nearly linear extrapolation towards the domain's boundary without the need to spend boundary points. Finally, we apply the hierarchical modified fundamental spline basis to gradient-based optimization with sparse grid surrogates.

1 Introduction

When dealing with real-valued functions $f\colon [0, 1]^d \to \mathbb{R}$ on higher-dimensional domains (such as interpolation and optimization), one must consider the *curse of dimensionality*. The curse states that the number of discretization points needed for a uniform sampling of $[0, 1]^d$ depends exponentially on d. As a result, conventional approaches are typically ruled out even for moderate dimensionalities, where $d \geq 4$. To tackle the curse, sparse grids have been used successfully for discretization [2, 4, 19].

J. Valentin · D. Pflüger (✉)
Simulation of Large Systems (SGS), Institute for Parallel and Distributed Systems (IPVS), University of Stuttgart, Stuttgart, Germany
e-mail: julian.valentin@ipvs.uni-stuttgart.de; dirk.pflueger@ipvs.uni-stuttgart.de

J. Garcke et al. (eds.), *Sparse Grids and Applications – Miami 2016*, Lecture Notes in Computational Science and Engineering 123,
https://doi.org/10.1007/978-3-319-75426-0_10

Sparse grids can be used with different types of hierarchical bases. Common types include the piecewise linear [22] and the piecewise polynomial functions of Bungartz [1]. All these 1D bases are continuous on [0, 1], but not continuously differentiable. This leads to problems when using sparse grid surrogates in applications that depend on derivatives, such as gradient-based optimization. Using hierarchical B-splines on sparse grids is a promising approach to overcome this problem. Their smoothness can be adjusted by their degree p, and their compact support and piecewise polynomial structure facilitate their handling both analytically and numerically. Moreover, the piecewise linear functions are a special case of B-splines, obtained by setting $p = 1$. B-spline sparse grid interpolants have been employed successfully for gradient-based optimization and topology optimization [19].

The main drawback of B-splines is their non-local support (even though compact) in the sense that a B-spline does not necessarily vanish at neighboring grid points. This greatly complicates the process of hierarchization (i.e., computing the interpolant): For basis functions with local support (i.e., $\operatorname{supp} \varphi_{l,i} = [x_{l,i} - h_l, x_{l,i} + h_l]$ with $x_{l,i} := ih_l$, $h_l := 2^{-l}$), hierarchization can be done efficiently via the unidirectional principle, which works for spatially adaptive grids in which direct ancestors exist in every coordinate direction (see [11, 12]). For B-splines, however, a linear system must be solved to perform hierarchization. The corresponding system matrix is generally non-symmetric (in contrast to regression tasks [11]) and may be either sparsely or densely populated. Consequently, considerable time and system memory are required to solve the linear system.

In this paper, we propose three transformations that convert a general hierarchical basis into another hierarchical basis that satisfies a specific *fundamental property*. The resulting transformed bases lead to easily solvable triangular hierarchization systems. One transformation leaves the spanned sparse grid space invariant, while the other two leave the spanned nodal spaces invariant. If we apply the third transformation to the hierarchical B-spline space, we obtain the hierarchical fundamental spline basis, whose analytical approximation quality on sparse grids has been previously studied [17]. We propose modifying the fundamental splines, analogously to the modified B-spline basis [11], to attain meaningful boundary values without having to "spend" sparse grid points on the boundary itself. We compare the quality of the hierarchical modified fundamental splines with the modified B-spline basis in an application-based setting. Here, we choose the gradient-based optimization of the corresponding sparse grid surrogates since optimization is one area in which B-splines have shown good results [19].

The paper is structured as follows: First, we introduce our notation of hierarchical B-splines on sparse grids in Sect. 2 and state the hierarchization problem. In Sect. 3, we study the behavior of general hierarchical basis functions which fulfill the fundamental property. We give definitions of the different transformations to transform an existing sparse grid basis into one that satisfies the fundamental property in Sect. 4. As already mentioned, one of them leads to the well-known fundamental splines, which we discuss in Sect. 5. We compare the performance of the fundamental splines versus the B-spline basis (both modified) in gradient-based optimization in Sect. 6. Finally, we conclude in Sect. 7.

2 B-Splines on Sparse Grids

In this section, we first repeat the definition of B-splines on sparse grids [11, 19] and then state the problems which arise for typical algorithms at the example of hierarchization.

2.1 *Definition and Properties*

Let $p \in \mathbb{N}_0$. The *cardinal B-spline* $b^p\colon \mathbb{R} \to \mathbb{R}$ *of degree* p is defined by

$$b^0(x) := \chi_{[0,1)}(x)\,, \qquad b^p(x) := \int_0^1 b^{p-1}(x-y)\,\mathrm{d}y\,.$$

Equivalently (cf. [5, 6]), b^p can be defined by the recursion

$$b^p(x) = \frac{x}{p} b^{p-1}(x) + \frac{p+1-x}{p} b^{p-1}(x-1)\,,$$

which can be used to prove simple properties such as boundedness ($0 \le b^p \le 1$) and compact support ($\operatorname{supp} b^p = [0, p+1]$). b^p is indeed a spline of degree p, i.e., a piecewise polynomial of degree $\le p$ and $(p-1)$ times continuously differentiable. The pieces are the so-called *knot intervals* $[k, k+1)$, $k = 0, \dots, p$, whereas the points $k = 0, \dots, p+1$ are called *knots* (places where the pth derivative is discontinuous). The translates $\{b^p(\cdot - k) \mid k \in \mathbb{Z}\}$ form a basis of the *spline space of degree* p (all functions on $\mathbb{R}$ that are, for every $k \in \mathbb{Z}$, polynomials of degree $\le p$ on $[k, k+1)$ and at least $(p-1)$ times continuously differentiable at $x = k$).

Hierarchical B-splines $\varphi^p_{l,i}\colon [0,1] \to \mathbb{R}$ of *level* $l \in \mathbb{N}$ and *index* $i \in I_l := \{1, 3, 5, \dots, 2^l - 1\}$ are defined by

$$\varphi^p_{l,i}(x) := b^p\left(\frac{x}{h_l} + \frac{p+1}{2} - i\right), \quad h_l := 2^{-l}\,.$$

We will consider only odd degree p (covers the common cases $p \in \{1, 3\}$), as otherwise *grid points* $x_{l,i} := ih_l$ and knots do not coincide. This leads to non-nested nodal spaces and prevents their necessary decomposition into hierarchical subspaces.

Multivariate hierarchical B-splines are defined by the usual tensor product approach, i.e., $\varphi^{\mathbf{p}}_{\mathbf{l},\mathbf{i}}\colon [0,1]^d \to \mathbb{R}$ for $\mathbf{p} \in \mathbb{N}_0^d$, $\mathbf{l} \in \mathbb{N}^d$, $\mathbf{i} \in I_{\mathbf{l}} := I_{l_1} \times \cdots \times I_{l_d}$, and

$$\varphi^{\mathbf{p}}_{\mathbf{l},\mathbf{i}}(\mathbf{x}) := \prod_{t=1}^{d} \varphi^{p_t}_{l_t,i_t}(x_t)\,.$$

Nodal and *hierarchical subspaces* are defined by

$$V_{\mathbf{l}}^{\mathbf{p}} := \operatorname{span}\{\varphi_{\mathbf{l},\mathbf{i}}^{\mathbf{p}} \mid \mathbf{1} \le \mathbf{i} \le \mathbf{2}^{\mathbf{l}} - \mathbf{1}\}\,, \qquad W_{\mathbf{l}}^{\mathbf{p}} := \operatorname{span}\{\varphi_{\mathbf{l},\mathbf{i}}^{\mathbf{p}} \mid \mathbf{i} \in I_{\mathbf{l}}\}\,,$$

where "$\mathbf{1} \le \mathbf{i} \le \mathbf{2}^{\mathbf{l}} - \mathbf{1}$" is to be read coordinate-wise. The basis functions that span these subspaces are shown in Fig. 1. The corresponding grid points are given by $\mathbf{x}_{\mathbf{l},\mathbf{i}} := (x_{l_1,i_1}, \dots, x_{l_d,i_d})$. If we restrict the domain of all functions to

$$D_{\mathbf{l}}^{\mathbf{p}} := D_{l_1}^{p_1} \times \cdots \times D_{l_d}^{p_d}\,, \quad D_{l_t}^{p_t} := \left[\frac{p_t+1}{2} h_{l_t}, 1 - \frac{p_t+1}{2} h_{l_t}\right],$$

we can infer

$$V_{\mathbf{l}}^{\mathbf{p}}\big|_{D_{\mathbf{l}}^{\mathbf{p}}} = \bigoplus_{\mathbf{l}' \le \mathbf{l}} W_{\mathbf{l}'}^{\mathbf{p}}\Big|_{D_{\mathbf{l}}^{\mathbf{p}}}\,,$$

where both sides coincide with the spline space of degree $\mathbf{p}$ corresponding to the full grid of level $\mathbf{l}$, as we have shown in [19]. The sparse grid space $V_n^{\mathbf{p},\mathrm{s}}$ of level $n \in \mathbb{N}$ can now be constructed as usual by

$$V_n^{\mathbf{p},\mathrm{s}} := \bigoplus_{\|\mathbf{l}\|_1 \le n+d-1} W_{\mathbf{l}}^{\mathbf{p}}\,.$$

Here and in the following, the superscript "s" denotes "sparse grid" (in contrast to full grids). Note that the well-known piecewise linear sparse grid space is a special case for $\mathbf{p} = \mathbf{1}$.

The basis functions of the common hierarchical bases (such as piecewise linear or piecewise polynomial ones) of level $\mathbf{l} \ge \mathbf{1}$ vanish on the boundary of the domain $[0, 1]^d$. As a result, interpolants $\tilde{f}$ of a function f will vanish there, too. Even though not all B-splines vanish on the boundary, interpolants decay unnaturally towards it, as necessary degrees of freedom are missing near the boundary.

One approach to represent functions with non-zero boundary values is to use sparse grids with boundary points, i.e., an additional 1D level $l = 0$ is introduced together with two basis functions ($i \in \{0, 1\}$). However, especially in higher dimensions, disproportionately many points (and, thus, function evaluations) have to be spent on the boundary. This severely limits the number of dimensions that can be dealt with.

An alternative approach is instead to modify the basis to extrapolate towards the boundary [11]: The 1D basis function of level 1 is set to the constant 1, while on higher levels l, the left-most and right-most basis functions (with index $i = 1$ or $i = 2^l - 1$) are modified such that the basis function is linear or almost linear at the boundary. This mimics linear extrapolation. Starting from the linear case $p = 1$, in which this is achieved by $\varphi_{l,1}^{1,\mathrm{mod}} := \varphi_{l,1}^{1} + 2\varphi_{l,0}^{1}$ and $\varphi_{l,2^l-1}^{1,\mathrm{mod}} := \varphi_{l,2^l-1}^{1} + 2\varphi_{l,2^l}^{1}$, this modification can be generalized for B-splines [11]:

$$\varphi_{l,i}^{p,\mathrm{mod}}(x) := \begin{cases} 1 & \text{if } l = 1, i = 1, \\ \psi_l^p(x) & \text{if } l > 1, i = 1, \\ \psi_l^p(1-x) & \text{if } l > 1, i = 2^l - 1, \\ \varphi_{l,i}^p(x) & \text{otherwise,} \end{cases} \qquad \psi_l^p := \sum_{k=0}^{\lceil (p+1)/2 \rceil} (k+1)\varphi_{l,1-k}^p \,.$$

One can prove with Marsden's identity [6] that for $1 \leq p \leq 4$, ψ_l^p is linear near the boundary:

$$\psi_l^p(x) = 2 - \frac{x}{h_l}\,, \quad x \in \left[0, \frac{5-p}{2} h_l\right]. \tag{1}$$

For higher degrees p, the deviation from $2 - x/h_l$ is very small [11]. Multivariate modified basis functions are formed via the tensor product approach. The one-dimensional modification is shown in Fig. 1 (right).

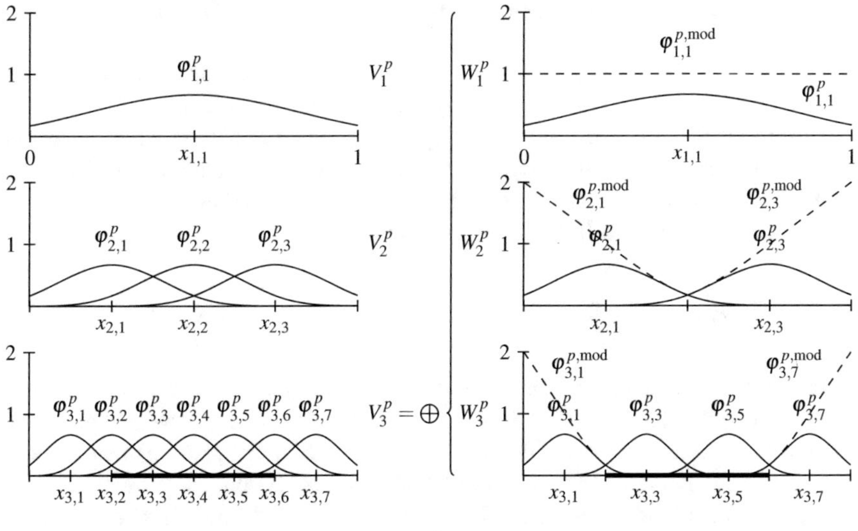

Fig. 1 *Left:* Nodal cubic B-splines, spanning V_l^p. *Right:* Hierarchical cubic B-splines, spanning W_l^p. On D_l^p *(thick)*, the span of $\bigoplus_{l'=1}^{l} W_{l'}^p$ equals V_l^p (here, $l = 3$). For reasonable boundary values, use the modified cubic B-splines *(dashed)*

2.2 *Hierarchization with B-Splines*

Hierarchical B-splines can be used for interpolation on sparse grids. In the following, we assume $\Omega^{\mathrm{s}} \subsetneq [0,1]^d$, $N := |\Omega^{\mathrm{s}}| < \infty$, to be a spatially adaptive sparse grid, i.e., consisting of grid points $\mathbf{x}_{\mathbf{l},\mathbf{i}}$ which have at least one ancestor point:

$$\forall_{\mathbf{x}_{\mathbf{l},\mathbf{i}} \in \Omega^{\mathrm{s}}} \exists_{\mathbf{x}_{\mathbf{l}',\mathbf{i}'} \in \Omega^{\mathrm{s}}} \exists_{t' \in \{1,\dots,d\}} \quad (l_t, i_t) = \begin{cases} (l'_t + 1, 2i'_t \pm 1) & \text{if } t = t', \\ (l'_t, i'_t) & \text{otherwise.} \end{cases} \tag{2}$$

We further assume that for every grid point $\mathbf{x}_{\mathbf{k},\mathbf{j}} \in \Omega^{\mathrm{s}}$ some data $f(\mathbf{x}_{\mathbf{k},\mathbf{j}}) \in \mathbb{R}$ is given. The task of computing coefficients $\alpha_{\mathbf{l},\mathbf{i}} \in \mathbb{R}$ with

$$\forall_{\mathbf{x}_{\mathbf{k},\mathbf{j}} \in \Omega^{\mathrm{s}}} \quad \tilde{f}(\mathbf{x}_{\mathbf{k},\mathbf{j}}) = f(\mathbf{x}_{\mathbf{k},\mathbf{j}}), \qquad \tilde{f} := \sum_{\mathbf{x}_{\mathbf{l},\mathbf{i}} \in \Omega^{\mathrm{s}}} \alpha_{\mathbf{l},\mathbf{i}} \varphi^{\mathbf{p}}_{\mathbf{l},\mathbf{i}},$$

is called *hierarchization*. The coefficients are called *hierarchical surpluses*, a name which stems from the piecewise linear and the piecewise polynomial functions. Hierarchization is equivalent to solving a system of linear equations,

$$A\boldsymbol{\alpha} = \mathbf{f}, \quad A := (\varphi^{\mathbf{p}}_s(\mathbf{x}_r))_{r,s}, \quad \boldsymbol{\alpha} := (\alpha_s)_s, \quad \mathbf{f} := (f(\mathbf{x}_r))_r, \tag{3}$$

where the row and column indices $r, s \in \{1, \dots, N\}$ are linearized indices (in arbitrary order) of the level-index pairs $(\mathbf{k}, \mathbf{j})$ and $(\mathbf{l}, \mathbf{i})$, respectively.

Unfortunately, using hierarchical B-splines as basis functions leads to linear systems which are hard to solve, since they are in general non-symmetric and densely populated: The matrix entry in the row corresponding to $(\mathbf{k}, \mathbf{j})$ and the column corresponding to $(\mathbf{l}, \mathbf{i})$ vanishes if and only if

$$\mathbf{x}_{\mathbf{k},\mathbf{j}} \notin \operatorname{supp}^0 \varphi^{\mathbf{p}}_{\mathbf{l},\mathbf{i}} \iff \exists_{t=1,\dots,d} \quad x_{k_t,j_t} \notin \left(x_{l_t,i_t} - \frac{p_t+1}{2} h_{l_t}, \; x_{l_t,i_t} + \frac{p_t+1}{2} h_{l_t} \right),$$

where supp^0 denotes the interior of the support. However, for low levels $\mathbf{l}$, the mesh width h_{l_t} is rather large in every dimension t, implying that only few grid points lie outside $\operatorname{supp}^0 \varphi^{\mathbf{p}}_{\mathbf{l},\mathbf{i}}$. For general degree $\mathbf{p}$, every matrix entry $\varphi^{\mathbf{p}}_{\mathbf{l},\mathbf{i}}(\mathbf{x}_{\mathbf{k},\mathbf{j}})$ can be non-zero due to the overlapping supports of the hierarchical B-splines. In contrast to the case $\mathbf{p} = \mathbf{1}$ of hat functions, the value of $\alpha_{\mathbf{l},\mathbf{i}}$ depends not only on $f(\mathbf{x}_{\mathbf{l},\mathbf{i}})$ and the data at the $3^d - 1$ neighboring grid points on the boundary of $\operatorname{supp} \varphi^{\mathbf{1}}_{\mathbf{l},\mathbf{i}}$. This makes the use of the so-called unidirectional principle, which hierarchizes with $\mathcal{O}(Nd)$ one-dimensional basis evaluations, only possible for dimensionally adaptive sparse grids.

3 Fundamental Property

Much of the root of the difficulties of hierarchization lies in the overlapping support of the B-splines, which the B-splines need for their good approximation quality. However, it would already suffice if we had 1D basis functions $\phi^{\mathrm{F}}_{l,i}\colon [0,1] \to \mathbb{R}$ that satisfy

$$\begin{aligned} \phi^{\mathrm{F}}_{l,i}(x_{k,j}) &= 0\,, & k < l\,,\quad & j \in I_k\,,\\ \phi^{\mathrm{F}}_{l,i}(x_{l,j}) &= \delta_{i,j}\,, & & j \in I_l\,. \end{aligned} \tag{4}$$

We call the property *fundamental property*. The first equation in (4) makes sure that every basis function vanishes at all grid points of coarser levels. The second equation says that the basis functions should additionally vanish at all other grid points of the same level. As a result, the corresponding 1D hierarchization system matrix $A = (\phi^{\mathrm{F}}_s(x_r))_{r,s}$ is in lower triangular form with ones on the diagonal if the row and column indices r and s are identically sorted by increasing level of the corresponding grid points and basis functions.

In the multidimensional tensor product case, the arguments are very similar: We observe that (4) implies

$$\phi^{\mathrm{F}}_{\mathbf{l},\mathbf{i}}(\mathbf{x}_{\mathbf{k},\mathbf{j}}) \neq 0 \quad\Longrightarrow\quad \forall_{t=1,\dots,d}\ \Big[(l_t < k_t) \vee ((l_t, i_t) = (k_t, j_t))\Big]\,, \qquad \mathbf{x}_{\mathbf{k},\mathbf{j}} \in \Omega^{\mathrm{s}}\,. \tag{5}$$

In other words, at a fixed grid point $\mathbf{x}_{\mathbf{k},\mathbf{j}}$, only those basis functions $\phi^{\mathrm{F}}_{\mathbf{l},\mathbf{i}}$ can be non-zero which, in each dimension, have coarser levels than the grid point $\mathbf{x}_{\mathbf{k},\mathbf{j}}$ or the same level-index pair. For a suitable ordering as in the 1D case (but sorting by level sum $\|\cdot\|_1$ instead), A will be in lower triangular form, which is the statement of the following forward substitution lemma. The following considerations hold for arbitrary tensor product bases, which we denote with $\phi_{\mathbf{l},\mathbf{i}}$ (as opposed to the B-splines $\varphi^{\mathbf{p}}_{\mathbf{l},\mathbf{i}}$). In each dimension, we could even employ different types of basis functions, which would enable dimensional p-adaptivity.

Lemma 1 *Let $\Omega^{\mathrm{s}} \subsetneq [0,1]^d$ be a spatially adaptive sparse grid with arbitrary tensor product basis functions $\phi^{\mathrm{F}}_{\mathbf{l},\mathbf{i}}$ consisting of 1D functions $\phi^{\mathrm{F}}_{l_t,i_t}$ which fulfill* (4). *Then the hierarchical surplus of a grid point $\mathbf{x}_{\mathbf{k},\mathbf{j}} \in \Omega^{\mathrm{s}}$ satisfies*

$$\alpha_{\mathbf{k},\mathbf{j}} = f(\mathbf{x}_{\mathbf{k},\mathbf{j}}) - \sum_{\substack{\mathbf{x}_{\mathbf{l},\mathbf{i}} \in \Omega^{\mathrm{s}},\\ \|\mathbf{l}\|_1 < \|\mathbf{k}\|_1}} \alpha_{\mathbf{l},\mathbf{i}} \phi^{\mathrm{F}}_{\mathbf{l},\mathbf{i}}(\mathbf{x}_{\mathbf{k},\mathbf{j}})\,.$$

Proof The row of the linear system (3) corresponding to the level-index pair $(\mathbf{k}, \mathbf{j})$ is

$$\sum_{\mathbf{x}_{\mathbf{l},\mathbf{i}} \in \Omega^{\mathrm{s}}} \alpha_{\mathbf{l},\mathbf{i}} \phi^{\mathrm{F}}_{\mathbf{l},\mathbf{i}}(\mathbf{x}_{\mathbf{k},\mathbf{j}}) = f(\mathbf{x}_{\mathbf{k},\mathbf{j}})\,. \tag{6}$$

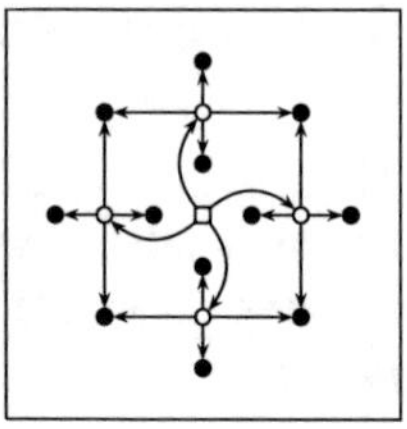

Fig. 2 Directed acyclic graph (DAG) of the two-dimensional regular sparse grid of level 3. The markers indicate the depth of the grid points, i.e., their distance from the "root" $\mathbf{x}_{\mathbf{1},\mathbf{1}}$

Algorithm 1: BFS for hierarchization with fundamental basis functions

---

Input : Sparse grid $\Omega^{\mathrm{s}} \subsetneq [0,1]^d$, vector $\mathbf{f} = (f(\mathbf{x}_{\mathbf{k},\mathbf{j}}))_{(\mathbf{k},\mathbf{j})}$ of function values
Output: Vector $\boldsymbol{\alpha} = (\alpha_{\mathbf{l},\mathbf{i}})_{(\mathbf{l},\mathbf{i})}$ of hierarchical surpluses

$\boldsymbol{\alpha} \leftarrow \mathbf{f}$;
$q \leftarrow$ empty FIFO queue;
`Push(`q, $\mathbf{x}_{\mathbf{1},\mathbf{1}}$`);` `// insert root point`
$\Omega^{\mathrm{s}}_{\mathrm{p}} \leftarrow \{\mathbf{x}_{\mathbf{1},\mathbf{1}}\}$; `// list of processed grid points`
while $q \neq \emptyset$ **do**
  $\mathbf{x}_{\mathbf{l},\mathbf{i}} \leftarrow$ `Pop(`q`);`
  foreach $\{\mathbf{x}_{\mathbf{k},\mathbf{j}} \in \Omega^{\mathrm{s}} \setminus \{\mathbf{x}_{\mathbf{l},\mathbf{i}}\} \mid \forall_{t=1,\dots,d}\ (k_t > l_t) \vee ((k_t, j_t) = (l_t, i_t))\}$ **do**
    $\alpha_{\mathbf{k},\mathbf{j}} \leftarrow \alpha_{\mathbf{k},\mathbf{j}} - \alpha_{\mathbf{l},\mathbf{i}} \phi^{\mathrm{F}}_{\mathbf{l},\mathbf{i}}(\mathbf{x}_{\mathbf{k},\mathbf{j}})$; `// update surpluses`
  foreach $\{\mathbf{x}_{\mathbf{k},\mathbf{j}} \in \Omega^{\mathrm{s}} \setminus \Omega^{\mathrm{s}}_{\mathrm{p}} \mid \mathbf{x}_{\mathbf{k},\mathbf{j}}$ *direct child of* $\mathbf{x}_{\mathbf{l},\mathbf{i}}\}$ **do**
    `Push(`q, $\mathbf{x}_{\mathbf{k},\mathbf{j}}$`);`
    $\Omega^{\mathrm{s}}_{\mathrm{p}} \leftarrow \Omega^{\mathrm{s}}_{\mathrm{p}} \cup \{\mathbf{x}_{\mathbf{k},\mathbf{j}}\}$; `// mark as processed`

In the case $\|\mathbf{l}\|_1 > \|\mathbf{k}\|_1$, we infer by (5) that $\phi^{\mathrm{F}}_{\mathbf{l},\mathbf{i}}(\mathbf{x}_{\mathbf{k},\mathbf{j}}) = 0$ as $l_t > k_t$ for some t. If $\|\mathbf{l}\|_1 = \|\mathbf{k}\|_1$, then either $\mathbf{l} = \mathbf{k}$ or $l_t > k_t$ for some t. In the latter case, $\phi^{\mathrm{F}}_{\mathbf{l},\mathbf{i}}(\mathbf{x}_{\mathbf{k},\mathbf{j}})$ vanishes as before, whereas in the former case, $\phi^{\mathrm{F}}_{\mathbf{l},\mathbf{i}}(\mathbf{x}_{\mathbf{k},\mathbf{j}}) = \prod_t \delta_{i_t, j_t}$ by (5). Hence, in (6), all summands with $\|\mathbf{l}\|_1 > \|\mathbf{k}\|_1$ vanish and for $\|\mathbf{l}\|_1 = \|\mathbf{k}\|_1$, only $\alpha_{\mathbf{k},\mathbf{j}} \phi^{\mathrm{F}}_{\mathbf{k},\mathbf{j}}(\mathbf{x}_{\mathbf{k},\mathbf{j}}) = \alpha_{\mathbf{k},\mathbf{j}}$ remains:

$$\sum_{\substack{\mathbf{x}_{\mathbf{l},\mathbf{i}} \in \Omega^{\mathrm{s}}, \\ \|\mathbf{l}\|_1 < \|\mathbf{k}\|_1}} \alpha_{\mathbf{l},\mathbf{i}} \phi^{\mathrm{F}}_{\mathbf{l},\mathbf{i}}(\mathbf{x}_{\mathbf{k},\mathbf{j}}) + \alpha_{\mathbf{k},\mathbf{j}} = f(\mathbf{x}_{\mathbf{k},\mathbf{j}}) ,$$

which implies the assertion. □

The lemma allows the hierarchical surpluses to be calculated level by level via a breadth-first search (BFS) in the directed acyclic graph (DAG) of Ω^{s} (see Fig. 2). We need here that every node is reachable from the root, which is implied by (2). Algorithm 1 shows an implementation of the BFS. Its correctness is somewhat obvious, but can be proved formally with the aid of Lemma 1.

With Algorithm 1, the hierarchical surpluses can be calculated relatively efficiently for basis functions which satisfy (4): The algorithm performs $\mathcal{O}(N^2 d)$ many 1D evaluations with $N := |\Omega^{\mathrm{s}}|$, and, more importantly, only needs linear space $\mathcal{O}(N)$. This is a significant gain compared to the usual space complexity $\mathcal{O}(N^2)$ for directly solving general (densely populated) linear systems of dimension N.

4 Sparse Grid Basis Transformations

We now consider the question of how to construct 1D basis functions $\phi^{\mathrm{F}}_{l,i}$ which fulfill (4) starting from some existing sparse grid basis $\phi_{l,i}$. The basis transformations are formulated for arbitrary tensor product basis functions $\phi_{l,i}$, but we always keep in mind our primary case of application, the B-splines $\varphi^p_{l,i}$. We first restrict ourselves to the 1D case, as the multivariate case is simply handled by the tensor product approach.

4.1 Hierarchical Fundamental Transformation

A canonical way to construct fundamental bases is to find coefficients $c^{l,i}_{l',i'} \in \mathbb{R}$ with

$$\phi^{\mathrm{hft}}_{l,i} := \sum_{l' \leq l} \sum_{i' \in I_{l'}} c^{l,i}_{l',i'} \phi_{l',i'} \quad \text{s.t.} \quad \forall_{k \leq l} \forall_{j \in I_k} \quad \phi^{\mathrm{hft}}_{l,i}(x_{k,j}) = \delta_{l,k}\delta_{i,j} \ , \tag{7}$$

where $\phi^{\mathrm{hft}}_{l,i}$ is the interpolant of the data $\{(x_{l,j}, \delta_{i,j}) \mid j = 1, \ldots, 2^l - 1\}$ using the hierarchical basis of the regular grid of level l. The coefficients $c^{l,i}_{l',i'}$ depend on level and index of the basis function $\phi^{\mathrm{hft}}_{l,i}$ and are (in general) different for each basis function. This makes precomputation and storage of the $2^l - 1$ coefficients (which are the solution of a linear system) cumbersome. In addition, to evaluate $\phi^{\mathrm{hft}}_{l,i}$ at a given x, $2^l - 1$ basis functions $\phi_{l',i'}$ have to be evaluated (when globally supported). If the $\phi_{l',i'}$ are B-splines $\varphi^p_{l',i'}$ of degree p, then we have to evaluate $\mathcal{O}(p)$ functions on each level $l' \leq l$ due to their local support, leading to a level-dependent complexity of $\mathcal{O}(l \cdot p)$. The resulting functions $\varphi^{p,\mathrm{hft}}_{l,i}$ are shown in Fig. 3a.

We call the transformation $\phi_{l,i} \mapsto \phi^{\mathrm{hft}}_{l,i}$ in (7) *hierarchical fundamental transformation*. The transition is only a change of basis, as we obtain the same interpolants $\tilde{f}$ on a regular sparse grid compared to the old basis functions $\phi_{l,i}$ (i.e., the regular sparse grid spaces spanned by the bases remain unchanged). Note that using a similar argument, the claim can be generalized to dimensionally adaptive sparse grids.

Lemma 2 *Let $n \in \mathbb{N}$ and let $\Omega^{\mathrm{s}}_n := \{\mathbf{x}_{\mathbf{l},\mathbf{i}} \mid \|\mathbf{l}\|_1 \leq n + d - 1, \mathbf{i} \in I_{\mathbf{l}}\}$ be the set of the grid points of the regular sparse grid of level n. Then $\{\phi_{\mathbf{l},\mathbf{i}}\}$ and $\{\phi^{\mathrm{hft}}_{\mathbf{l},\mathbf{i}}\}$ span the same regular sparse grid space of level n (grid point set Ω^{s}_n):*

$$\mathrm{span}\{\phi_{\mathbf{l},\mathbf{i}} \mid \mathbf{x}_{\mathbf{l},\mathbf{i}} \in \Omega^{\mathrm{s}}_n\} =: V^{\mathrm{s}}_n = V^{\mathrm{hft,s}}_n := \mathrm{span}\{\phi^{\mathrm{hft}}_{\mathbf{l},\mathbf{i}} \mid \mathbf{x}_{\mathbf{l},\mathbf{i}} \in \Omega^{\mathrm{s}}_n\} \ .$$

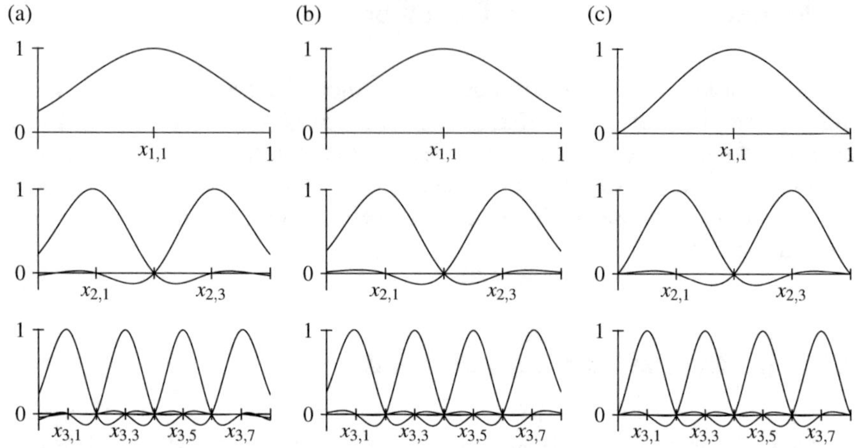

Fig. 3 Hierarchical fundamental bases illustrating the general fundamental transformations for the hierarchical cubic B-spline basis. (**a**) Hierarchical fund. transformation ($\varphi_{l,i}^{p,\mathrm{hft}}$). (**b**) Nodal fund. transformation ($\varphi_{l,i}^{p,\mathrm{nft}}$). (**c**) Translation-invariant fund. transformation ($\varphi_{l,i}^{p,\mathrm{tift}} = \varphi_{l,i}^{p,\mathrm{fs}}$)

Proof Clearly, $V_n^{\mathrm{s}} \supseteq V_n^{\mathrm{hft,s}}$ holds as $\phi_{\mathbf{l},\mathbf{i}}^{\mathrm{hft}} \in V_n^{\mathrm{s}}$ for all $\|\mathbf{l}\|_1 \leq n + d - 1, \mathbf{i} \in I_{\mathbf{l}}$:

$$\phi_{\mathbf{l},\mathbf{i}}^{\mathrm{hft}} = \prod_{t=1}^{d} \sum_{l'_t \leq l_t} \sum_{i'_t \in I_{l'_t}} c_{l'_t,i'_t}^{l_t,i_t} \phi_{l'_t,i'_t} = \sum_{\mathbf{l}' \leq \mathbf{l}} \sum_{\mathbf{i}' \in I_{\mathbf{l}'}} c_{\mathbf{l}',\mathbf{i}'}^{\mathbf{l},\mathbf{i}} \phi_{\mathbf{l}',\mathbf{i}'} \in V_n^{\mathrm{s}}, \quad c_{\mathbf{l}',\mathbf{i}'}^{\mathbf{l},\mathbf{i}} := \prod_{t=1}^{d} c_{l'_t,i'_t}^{l_t,i_t}.$$

For $V_n^{\mathrm{s}} \subseteq V_n^{\mathrm{hft,s}}$, it suffices to check that the dimensions of both spaces match. We show that $\{\phi_{\mathbf{l},\mathbf{i}}^{\mathrm{hft}} \mid \mathbf{x}_{\mathbf{l},\mathbf{i}} \in \Omega_n^{\mathrm{s}}\}$ is linearly independent. For arbitrary coefficients $\alpha_{\mathbf{l},\mathbf{i}} \in \mathbb{R}$ with $\sum_{\mathbf{x}_{\mathbf{l},\mathbf{i}} \in \Omega_n^{\mathrm{s}}} \alpha_{\mathbf{l},\mathbf{i}} \phi_{\mathbf{l},\mathbf{i}}^{\mathrm{hft}} = 0$, we have

$$\forall_{\mathbf{x}_{\mathbf{k},\mathbf{j}} \in \Omega_n^{\mathrm{s}}} \quad \sum_{\mathbf{x}_{\mathbf{l},\mathbf{i}} \in \Omega_n^{\mathrm{s}}} \alpha_{\mathbf{l},\mathbf{i}} \phi_{\mathbf{l},\mathbf{i}}^{\mathrm{hft}}(\mathbf{x}_{\mathbf{k},\mathbf{j}}) = 0\,.$$

This is a lower triangular linear system with ones on the diagonal if we arrange rows and columns appropriately as before. This implies regularity of the linear system and thus $\alpha_{\mathbf{l},\mathbf{i}} = 0$ for all $\mathbf{x}_{\mathbf{l},\mathbf{i}} \in \Omega_n^{\mathrm{s}}$. Therefore, $\{\phi_{\mathbf{l},\mathbf{i}}^{\mathrm{hft}} \mid \mathbf{x}_{\mathbf{l},\mathbf{i}} \in \Omega_n^{\mathrm{s}}\}$ is linearly independent and $\dim V_n^{\mathrm{hft,s}} = |\Omega_n^{\mathrm{s}}| = \dim V_n^{\mathrm{s}}$ (the latter equation holds as $\{\phi_{\mathbf{l},\mathbf{i}}\}$ is assumed to be a sparse grid basis). Together with $V_n^{\mathrm{s}} \supseteq V_n^{\mathrm{hft,s}}$, this implies $V_n^{\mathrm{s}} = V_n^{\mathrm{hft,s}}$. □

4.2 Nodal Fundamental Transformation

The hierarchical fundamental transformation has the drawback of the level-dependent complexity for evaluations ($\mathcal{O}(l \cdot p)$ in the B-spline case). This is not the case if we use the nodal basis of level l instead of the hierarchical functions of level $\leq l$ to interpolate the same 1D data $\{(x_{l,j}, \delta_{i,j}) \mid j = 1, \ldots, 2^l - 1\}$ for constructing the new basis function of level l:

$$\varphi_{l,i}^{\text{nft}} := \sum_{i'=1}^{2^l-1} c_{i'}^{l,i} \varphi_{l,i'} \quad \text{s.t.} \quad \forall_{j=1,\ldots,2^l-1} \quad \varphi_{l,i}^{\text{nft}}(x_{l,j}) = \delta_{i,j} \, .$$

We call this transformation *nodal fundamental transformation*. The evaluation complexity then only depends on the nodal basis support; for B-splines, only $\mathcal{O}(p)$ basis functions can be non-zero at a certain point. The resulting fundamental basis is shown in Fig. 3b. Nevertheless, if we compare again the spaces spanned by the original and the nodal fundamental basis, a sparse-grid-based comparison as for the hierarchical fundamental transformation is difficult: The multivariate fundamental functions $\varphi_{\mathbf{l},\mathbf{i}}^{\text{nft}} := \prod_{t=1}^{d} \varphi_{l_t,i_t}^{\text{nft}}$ are tensor products of nodal 1D basis functions $\varphi_{l_t,i_t}^{\text{nft}}$ (some with even index), leading to linear combinations of nodal multivariate basis functions, which, in general, cannot be represented by the original functions in the sparse grid space. On full grids, however, the spanned spaces coincide.

Lemma 3 $\{\varphi_{\mathbf{l},\mathbf{i}}\}$ *and* $\{\varphi_{\mathbf{l},\mathbf{i}}^{\text{nft}}\}$ *span the same nodal space:*

$$\operatorname{span}\left\{\varphi_{\mathbf{l},\mathbf{i}} \mid \mathbf{1} \leq \mathbf{i} \leq \mathbf{2}^{\mathbf{l}} - \mathbf{1}\right\} =: V_{\mathbf{l}} = V_{\mathbf{l}}^{\text{nft}} := \operatorname{span}\left\{\varphi_{\mathbf{l},\mathbf{i}}^{\text{nft}} \mid \mathbf{1} \leq \mathbf{i} \leq \mathbf{2}^{\mathbf{l}} - \mathbf{1}\right\}, \quad \mathbf{l} \in \mathbb{N}^d \, .$$

This statement can be proved analogously to Lemma 2.

4.3 Translation-Invariant Fundamental Transformation

Both the hierarchical and the nodal fundamental transformation do not preserve the translation invariance ($\varphi_{l,i}(x) = \varphi(x/h_l - i)$, $i \in I_l$, for some mother function $\varphi\colon \mathbb{R} \to \mathbb{R}$) of the original basis. From a computational point of view, this is a major disadvantage, since the coefficients $c_{l',i'}^{l,i}$ or $c_{i'}^{l,i}$ of the basis function of level l and index i depend on l, i and would have to be recalculated for every basis function. To tackle this problem, we use the nodal fundamental transformation for the construction, but now allow for general integer indices:

$$\varphi_{l,i}^{\text{tift}} := \sum_{i' \in \mathbb{Z}} \tilde{c}_{i'} \varphi_{l,i'} \quad \text{s.t.} \quad \forall_{j \in \mathbb{Z}} \quad \varphi_{l,i}^{\text{tift}}(x_{l,j}) = \delta_{i,j} \, . \tag{8}$$

This transformation is invariant under translation-invariance, i.e., if the original basis functions have the form $\phi_{l,i'}(x) = \phi(x/h_l - i')$, then

$$\phi_{l,i}^{\text{tift}}(x) = \sum_{i'\in\mathbb{Z}} \tilde{c}_{i'}\phi\left(\frac{x}{h_l} - i'\right) = \sum_{k\in\mathbb{Z}} \tilde{c}_{k+i}\phi\left(\frac{x}{h_l} - i - k\right) = \phi^{\text{tift}}\left(\frac{x}{h_l} - i\right),$$

where ϕ^{tift} can be obtained by setting $c_k := \tilde{c}_{k+i}$ in

$$\phi^{\text{tift}} := \sum_{k\in\mathbb{Z}} c_k\phi(\cdot - k) \quad \text{s.t.} \quad \forall_{j\in\mathbb{Z}} \quad \phi^{\text{tift}}(j) = \delta_{j,0}\,. \tag{9}$$

We therefore call this the *translation-invariant fundamental transformation*. We assume that the relevant index set

$$J_l := \{i' \in \mathbb{Z} \mid \phi_{l,i'}|_{[0,1]} \not\equiv 0\}$$

is finite, which implies that in each point $x \in [0, 1]$, almost all functions $\phi_{l,i'}$ of a level l vanish. This condition is satisfied for compactly supported, translation-invariant (continuous) basis functions such as the B-splines, and this allows us to replace the series over $i' \in \mathbb{Z}$ in (8) and (9) with sums over $i' \in J_l$, if we evaluate $\phi_{l,i}^{\text{tift}}$ on $[0, 1]$. In d dimensions, we write $J_{\mathbf{l}} := J_{l_1} \times \cdots \times J_{l_d}$. The resulting basis when using B-splines $\varphi_{l,i'}^{p}$ as $\phi_{l,i'}$ is shown in Fig. 3c; it will be studied in the next section.

As for the (non-translation-invariant) nodal fundamental transformation, a comparison of the spanned subspaces of original and transformed basis functions can, in general, only be made for the nodal bases. This time, we have to additionally include exterior basis functions $\phi_{\mathbf{l},\mathbf{i}}$ where $i_t < 1$ or $i_t > 2^{l_t} - 1$ for some t. These are the basis functions which have supports that extend into $[0, 1]^d$.

Lemma 4 $\{\phi_{\mathbf{l},\mathbf{i}}\}$ *and* $\{\phi_{\mathbf{l},\mathbf{i}}^{\text{tift}}\}$ *span the same nodal space, if we include exterior basis functions:*

$$\text{span}\{\phi_{\mathbf{l},\mathbf{i}} \mid \mathbf{i} \in J_{\mathbf{l}}\} =: V_{\mathbf{l}}^{\text{ext}} = V_{\mathbf{l}}^{\text{tift,ext}} := \text{span}\left\{\phi_{\mathbf{l},\mathbf{i}}^{\text{tift}} \mid \mathbf{i} \in J_{\mathbf{l}}\right\}, \quad \mathbf{l} \in \mathbb{N}^d\,. \tag{10}$$

The proof is analogous to Lemmas 2 and 3.

5 Hierarchical Fundamental Splines

5.1 Definition and Properties

In the previous section, we formulated the translation-invariant fundamental transformation for general hierarchical bases. In the following, we will apply the

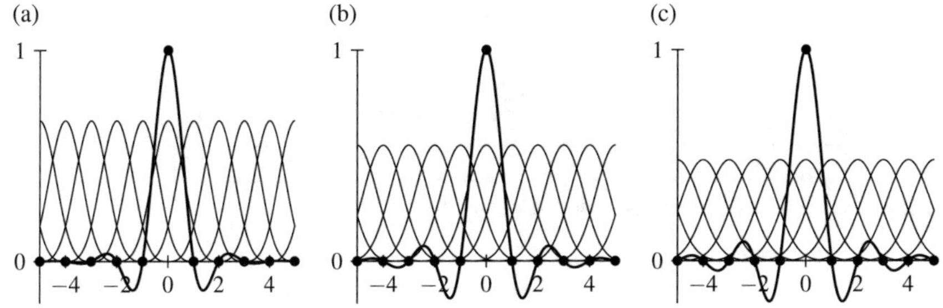

Fig. 4 The fundamental spline $\varphi^{p,\text{fs}}$ *(thick line)* as a linear combination of the B-splines $\varphi^p(\cdot - k)$, $k \in \mathbb{Z}$ *(thin lines)*, for degrees. (**a**) $p = 3$. (**b**) $p = 5$. (**c**) $p = 7$

transformation to the case of B-splines, i.e., in the notation of Sect. 4.3,

$$\varphi^p(x) = b^p\left(x + \frac{p+1}{2}\right), \quad J_l^p = \left\{-\frac{p-1}{2}, -\frac{p-1}{2} + 1, \ldots, 2^l + \frac{p-1}{2}\right\}.$$

Equation (9) then leads to a bi-infinite-dimensional linear system for the coefficients c_k^p of $\varphi^{p,\text{tift}}$:

$$\begin{pmatrix} \ddots & & \ddots & & \ddots & \\ \ddots & b^p(\frac{p+1}{2}) & b^p(\frac{p+1}{2} - 1) & b^p(\frac{p+1}{2} - 2) & & \\ \ddots & b^p(\frac{p+1}{2} + 1) & b^p(\frac{p+1}{2}) & b^p(\frac{p+1}{2} - 1) & \ddots & \\ & b^p(\frac{p+1}{2} + 2) & b^p(\frac{p+1}{2} + 1) & b^p(\frac{p+1}{2}) & \ddots & \\ & & \ddots & \ddots & \ddots \end{pmatrix} \cdot \begin{pmatrix} \vdots \\ c_{-1}^p \\ c_0^p \\ c_1^p \\ \vdots \end{pmatrix} = \begin{pmatrix} \vdots \\ 0 \\ 1 \\ 0 \\ \vdots \end{pmatrix}. \tag{11}$$

The system matrix is a symmetric banded Toeplitz matrix, since in each row only p entries are non-zero. The linear system has indeed a unique solution:

Theorem 1 *The linear system* (11) *has a unique solution* $(c_k^p)_{k\in\mathbb{Z}}$, *and the corresponding mother function* $\varphi^{p,\text{fs}} := \sum_{k\in\mathbb{Z}} c_k^p \varphi^p(\cdot - k)$ *satisfies*

$$\exists_{\beta_p,\gamma_p \in \mathbb{R}^+} \forall_{x\in\mathbb{R}} \quad |\varphi^{p,\text{fs}}(x)| \le \beta_p \exp(-\ln(\gamma_p)|x|).$$

A proof of this theorem can be found in [15]. The corresponding mother function $\varphi^{p,\text{fs}}$ is well-known as the *fundamental spline of degree p* [15, 16], see Fig. 4. Fundamental splines can be used, for example, to define spline wavelets [3].

The resulting hierarchical basis for cubic degree can be seen in Fig. 3c. The common nodal space $V_{\mathbf{l}}^{\mathbf{p},\text{fs,ext}}$ (defined by applying fundamental splines to (10)) of the hierarchical B-spline and the hierarchical fundamental spline basis is, using the notation of Sect. 4.3 and [6], exactly the tensor product spline space $S_{\Xi}^{\mathbf{p}}$ with

Table 1 Best-approximated values of β_p and γ_p for different fundamental spline degrees p

p	1	3	5	7	9	11	13	15
β_p	1	1.241	1.104	1.058	1.037	1.026	1.019	1.015
γ_p	e	3.732	2.322	1.868	1.645	1.512	1.425	1.363
n_p	1	18	29	40	52	64	77	90

n_p is the smallest positive integer such that $\forall_{|k|\geq n_p}\ |c_k^p| < 10^{-10}$

knots

$$\Xi := \boldsymbol{\xi}_1 \times \cdots \times \boldsymbol{\xi}_d\,,$$

$$\boldsymbol{\xi}_t := (\xi_{t,k_t})_{k_t=0}^{m_t+p_t}\,,\quad \xi_{t,k_t} := (k_t - p_t)h_{l_t}\,,\quad m_t := 2^{l_t} + p_t\,,\quad t = 1, \ldots, d\,.$$

This is the space of all functions on $[0, 1]^d$ which are d-variate polynomials of coordinate degree $\leq \mathbf{p}$ on every knot hyperrectangle $[\xi_{1,k_1}, \xi_{1,k_1+1}] \times \cdots \times [\xi_{d,k_d}, \xi_{d,k_d+1}]$ ($\mathbf{p} \leq \mathbf{k} \leq \mathbf{m} - \mathbf{1}$) and at least ($\mathbf{p} - \mathbf{1}$) times continuously partially differentiable at every knot $(\xi_{1,k_1}, \ldots, \xi_{d,k_d})$ in the interior of $[0, 1]^d$ ($\mathbf{p} + \mathbf{1} \leq \mathbf{k} \leq \mathbf{m} - \mathbf{1}$).

Due to the stability of the B-spline basis [6], Theorem 1 implies that the coefficients c_k^p of $\varphi^{p,\mathrm{fs}}$ observe the same exponential decay as the fundamental spline itself, i.e.,

$$|c_k^p| \leq \tilde{\beta}_p \exp(-\ln(\gamma_p)|k|)$$

with some $\tilde{\beta}_p > 0$ independent of k. Because of this inequality, we may solve (11) approximately if we choose a large enough $n_p \in \mathbb{N}$, truncate the linear system to $2n_p - 1$ dimensions, and set $c_k^p = 0$ for all $|k| \geq n_p$. It is an interesting fact that the optimal decay rate γ_p is algebraically determined as the absolute value of the largest root smaller than -1 of the polynomial $\sum_{k=1}^{p} b^p(k)x^{k-1}$ whose coefficients are the values of the cardinal B-spline b^p at its knots, see [3, 16]. In Table 1, we give approximations for the value of β_p and γ_p in addition to the truncation index n_p that we use. Note that due to the local support of the B-splines, we do not have to perform $2n_p + 1$ cardinal B-spline evaluations to evaluate the fundamental spline once, but only $p + 1$ many.

5.2 *Modified Fundamental Splines*

The modifications of the basis functions in Sect. 2.1 were motivated by non-zero boundary values: Without modification and without boundary grid points, every linear combination of the hierarchical fundamental spline basis vanishes on the boundary of $[0, 1]^d$.

As for the B-splines [11, 19], we define modified fundamental splines by modifying the first and last basis function of each level,

$$\varphi_{l,i}^{p,\mathrm{fs,mod}}(x) := \begin{cases} 1 & \text{if } l = 1, i = 1, \\ \psi_l^{p,\mathrm{fs}}(x) & \text{if } l > 1, i = 1, \\ \psi_l^{p,\mathrm{fs}}(1-x) & \text{if } l > 1, i = 2^l - 1, \\ \varphi_{l,i}^{p,\mathrm{fs}}(x) & \text{otherwise.} \end{cases}$$

Here, $\psi_l^{p,\mathrm{fs}}$ is defined by

$$\psi_l^{p,\mathrm{fs}} := \sum_{i=1-(p+1)/2}^{\infty} c_i^{p,\mathrm{mod}} \varphi_{l,i}^p \quad \text{s.t.} \quad \begin{cases} \psi_l^{p,\mathrm{fs}}(x_{l,j}) = \delta_{j,1}, & j \in \mathbb{N}, \\ (\psi_l^{p,\mathrm{fs}})''(x_{l,1}) = 0, & \\ (\psi_l^{p,\mathrm{fs}})^{(j)}(0) = 0, & j = 2, 3, \ldots, (p+1)/2, \end{cases} \tag{12}$$

for $p > 1$, where $(\psi_l^{p,\mathrm{fs}})''$ and $(\psi_l^{p,\mathrm{fs}})^{(j)}$ denote the second and the jth derivative, respectively, and $\psi_l^{p,\mathrm{fs}} := \psi_l^p$ for $p = 1$. Note that the coefficients $c_i^{p,\mathrm{mod}}$ do not depend on the level l, as $\psi_l^{p,\mathrm{fs}}(x) = \psi_0^{p,\mathrm{fs}}(x/h_l)$. The resulting functions are depicted in Fig. 5. The modification coefficients $c_i^{p,\mathrm{mod}}$ experience the same exponential decay as c_i^p and can thus be approximated as the solution of a truncated linear system.

The choice of the conditions (12) is motivated by the cubic case $p = 3$ (Fig. 5a). As the modified function $\psi_l^{p,\mathrm{fs}}$ will not be evaluated for $x < 0$, it suffices to start the summation at the index $1 - (p+1)/2$ of the first relevant B-spline (-1 in the cubic case). We want $\psi_l^{p,\mathrm{fs}}$ to satisfy the fundamental property at inner grid points $x_{l,i}$, which gives one condition for each B-spline $\varphi_{l,i}^p$, $i \in \mathbb{N}$, resulting in exactly two remaining degrees of freedom in the cubic case (namely $i = -1$ and $i = 0$). With these conditions, we want to extrapolate linearly on $[0, x_{l,1}]$ as we did with the modification (1) of the cubic hierarchical B-splines, which is suitable for financial

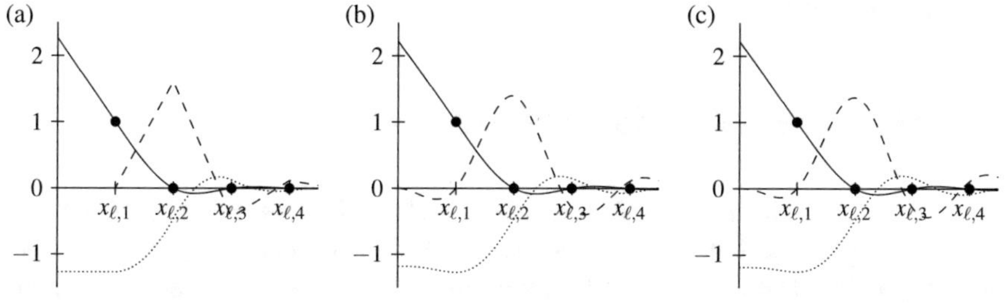

Fig. 5 Modified fundamental spline $\psi_l^{p,\mathrm{fs}}$ *(solid line)* together with its first *(dotted)* and second derivative *(dashed)*. For illustrative purposes, the first and second derivatives are scaled to $l = 0$. **(a)** $p = 3$. **(b)** $p = 5$. **(c)** $p = 7$

applications [11], for instance. Thus, it seems natural to set the second derivative of $\psi_l^{p,\mathrm{fs}}$ to zero in $x = 0$ and $x = x_{l,1}$, since this leads to $(\psi_l^{p,\mathrm{fs}})''(x) \equiv 0$ on $[0, x_{l,1}]$ in the cubic case as $(\psi_l^{p,\mathrm{fs}})''$ is piecewise linear. For higher degrees p, we use the additional degrees of freedom to increase the multiplicity of the root of $(\psi_l^{p,\mathrm{fs}})''$ in $x = 0$, making sure that $(\psi_l^{p,\mathrm{fs}})''$ is "as linear as possible" near $x = 0$. Note that we cannot maintain $(\psi_l^{p,\mathrm{fs}})''(x) \equiv 0$ on $[0, x_{l,1}]$ for higher degrees. This would require $p - 1$ conditions, but after taking the fundamental conditions into account, there are only $(p + 1)/2$ degrees of freedom left.

6 Application to Gradient-Based Optimization

In the following, we compare the modified fundamental spline basis with the modified B-spline basis in an application-based setting. We do not consider the hierarchical fundamental transformation and the nodal fundamental transformation, as the resulting bases are not translation-invariant. As the application, we choose gradient-based, global optimization: Given an *objective function* $f\colon [0, 1]^d \to \mathbb{R}$, the problem is to find $\mathbf{x}_{\mathrm{opt}} := \arg\min_{\mathbf{x}\in[0,1]^d} f(\mathbf{x})$ or a good approximation $\mathbf{x}^*_{\mathrm{opt}} \in [0, 1]^d$ (i.e., $f(\mathbf{x}^*_{\mathrm{opt}}) \approx f(\mathbf{x}_{\mathrm{opt}})$) with as few evaluations of f as possible. We assume that each evaluation can trigger a lengthy simulation or a time-consuming experiment (e.g., in inverse problems). Hence, an upper bound $N_{\mathrm{ub}} \in \mathbb{N}$ on the number of allowed evaluations of f is given.

We solve the optimization problem as in [19] with a surrogate-based approach: First, we generate a spatially adaptive sparse grid, in which we aim to spend more grid points close to the global minimum while exploring the whole domain $[0, 1]^d$ to identify the global minimum. Then, we interpolate at the grid points using either the modified B-spline or the modified fundamental spline basis, resulting in an interpolant $\tilde{f}\colon [0, 1]^d \to \mathbb{R}$. Finally, we apply gradient-based methods to $\tilde{f}$ to find $\mathbf{x}^*_{\mathrm{opt}}$. Note that one evaluation of the surrogate $\tilde{f}$ is, in most applications, orders of magnitude cheaper than an evaluation of the objective function f. Additionally, we can exploit the existence of gradients and Hessians of $\tilde{f}$, even though we only used function evaluations of f.

6.1 Generation of the Spatially Adaptive Grid

For the spatially adaptive grid generation, we employ the method of Novak–Ritter [10, 19]. In each iteration, it refines exactly one grid point $\mathbf{x}_{\mathbf{l}^*,\mathbf{i}^*}$ of the current sparse grid Ω^{s}. The refinement of a point $\mathbf{x}_{\mathbf{l}^*,\mathbf{i}^*} \in \Omega^{\mathrm{s}}$ is done by augmenting the sparse grid with $2d$ so-called *mth order children* $\mathbf{x}_{\mathbf{l},\mathbf{i}}$ which satisfy

$$\exists_{t'\in\{1,\ldots,d\}} \quad (l_t, i_t) = \begin{cases} (l_t^* + m, 2^m i_t^* \pm 1) & \text{if } t = t', \\ (l_t^*, i_t^*) & \text{otherwise.} \end{cases} \tag{13}$$

Since, e.g., a first-order child of $\mathbf{x}_{\mathbf{l}^*,\mathbf{i}^*}$ might already exist in the grid, m is individually chosen for each of the $2d$ points as the lowest number such that the point to be inserted is not yet in the grid. Thus, in each iteration, exactly $2d$ points are inserted. We do not automatically insert the hierarchical ancestors of (2) in all directions. Their existence is an assumption needed for the unidirectional principle (UP [11]), but the UP only works for hat functions and piecewise polynomials, not for B-splines in general. In addition, due to the recursive insertion of the hierarchical ancestors, this would lead to many grid points at uninteresting places [19]. Note that we are not interested in a low L^2 error in the whole domain, but rather in a good representation locally around the global minimum.

The point to be refined is selected as the point $\mathbf{x}_{\mathbf{l}_k,\mathbf{i}_k} \in \Omega^{\mathrm{s}}$ with the smallest *quality*

$$\beta_k := (\|\mathbf{l}_k\|_1 + d_k + 1)^\gamma \cdot r_k^{1-\gamma} \,.$$

Here, $\|\mathbf{l}_k\|_1$ is the level sum of the grid point, d_k is the number of previous refinements of $\mathbf{x}_{\mathbf{l}_k,\mathbf{i}_k}$, $r_k := |\{j = 1, \ldots, N \mid f(\mathbf{x}_{\mathbf{l}_j,\mathbf{i}_j}) \leq f(\mathbf{x}_{\mathbf{l}_k,\mathbf{i}_k})\}|$ is the *rank* of $\mathbf{x}_{\mathbf{l}_k,\mathbf{i}_k}$ in Ω^{s}, and $\gamma \in [0, 1]$ the adaptivity of the method. In the case $\gamma = 0$, the method always refines the point of Ω^{s} with the best function value, without exploring the whole domain. For $\gamma = 1$, the function values have no influence on the grid generation anymore and the method generates more or less regular sparse grids. For a visual example of the influence of γ, see Fig. 6. The choice of the adaptivity parameter is of course important, but we choose a priori a fixed adaptivity $\gamma = 0.85$ for the following experiments. In our experience, this value seems to give a reasonable trade-off between exploration and exploitation.

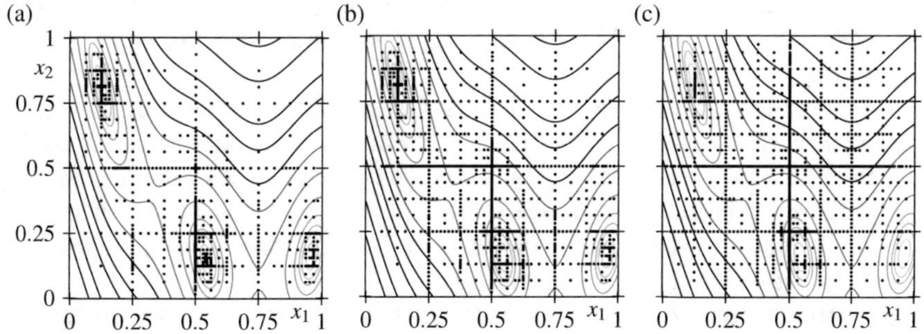

Fig. 6 Spatially adaptive sparse grids with 1000 grid points generated with Novak–Ritter's method for the bivariate Branin test function and different values of the adaptivity parameter (dark contour lines correspond to large function values). The function has three global minima (see Table 2). (**a**) $\gamma = 0.6$. (**b**) $\gamma = 0.85$. (**c**) $\gamma = 0.95$

6.2 Optimization Procedure

After generating the spatially adaptive sparse grid $\Omega^{\mathrm{s}} = \{\mathbf{x}_{\mathbf{l}_k,\mathbf{i}_k} \mid k = 1, \ldots, N\}$, $N := |\Omega^{\mathrm{s}}| \leq N_{\mathrm{ub}}$, we choose the type of hierarchical basis functions (either modified B-splines or modified fundamental splines of degree $\mathbf{p}$) and perform the hierarchization of the function values $\mathbf{f} = (f(\mathbf{x}_{\mathbf{l}_k,\mathbf{i}_k}))_{k=1}^{N}$, which are already known from the grid generation phase. In the B-spline case, hierarchization is done by solving the linear system (3) using various solver libraries, depending on the sparsity structure. For fundamental splines, the system is solved more efficiently using Algorithm 1.

The resulting hierarchical surpluses $\boldsymbol{\alpha} = (\alpha_k)_{k=1}^{N}$ determine the spline interpolant $\tilde{f}\colon [0,1]^d \to \mathbb{R}$, $\tilde{f} := \sum_{k=1}^{N} \alpha_k \varphi_{\mathbf{l}_k,\mathbf{i}_k}$, where the $\varphi_{\mathbf{l}_k,\mathbf{i}_k}$ are either the modified B-splines $\varphi_{\mathbf{l}_k,\mathbf{i}_k}^{\mathbf{p},\mathrm{mod}}$ (see [11, 19]) or the modified fundamental splines $\varphi_{\mathbf{l}_k,\mathbf{i}_k}^{\mathbf{p},\mathrm{fs},\mathrm{mod}}$ of degree $\mathbf{p}$. The interpolant $\tilde{f}$ is $(\mathbf{p} - \mathbf{1})$ times continuously partially differentiable, hence, gradient-based optimization methods can be used to minimize $\tilde{f}$. As in [19], we used local methods [9] such as simple gradient descent, nonlinear conjugate gradients [13], Newton's method, BFGS, and Rprop [14]. Additionally, we use the gradient-free methods Nelder–Mead (NM [8]) and Differential Evolution (DE [18]). All methods have additionally been implemented in a globalized version using $m := \min(10d, 100)$ uniformly distributed random starting points in a multi-start approach.

Like in [19], we optimize not only the spline interpolant $\tilde{f}$, but, for comparison reasons, also the piecewise linear interpolant and the objective function directly:

(S) Apply all methods to $\tilde{f}$ with $\mathbf{y}_0 := \mathbf{x}_{\mathbf{l}_{k^*},\mathbf{i}_{k^*}}$, $k^* := \arg\min_k f(\mathbf{x}_{\mathbf{l}_k,\mathbf{i}_k})$, as starting point. Additionally, we apply the globalized versions of the algorithms to $\tilde{f}$. We evaluate f at all resulting points and take the point as final result $\mathbf{x}_{\mathrm{opt}}^*$ with the smallest f value, falling back to $\mathbf{y}_0$ if we somehow managed to become worse.

(L) Same as (S), but optimize the piecewise linear interpolant instead. The piecewise linear interpolant is constructed by using the same sparse grid Ω^{s} as for the spline interpolant, but interpolating the function values with the common hat function basis by setting $\mathbf{p} = \mathbf{1}$. Since the piecewise linear interpolant is not continuously partially differentiable, only the gradient-free methods NM and DE are used for optimization.

(O) Apply the globalized NM directly to the objective function, always making sure that the number of evaluations of f does not exceed N_{ub}.

6.3 Test Functions and Results

We use the set of six test functions which we used in [19] and which are mentioned in Table 2. The domains of some of the functions were translated to make sure that the optima do not lie in the centers of the domains, as this would give sparse-grid-

Table 2 Test functions of two and arbitrary number of variables before scaling to $[0, 1]^d$

Name	Domain	$\mathbf{x}_{\text{opt}}$	$f(\mathbf{x}_{\text{opt}})$	Reference
Branin	$[-5, 10] \times [0, 15]$	$(-\pi, 12.275)$, $(\pi, 2.275)$, $(9.42478, 2.475)$	0.397887	[7, Branin RCOS]
Eggholder	$[-512, 512]^2$	$(512, 404.2319)$	−959.6407	[20, $F101$]
Rosenbrock	$[-5, 10]^2$	$(1, 1)$	0	[21]
Ackley	$[-1, 9]^d$	$\mathbf{0}$	0	[21]
Rastrigin	$[-2, 8]^d$	$\mathbf{0}$	0	[21]
Schwefel	$[-500, 500]^d$	$420.9687 \cdot \mathbf{1}$	$-418.9829d$	[21]

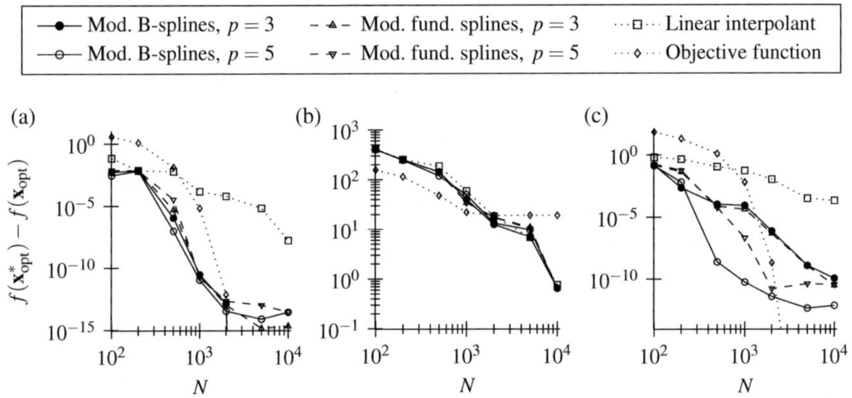

Fig. 7 Optimization results for two-dimensional test functions. The log–log plots depict the difference between the optimization error in terms of the function value versus the maximal number N of objective function evaluations. Shown are the results of (S) with modified B-splines *(solid lines)* and modified fundamental splines *(dashed)*. For comparison, the performance of the hat function basis ((L), *square markers*) and of the direct optimization of the objective function ((O), *diamonds*) can be seen. (**a**) Branin, $d = 2$. (**b**) Eggholder, $d = 2$. (**c**) Rosenbrock, $d = 2$

based approaches an advantage. After scaling the domains to the unit hypercube $[0, 1]^d$, we perturbed all functions with a Gaussian displacement (standard deviation 0.01). All results shown are means of five passes with different perturbations. Each plot displays three types of results. First, the results of (S), optimizing modified B-splines and modified fundamental B-splines (each for $p = 3$ and $p = 5$). Second, the results of (L), optimizing the piecewise linear interpolant. Third, the results of (O), directly optimizing the objective function.

The results for the three bivariate functions can be seen in Fig. 7. Clearly, for relatively “smooth” functions such as Branin and Rosenbrock, both B-splines and fundamental splines gain a significant advantage compared to the hat function basis. The B-splines perform slightly better than the fundamental splines for both functions. With its many oscillations and its optimum lying on the domain’s boundary, the Eggholder function is much harder to optimize for all approaches shown (cf. Fig. 7b).

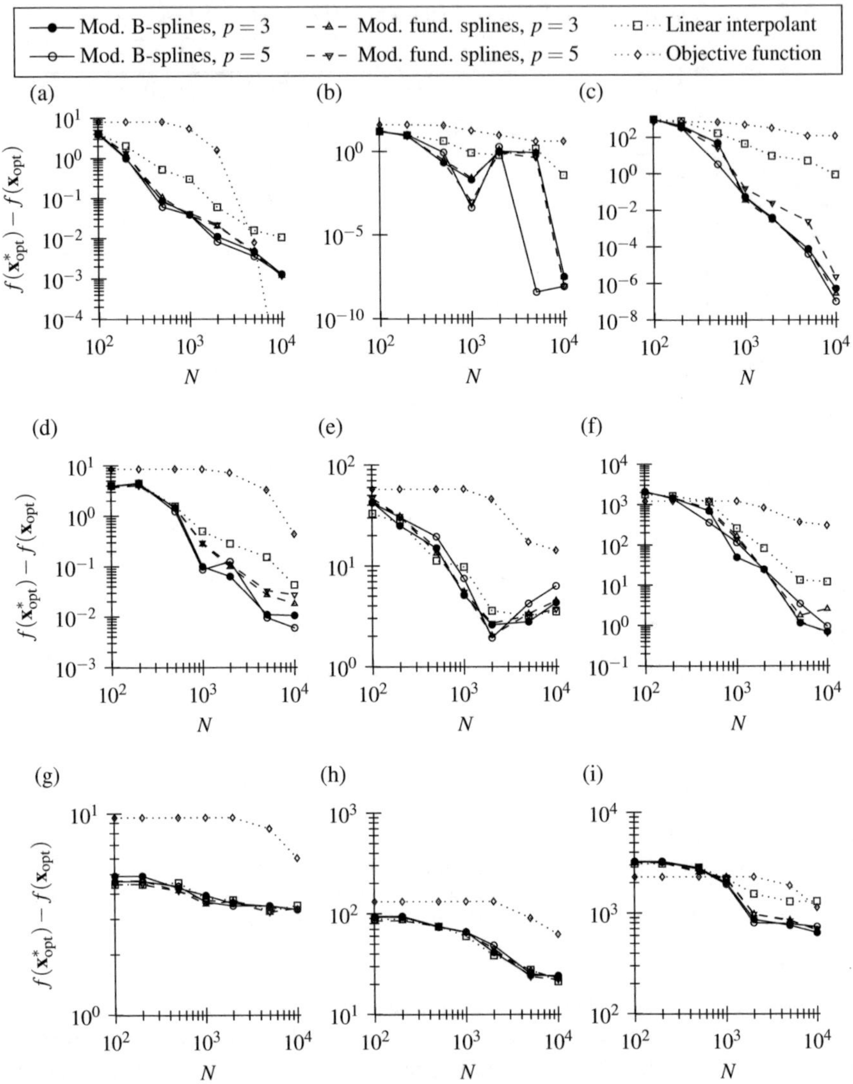

Fig. 8 Optimization results for higher-dimensional test functions. See Fig. 7 for details. (**a**) Ackley, $d = 4$. (**b**) Rastrigin, $d = 4$. (**c**) Schwefel, $d = 4$. (**d**) Ackley, $d = 6$. (**e**) Rastrigin, $d = 6$. (**f**) Schwefel, $d = 6$. (**g**) Ackley, $d = 10$. (**h**) Rastrigin, $d = 10$. (**i**) Schwefel, $d = 10$

Figure 8 shows the convergence behavior for the three d-variate test functions. It can be said that for a moderate number of dimensions, both B-splines and fundamental splines perform roughly equally well, while for higher dimensionalities ($d = 10$), convergence slows down evidently. Note that all three d-variate test functions Ackley, Rastrigin, and Schwefel have many local minima, which makes it computationally hard to find the global minimum. Especially for coarse sparse grids, it is possible that the minima of the interpolant and the objective function coincide by chance. This means that the error curves do not need to be strictly monotonically decreasing, which can be clearly seen for the Rastrigin function in $d = 4$ or $d = 6$.

6.4 Comparison of Runtime and Memory Consumption

Figure 9 visualizes the runtime and memory needed by the hierarchization step for the sparse grids which were generated for the optimization of the test functions in Fig. 8 (averaged over all objective functions and dimensionalities), measured on a laptop with Intel Core i5-4300U. As already mentioned in Sect. 6.2, hierarchization was done by linear system solvers in the B-spline case and with Algorithm 1 in the fundamental spline case. Note that the effort for degree $p = 3$ and $p = 5$ is nearly identical such that the corresponding lines almost overlap. The runtime of our implementation is both asymptotically and in absolute numbers much better for fundamental splines (factor 20 for $N = 20{,}000$), although slightly worse than the theoretical complexity of $\mathcal{O}(N^2)$. The memory consumption is of course considerably lower for fundamental splines, since Algorithm 1 needs only a linear amount $\mathcal{O}(N)$ of memory, while the linear system solvers can allocate up to $\mathcal{O}(N^2)$ bytes, depending on the type of solver (dense/sparse). This means for $N = 20{,}000$ points already a memory usage of up to 7.2 GB, in contrast to the fundamental splines, which merely need 25 MB.

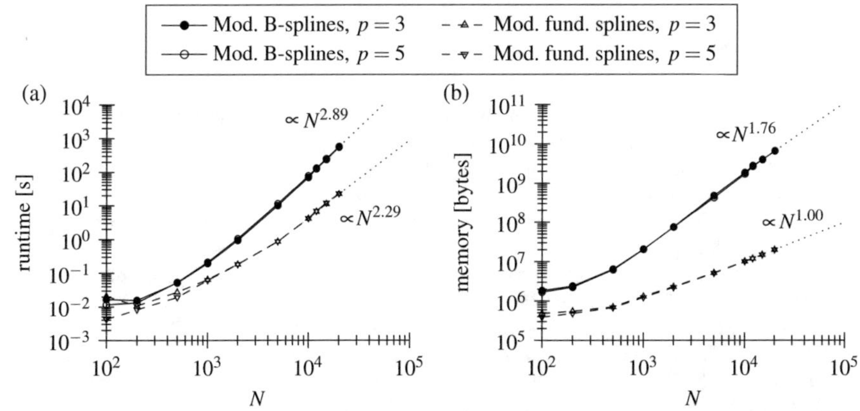

Fig. 9 Average runtime *(left)* and memory consumption *(right)* of the hierarchization of the spatially adaptive sparse grids used for Fig. 8 vs. the number of sparse grid points. The respectively last two data points were extrapolated exponentially *(dotted)*, using the mean of $p = 3$ and $p = 5$. (**a**) Runtime. (**b**) Memory consumption

7 Conclusion

In this paper, we have treated the problem of hierarchization with hierarchical bases which have larger supports than the usual hat functions. We have given a simple breadth-first search (BFS) algorithm for the hierarchization of bases which fulfill the fundamental property. Furthermore, we have studied three different basis transformations which transform a general basis to a basis satisfying the fundamental property. One of the transformations has led to the well-known definition of fundamental splines, which we compared with the conventional B-spline basis in the application of gradient-based optimization. We have seen that in this setting, the modified fundamental spline basis exhibits comparable results as the modified B-splines with significant improvements in runtime and memory consumption. Thus, for practical applications, the modified fundamental splines can be preferred over the modified B-spline basis.

Future work should try to improve the implementation of the BFS algorithm to reduce the runtime to the theoretical complexity of $\mathcal{O}(N^2)$. In addition, the performance of the modified fundamental spline basis should be studied if applied to the optimization of objective functions stemming from real-world problems, in contrast to the test functions with explicit formulas, as we did in this work.

Acknowledgements This work was financially supported by the Ministry of Science, Research and the Arts of the State of Baden-Württemberg. We thank the referees for their valuable comments.

References

1. H.-J. Bungartz, Finite elements of higher order on sparse grids. Habilitationsschrift, Institut für Informatik, TU München, 1998
2. H.-J. Bungartz, M. Griebel, Sparse grids. Acta Numer. **13**, 147–269 (2004)
3. C.K. Chui, *An Introduction to Wavelets* (Academic, San Diego, 1992)
4. F. Franzelin, D. Pflüger, From data to uncertainty: an efficient integrated data-driven sparse grid approach to propagate uncertainty, in *Sparse Grids and Applications – Stuttgart 2014*, ed. by J. Garcke, D. Pflüger. Lecture Notes in Computational Science and Engineering, vol. 109 (Springer, Cham, 2016), pp. 29–49
5. K. Höllig, *Finite Element Methods with B-Splines* (SIAM, Philadelphia, 2003)
6. K. Höllig, J. Hörner, *Approximation and Modeling with B-Splines* (SIAM, Philadelphia, 2013)
7. M. Jamil, X.-S. Yang, A literature survey of benchmark functions for global optimisation problems. Int. J. Math. Model. Numer. Optim. **4**(2), 150–194 (2013)
8. J.A. Nelder, R. Mead, A simplex method for function minimization. Comput. J. **7**(4), 308–313 (1965)
9. J. Nocedal, S.J. Wright, *Numerical Optimization* (Springer, New York, 1999)
10. E. Novak, K. Ritter, Global optimization using hyperbolic cross points, in *State of the Art in Global Optimization*, ed. by C.A. Floudas, P.M. Pardalos. Nonconvex Optimization and Its Applications, vol. 7 (Springer, Boston, 1996), pp. 19–33
11. D. Pflüger, *Spatially Adaptive Sparse Grids for High-Dimensional Problems* (Verlag Dr. Hut, Munich, 2010)

12. D. Pflüger, Spatially adaptive refinement, in *Sparse Grids and Applications*, ed. by J. Garcke, M. Griebel. Lecture Notes in Computational Science and Engineering (Springer, Berlin, 2012), pp. 243–262
13. E. Polak, G. Ribière, Note sur la convergence de méthodes de directions conjuguées. Rev. Fr. Inf. Rech. Oper. **3**(1), 35–43 (1969)
14. M. Riedmiller, H. Braun, A direct adaptive method for faster backpropagation learning: the RPROP algorithm, in *Proceedings of IEEE International Conference on Neural Networks*, vol. 1 (1993), pp. 586–591
15. I.J. Schoenberg, Cardinal interpolation and spline functions: II. Interpolation of data of power growth. J. Approx. Theory **6**, 404–420 (1972)
16. I.J. Schoenberg, *Cardinal Spline Interpolation* (SIAM, Philadelphia, 1973)
17. W. Sickel, T. Ullrich, Spline interpolation on sparse grids. Appl. Anal. **90**(3–4), 337–383 (2011)
18. R. Storn, K. Price, Differential Evolution – a simple and efficient heuristic for global optimization over continuous spaces. J. Global Optim. **11**(4), 341–359 (1997)
19. J. Valentin, D. Pflüger, Hierarchical gradient-based optimization with B-splines on sparse grids, in *Sparse Grids and Applications – Stuttgart 2014*, ed. by J. Garcke, D. Pflüger. Lecture Notes in Computational Science and Engineering, vol. 109 (Springer, Cham, 2016), pp. 315–336
20. D. Whitley, S. Rana, J. Dzubera, K.E. Mathias, Evaluating evolutionary algorithms. Artif. Intell. **85**(1–2), 245–276 (1996)
21. X.-S. Yang, *Engineering Optimization* (Wiley, Hoboken, 2010)
22. C. Zenger, *Sparse Grids*. Notes on Numerical Fluid Mechanics, vol. 31 (Vieweg, Braunschweig, 1991), pp. 241–251

Editorial Policy

1. Volumes in the following three categories will be published in LNCSE:

i) Research monographs
ii) Tutorials
iii) Conference proceedings

Those considering a book which might be suitable for the series are strongly advised to contact the publisher or the series editors at an early stage.

2. Categories i) and ii). Tutorials are lecture notes typically arising via summer schools or similar events, which are used to teach graduate students. These categories will be emphasized by Lecture Notes in Computational Science and Engineering. **Submissions by interdisciplinary teams of authors are encouraged.** The goal is to report new developments – quickly, informally, and in a way that will make them accessible to non-specialists. In the evaluation of submissions timeliness of the work is an important criterion. Texts should be well-rounded, well-written and reasonably self-contained. In most cases the work will contain results of others as well as those of the author(s). In each case the author(s) should provide sufficient motivation, examples, and applications. In this respect, Ph.D. theses will usually be deemed unsuitable for the Lecture Notes series. Proposals for volumes in these categories should be submitted either to one of the series editors or to Springer-Verlag, Heidelberg, and will be refereed. A provisional judgement on the acceptability of a project can be based on partial information about the work: a detailed outline describing the contents of each chapter, the estimated length, a bibliography, and one or two sample chapters – or a first draft. A final decision whether to accept will rest on an evaluation of the completed work which should include

– at least 100 pages of text;
– a table of contents;
– an informative introduction perhaps with some historical remarks which should be accessible to readers unfamiliar with the topic treated;
– a subject index.

3. Category iii). Conference proceedings will be considered for publication provided that they are both of exceptional interest and devoted to a single topic. One (or more) expert participants will act as the scientific editor(s) of the volume. They select the papers which are suitable for inclusion and have them individually refereed as for a journal. Papers not closely related to the central topic are to be excluded. Organizers should contact the Editor for CSE at Springer at the planning stage, see *Addresses* below.

In exceptional cases some other multi-author-volumes may be considered in this category.

4. Only works in English will be considered. For evaluation purposes, manuscripts may be submitted in print or electronic form, in the latter case, preferably as pdf- or zipped ps-files. Authors are requested to use the LaTeX style files available from Springer at http://www.springer.com/gp/authors-editors/book-authors-editors/manuscript-preparation/5636 (Click on LaTeX Template → monographs or contributed books).

For categories ii) and iii) we strongly recommend that all contributions in a volume be written in the same LaTeX version, preferably LaTeX2e. Electronic material can be included if appropriate. Please contact the publisher.

Careful preparation of the manuscripts will help keep production time short besides ensuring satisfactory appearance of the finished book in print and online.

5. The following terms and conditions hold. Categories i), ii) and iii):

Authors receive 50 free copies of their book. No royalty is paid.
Volume editors receive a total of 50 free copies of their volume to be shared with authors, but no royalties.

Authors and volume editors are entitled to a discount of 33.3 % on the price of Springer books purchased for their personal use, if ordering directly from Springer.

6. Springer secures the copyright for each volume.

Addresses:

Timothy J. Barth
NASA Ames Research Center
NAS Division
Moffett Field, CA 94035, USA
barth@nas.nasa.gov

Michael Griebel
Institut für Numerische Simulation
der Universität Bonn
Wegelerstr. 6
53115 Bonn, Germany
griebel@ins.uni-bonn.de

David E. Keyes
Mathematical and Computer Sciences
and Engineering
King Abdullah University of Science
and Technology
P.O. Box 55455
Jeddah 21534, Saudi Arabia
david.keyes@kaust.edu.sa

and

Department of Applied Physics
and Applied Mathematics
Columbia University
500 W. 120 th Street
New York, NY 10027, USA
kd2112@columbia.edu

Risto M. Nieminen
Department of Applied Physics
Aalto University School of Science
and Technology
00076 Aalto, Finland
risto.nieminen@aalto.fi

Dirk Roose
Department of Computer Science
Katholieke Universiteit Leuven
Celestijnenlaan 200A
3001 Leuven-Heverlee, Belgium
dirk.roose@cs.kuleuven.be

Tamar Schlick
Department of Chemistry
and Courant Institute
of Mathematical Sciences
New York University
251 Mercer Street
New York, NY 10012, USA
schlick@nyu.edu

Editor for Computational Science
and Engineering at Springer:
Martin Peters
Springer-Verlag
Mathematics Editorial IV
Tiergartenstrasse 17
69121 Heidelberg, Germany
martin.peters@springer.com

Lecture Notes in Computational Science and Engineering

1. D. Funaro, *Spectral Elements for Transport-Dominated Equations.*
2. H.P. Langtangen, *Computational Partial Differential Equations.* Numerical Methods and Diffpack Programming.
3. W. Hackbusch, G. Wittum (eds.), *Multigrid Methods V.*
4. P. Deuflhard, J. Hermans, B. Leimkuhler, A.E. Mark, S. Reich, R.D. Skeel (eds.), *Computational Molecular Dynamics: Challenges, Methods, Ideas.*
5. D. Kröner, M. Ohlberger, C. Rohde (eds.), *An Introduction to Recent Developments in Theory and Numerics for Conservation Laws.*
6. S. Turek, *Efficient Solvers for Incompressible Flow Problems.* An Algorithmic and Computational Approach.
7. R. von Schwerin, ***M**ulti **B**ody **S**ystem **SIM**ulation.* Numerical Methods, Algorithms, and Software.
8. H.-J. Bungartz, F. Durst, C. Zenger (eds.), *High Performance Scientific and Engineering Computing.*
9. T.J. Barth, H. Deconinck (eds.), *High-Order Methods for Computational Physics.*
10. H.P. Langtangen, A.M. Bruaset, E. Quak (eds.), *Advances in Software Tools for Scientific Computing.*
11. B. Cockburn, G.E. Karniadakis, C.-W. Shu (eds.), *Discontinuous Galerkin Methods.* Theory, Computation and Applications.
12. U. van Rienen, *Numerical Methods in Computational Electrodynamics.* Linear Systems in Practical Applications.
13. B. Engquist, L. Johnsson, M. Hammill, F. Short (eds.), *Simulation and Visualization on the Grid.*
14. E. Dick, K. Riemslagh, J. Vierendeels (eds.), *Multigrid Methods VI.*
15. A. Frommer, T. Lippert, B. Medeke, K. Schilling (eds.), *Numerical Challenges in Lattice Quantum Chromodynamics.*
16. J. Lang, *Adaptive Multilevel Solution of Nonlinear Parabolic PDE Systems.* Theory, Algorithm, and Applications.
17. B.I. Wohlmuth, *Discretization Methods and Iterative Solvers Based on Domain Decomposition.*
18. U. van Rienen, M. Günther, D. Hecht (eds.), *Scientific Computing in Electrical Engineering.*
19. I. Babuška, P.G. Ciarlet, T. Miyoshi (eds.), *Mathematical Modeling and Numerical Simulation in Continuum Mechanics.*
20. T.J. Barth, T. Chan, R. Haimes (eds.), *Multiscale and Multiresolution Methods.* Theory and Applications.
21. M. Breuer, F. Durst, C. Zenger (eds.), *High Performance Scientific and Engineering Computing.*
22. K. Urban, *Wavelets in Numerical Simulation.* Problem Adapted Construction and Applications.
23. L.F. Pavarino, A. Toselli (eds.), *Recent Developments in Domain Decomposition Methods.*

24. T. Schlick, H.H. Gan (eds.), *Computational Methods for Macromolecules: Challenges and Applications.*

25. T.J. Barth, H. Deconinck (eds.), *Error Estimation and Adaptive Discretization Methods in Computational Fluid Dynamics.*

26. M. Griebel, M.A. Schweitzer (eds.), *Meshfree Methods for Partial Differential Equations.*

27. S. Müller, *Adaptive Multiscale Schemes for Conservation Laws.*

28. C. Carstensen, S. Funken, W. Hackbusch, R.H.W. Hoppe, P. Monk (eds.), *Computational Electromagnetics.*

29. M.A. Schweitzer, *A Parallel Multilevel Partition of Unity Method for Elliptic Partial Differential Equations.*

30. T. Biegler, O. Ghattas, M. Heinkenschloss, B. van Bloemen Waanders (eds.), *Large-Scale PDE-Constrained Optimization.*

31. M. Ainsworth, P. Davies, D. Duncan, P. Martin, B. Rynne (eds.), *Topics in Computational Wave Propagation.* Direct and Inverse Problems.

32. H. Emmerich, B. Nestler, M. Schreckenberg (eds.), *Interface and Transport Dynamics.* Computational Modelling.

33. H.P. Langtangen, A. Tveito (eds.), *Advanced Topics in Computational Partial Differential Equations.* Numerical Methods and Diffpack Programming.

34. V. John, *Large Eddy Simulation of Turbulent Incompressible Flows.* Analytical and Numerical Results for a Class of LES Models.

35. E. Bänsch (ed.), *Challenges in Scientific Computing - CISC 2002.*

36. B.N. Khoromskij, G. Wittum, *Numerical Solution of Elliptic Differential Equations by Reduction to the Interface.*

37. A. Iske, *Multiresolution Methods in Scattered Data Modelling.*

38. S.-I. Niculescu, K. Gu (eds.), *Advances in Time-Delay Systems.*

39. S. Attinger, P. Koumoutsakos (eds.), *Multiscale Modelling and Simulation.*

40. R. Kornhuber, R. Hoppe, J. Périaux, O. Pironneau, O. Wildlund, J. Xu (eds.), *Domain Decomposition Methods in Science and Engineering.*

41. T. Plewa, T. Linde, V.G. Weirs (eds.), *Adaptive Mesh Refinement – Theory and Applications.*

42. A. Schmidt, K.G. Siebert, *Design of Adaptive Finite Element Software.* The Finite Element Toolbox ALBERTA.

43. M. Griebel, M.A. Schweitzer (eds.), *Meshfree Methods for Partial Differential Equations II.*

44. B. Engquist, P. Lötstedt, O. Runborg (eds.), *Multiscale Methods in Science and Engineering.*

45. P. Benner, V. Mehrmann, D.C. Sorensen (eds.), *Dimension Reduction of Large-Scale Systems.*

46. D. Kressner, *Numerical Methods for General and Structured Eigenvalue Problems.*

47. A. Boriçi, A. Frommer, B. Joó, A. Kennedy, B. Pendleton (eds.), *QCD and Numerical Analysis III.*

48. F. Graziani (ed.), *Computational Methods in Transport.*

49. B. Leimkuhler, C. Chipot, R. Elber, A. Laaksonen, A. Mark, T. Schlick, C. Schütte, R. Skeel (eds.), *New Algorithms for Macromolecular Simulation.*

50. M. Bücker, G. Corliss, P. Hovland, U. Naumann, B. Norris (eds.), *Automatic Differentiation: Applications, Theory, and Implementations.*

51. A.M. Bruaset, A. Tveito (eds.), *Numerical Solution of Partial Differential Equations on Parallel Computers.*

52. K.H. Hoffmann, A. Meyer (eds.), *Parallel Algorithms and Cluster Computing.*

53. H.-J. Bungartz, M. Schäfer (eds.), *Fluid-Structure Interaction.*

54. J. Behrens, *Adaptive Atmospheric Modeling.*

55. O. Widlund, D. Keyes (eds.), *Domain Decomposition Methods in Science and Engineering XVI.*

56. S. Kassinos, C. Langer, G. Iaccarino, P. Moin (eds.), *Complex Effects in Large Eddy Simulations.*

57. M. Griebel, M.A Schweitzer (eds.), *Meshfree Methods for Partial Differential Equations III.*

58. A.N. Gorban, B. Kégl, D.C. Wunsch, A. Zinovyev (eds.), *Principal Manifolds for Data Visualization and Dimension Reduction.*

59. H. Ammari (ed.), *Modeling and Computations in Electromagnetics: A Volume Dedicated to Jean-Claude Nédélec.*

60. U. Langer, M. Discacciati, D. Keyes, O. Widlund, W. Zulehner (eds.), *Domain Decomposition Methods in Science and Engineering XVII.*

61. T. Mathew, *Domain Decomposition Methods for the Numerical Solution of Partial Differential Equations.*

62. F. Graziani (ed.), *Computational Methods in Transport: Verification and Validation.*

63. M. Bebendorf, *Hierarchical Matrices.* A Means to Efficiently Solve Elliptic Boundary Value Problems.

64. C.H. Bischof, H.M. Bücker, P. Hovland, U. Naumann, J. Utke (eds.), *Advances in Automatic Differentiation.*

65. M. Griebel, M.A. Schweitzer (eds.), *Meshfree Methods for Partial Differential Equations IV.*

66. B. Engquist, P. Lötstedt, O. Runborg (eds.), *Multiscale Modeling and Simulation in Science.*

67. I.H. Tuncer, Ü. Gülcat, D.R. Emerson, K. Matsuno (eds.), *Parallel Computational Fluid Dynamics 2007.*

68. S. Yip, T. Diaz de la Rubia (eds.), *Scientific Modeling and Simulations.*

69. A. Hegarty, N. Kopteva, E. O'Riordan, M. Stynes (eds.), *BAIL* 2008 – *Boundary and Interior Layers.*

70. M. Bercovier, M.J. Gander, R. Kornhuber, O. Widlund (eds.), *Domain Decomposition Methods in Science and Engineering XVIII.*

71. B. Koren, C. Vuik (eds.), *Advanced Computational Methods in Science and Engineering.*

72. M. Peters (ed.), *Computational Fluid Dynamics for Sport Simulation.*

73. H.-J. Bungartz, M. Mehl, M. Schäfer (eds.), *Fluid Structure Interaction II - Modelling, Simulation, Optimization.*

74. D. Tromeur-Dervout, G. Brenner, D.R. Emerson, J. Erhel (eds.), *Parallel Computational Fluid Dynamics 2008.*

75. A.N. Gorban, D. Roose (eds.), *Coping with Complexity: Model Reduction and Data Analysis.*

76. J.S. Hesthaven, E.M. Rønquist (eds.), *Spectral and High Order Methods for Partial Differential Equations.*

77. M. Holtz, *Sparse Grid Quadrature in High Dimensions with Applications in Finance and Insurance.*

78. Y. Huang, R. Kornhuber, O.Widlund, J. Xu (eds.), *Domain Decomposition Methods in Science and Engineering XIX.*

79. M. Griebel, M.A. Schweitzer (eds.), *Meshfree Methods for Partial Differential Equations V.*

80. P.H. Lauritzen, C. Jablonowski, M.A. Taylor, R.D. Nair (eds.), *Numerical Techniques for Global Atmospheric Models.*

81. C. Clavero, J.L. Gracia, F.J. Lisbona (eds.), *BAIL 2010 – Boundary and Interior Layers, Computational and Asymptotic Methods.*

82. B. Engquist, O. Runborg, Y.R. Tsai (eds.), *Numerical Analysis and Multiscale Computations.*

83. I.G. Graham, T.Y. Hou, O. Lakkis, R. Scheichl (eds.), *Numerical Analysis of Multiscale Problems.*

84. A. Logg, K.-A. Mardal, G. Wells (eds.), *Automated Solution of Differential Equations by the Finite Element Method.*

85. J. Blowey, M. Jensen (eds.), *Frontiers in Numerical Analysis - Durham 2010.*

86. O. Kolditz, U.-J. Gorke, H. Shao, W. Wang (eds.), *Thermo-Hydro-Mechanical-Chemical Processes in Fractured Porous Media - Benchmarks and Examples.*

87. S. Forth, P. Hovland, E. Phipps, J. Utke, A. Walther (eds.), *Recent Advances in Algorithmic Differentiation.*

88. J. Garcke, M. Griebel (eds.), *Sparse Grids and Applications.*

89. M. Griebel, M.A. Schweitzer (eds.), *Meshfree Methods for Partial Differential Equations VI.*

90. C. Pechstein, *Finite and Boundary Element Tearing and Interconnecting Solvers for Multiscale Problems.*

91. R. Bank, M. Holst, O. Widlund, J. Xu (eds.), *Domain Decomposition Methods in Science and Engineering XX.*

92. H. Bijl, D. Lucor, S. Mishra, C. Schwab (eds.), *Uncertainty Quantification in Computational Fluid Dynamics.*

93. M. Bader, H.-J. Bungartz, T. Weinzierl (eds.), *Advanced Computing.*

94. M. Ehrhardt, T. Koprucki (eds.), *Advanced Mathematical Models and Numerical Techniques for Multi-Band Effective Mass Approximations.*

95. M. Azaïez, H. El Fekih, J.S. Hesthaven (eds.), *Spectral and High Order Methods for Partial Differential Equations ICOSAHOM 2012.*

96. F. Graziani, M.P. Desjarlais, R. Redmer, S.B. Trickey (eds.), *Frontiers and Challenges in Warm Dense Matter.*

97. J. Garcke, D. Pflüger (eds.), *Sparse Grids and Applications – Munich 2012.*

98. J. Erhel, M. Gander, L. Halpern, G. Pichot, T. Sassi, O. Widlund (eds.), *Domain Decomposition Methods in Science and Engineering XXI.*

99. R. Abgrall, H. Beaugendre, P.M. Congedo, C. Dobrzynski, V. Perrier, M. Ricchiuto (eds.), *High Order Nonlinear Numerical Methods for Evolutionary PDEs - HONOM 2013.*

100. M. Griebel, M.A. Schweitzer (eds.), *Meshfree Methods for Partial Differential Equations VII.*

101. R. Hoppe (ed.), *Optimization with PDE Constraints - OPTPDE 2014.*

102. S. Dahlke, W. Dahmen, M. Griebel, W. Hackbusch, K. Ritter, R. Schneider, C. Schwab, H. Yserentant (eds.), *Extraction of Quantifiable Information from Complex Systems.*

103. A. Abdulle, S. Deparis, D. Kressner, F. Nobile, M. Picasso (eds.), *Numerical Mathematics and Advanced Applications - ENUMATH 2013.*

104. T. Dickopf, M.J. Gander, L. Halpern, R. Krause, L.F. Pavarino (eds.), *Domain Decomposition Methods in Science and Engineering XXII.*

105. M. Mehl, M. Bischoff, M. Schäfer (eds.), *Recent Trends in Computational Engineering - CE2014.* Optimization, Uncertainty, Parallel Algorithms, Coupled and Complex Problems.

106. R.M. Kirby, M. Berzins, J.S. Hesthaven (eds.), *Spectral and High Order Methods for Partial Differential Equations - ICOSAHOM'14.*

107. B. Jüttler, B. Simeon (eds.), *Isogeometric Analysis and Applications 2014.*

108. P. Knobloch (ed.), *Boundary and Interior Layers, Computational and Asymptotic Methods – BAIL 2014.*

109. J. Garcke, D. Pflüger (eds.), *Sparse Grids and Applications – Stuttgart 2014.*

110. H. P. Langtangen, *Finite Difference Computing with Exponential Decay Models.*

111. A. Tveito, G.T. Lines, *Computing Characterizations of Drugs for Ion Channels and Receptors Using Markov Models.*

112. B. Karazösen, M. Manguoğlu, M. Tezer-Sezgin, S. Göktepe, Ö. Uğur (eds.), *Numerical Mathematics and Advanced Applications - ENUMATH 2015.*

113. H.-J. Bungartz, P. Neumann, W.E. Nagel (eds.), *Software for Exascale Computing - SPPEXA 2013-2015.*

114. G.R. Barrenechea, F. Brezzi, A. Cangiani, E.H. Georgoulis (eds.), *Building Bridges: Connections and Challenges in Modern Approaches to Numerical Partial Differential Equations.*

115. M. Griebel, M.A. Schweitzer (eds.), *Meshfree Methods for Partial Differential Equations VIII.*

116. C.-O. Lee, X.-C. Cai, D.E. Keyes, H.H. Kim, A. Klawonn, E.-J. Park, O.B. Widlund (eds.), *Domain Decomposition Methods in Science and Engineering XXIII.*

117. T. Sakurai, S.-L. Zhang, T. Imamura, Y. Yamamoto, Y. Kuramashi, T. Hoshi (eds.), *Eigenvalue Problems: Algorithms, Software and Applications in Petascale Computing.* EPASA 2015, Tsukuba, Japan, September 2015.

118. T. Richter (ed.), *Fluid-structure Interactions.* Models, Analysis and Finite Elements.

119. M.L. Bittencourt, N.A. Dumont, J.S. Hesthaven (eds.), *Spectral and High Order Methods for Partial Differential Equations ICOSAHOM 2016.* Selected Papers from the ICOSAHOM Conference, June 27-July 1, 2016, Rio de Janeiro, Brazil.

120. Z. Huang, M. Stynes, Z. Zhang (eds.), *Boundary and Interior Layers, Computational and Asymptotic Methods BAIL 2016.*

121. S.P.A. Bordas, E.N. Burman, M.G. Larson, M.A. Olshanskii (eds.), *Geometrically Unfitted Finite Element Methods and Applications.* Proceedings of the UCL Workshop 2016.

122. A. Gerisch, R. Penta, J. Lang (eds.), *Multiscale Models in Mechano and Tumor Biology*. Modeling, Homogenization, and Applications.

123. J. Garcke, D. Pflüger, C.G. Webster, G. Zhang (eds.), *Sparse Grids and Applications - Miami 2016.*

For further information on these books please have a look at our mathematics catalogue at the following URL: www.springer.com/series/3527

Monographs in Computational Science and Engineering

1. J. Sundnes, G.T. Lines, X. Cai, B.F. Nielsen, K.-A. Mardal, A. Tveito, *Computing the Electrical Activity in the Heart.*

For further information on this book, please have a look at our mathematics catalogue at the following URL: www.springer.com/series/7417

Texts in Computational Science and Engineering

1. H. P. Langtangen, *Computational Partial Differential Equations.* Numerical Methods and Diffpack Programming. 2nd Edition
2. A. Quarteroni, F. Saleri, P. Gervasio, *Scientific Computing with MATLAB and Octave.* 4th Edition
3. H. P. Langtangen, *Python Scripting for Computational Science.* 3rd Edition
4. H. Gardner, G. Manduchi, *Design Patterns for e-Science.*
5. M. Griebel, S. Knapek, G. Zumbusch, *Numerical Simulation in Molecular Dynamics.*
6. H. P. Langtangen, *A Primer on Scientific Programming with Python.* 5th Edition
7. A. Tveito, H. P. Langtangen, B. F. Nielsen, X. Cai, *Elements of Scientific Computing.*
8. B. Gustafsson, *Fundamentals of Scientific Computing.*
9. M. Bader, *Space-Filling Curves.*
10. M. Larson, F. Bengzon, *The Finite Element Method: Theory, Implementation and Applications.*
11. W. Gander, M. Gander, F. Kwok, *Scientific Computing: An Introduction using Maple and MATLAB.*
12. P. Deuflhard, S. Röblitz, *A Guide to Numerical Modelling in Systems Biology.*
13. M. H. Holmes, *Introduction to Scientific Computing and Data Analysis.*
14. S. Linge, H. P. Langtangen, *Programming for Computations* - A Gentle Introduction to Numerical Simulations with MATLAB/Octave.
15. S. Linge, H. P. Langtangen, *Programming for Computations* - A Gentle Introduction to Numerical Simulations with Python.
16. H.P. Langtangen, S. Linge, *Finite Difference Computing with PDEs* - A Modern Software Approach.
17. B. Gustafsson, *Scientific Computing from a Historical Perspective.*
18. J. A. Trangenstein, *Scientific Computing.* Volume I - Linear and Nonlinear Equations.

19. J. A. Trangenstein, *Scientific Computing*. Volume II - Eigenvalues and Optimization.

20. J. A. Trangenstein, *Scientific Computing*. Volume III - Approximation and Integration.

For further information on these books please have a look at our mathematics catalogue at the following URL: www.springer.com/series/5151

Printed in the United States
By Bookmasters